GROUP THEORY
FOR ATOMS,
MOLECULES AND SOLIDS

To Shirley

GROUP THEORY FOR ATOMS, MOLECULES AND SOLIDS

Brian S. Wherrett
Department of Physics,
Heriot-Watt University

Prentice/Hall International

Englewood Cliffs, New Jersey London New Delhi Rio de Janeiro
Singapore Sydney Tokyo Toronto Wellington

Library of Congress Cataloging in Publication Data

Wherrett, Brian S.
Group theory for atoms, molecules, and solids.

Bibliography: p.
Includes index.
1. Groups, Theory of. 2. Energy-band theory of solids.
3. Energy levels (Quantum mechanics) 4. Symmetry
(Physics) I. Title.
QC174.17G7W44 1986 530.4′1 85–19396
ISBN 0–13–365461–3

British Library Cataloguing in Publication Data

Wherrett, B.S.
Group theory for atoms, molecules and solids.
1. Groups, Theory of 2. Mathematical physics I. Title
530.1′5122 QC20.7.G76

ISBN 0–13–365461–3

Prentice-Hall Inc., *Englewood Cliffs, New Jersey*
Prentice-Hall International (UK) Ltd, *London*
Prentice-Hall of Australia Pty Ltd, *Sydney*
Prentice-Hall Canada Inc., *Toronto*
Prentice-Hall Hispanoamericana S.A., *Mexico*
Prentice-Hall of India Private Ltd, *New Delhi*
Prentice-Hall of Japan Inc., *Tokyo*
Prentice-Hall of Southeast Asia Pte Ltd, *Singapore*
Editora Prentice-Hall do Brasil Ltda, *Rio de Janeiro*
Whitehall Books Ltd, *Wellington, New Zealand*

Printed and bound in Great Britain for
Prentice-Hall International (UK) Ltd,
66 Wood Lane End, Hemel Hempstead,
Hertfordshire, HP2 4RG
by A. Wheaton & Co. Ltd, Exeter

1 2 3 4 5 90 89 88 87 86

ISBN 0-13-365461-3

Contents

PART TWO

APPLICATION OF SYMMETRY TO NORMAL MODES

PART THREE

SYMMETRY APPLIED TO QUANTUM ENERGY LEVELS

PART FOUR

FURTHER SYMMETRY IDEAS AND APPLICATION TO THE RADIATION–MATTER INTERACTION

Preface

This text has arisen out of a course of lectures on the physical applications of group theory, given for several years at Heriot-Watt University, and a longer course on group theory and semiconductor energy-band theory given at North Texas State University.

The aim is to establish, at a level suitable for those new to the ideas of group theory, the basic requirements for understanding the electronic band structure of solids and those transitions between the energy levels that can be excited by incident electromagnetic radiation. In so doing the groundwork is developed for an understanding of absorption selection rules in atoms and molecules as well as solids, of the vibrational properties of molecules and solids, and of non-radiative scattering processes in solids.

The text is divided into five parts. Part One contains chapters on the basic ideas of group theory, in Part Two group theory is used to solve the classical problem of the oscillations of a harmonically coupled system, and the electronic quantum states of atoms and molecules are discussed in Part Three. Part Four is concerned primarily with selection rules for transitions between electronic states and between states of the quantized coupled oscillators. Finally in Part Five the symmetries of crystalline materials are described and applied to the modelling of electron energy bands of solids.

In general each of the nineteen chapters of the text relies on material presented in earlier chapters. Rather than present all of the concepts of group theory in an initial section, they are introduced as they are needed, in the context of the problem being tackled. Thus, for example, direct product groups do not appear until Chapter 13. Equally Wigner's theorem is introduced in Chapter 7, to bridge the gap between symmetry ideas and the physical problems of interest. This theorem is introduced for use in the normal-mode problem, which is considered to be more intuitive than the quantum problems tackled in Chapters 9–19.

The course has been given as a final year course, corresponding in length to a US one-semester course, and as part of a course for experimental post-graduate students wishing to become familiar with the notation and the essential techniques of group theory. However, very little background knowledge is demanded. Success in use of the text demands a willingness to think in three dimensions and patience in bearing with the new vocabulary until it is used in physical situations. A small amount of previous experience of matrices, harmonic oscillation and semiconductor physics is useful but not essential. The reader will need to be familiar with the basic ideas of quantum mechanics.

My thanks are due to many people for helping me put this text together; to my family for putting up with my absences, to Art Smirl who conned me into writing up my notes in the first place, to the students who suffered the growing pains of the text, and to my typists – Becky, June and Janice.

Brian S. Wherrett

PART ONE

SYMMETRY IDEAS

1

An introduction to group theory

Group theory, in its application to physics, is concerned with the necessary consequences of symmetry.

By use of group theory one obtains physically significant results that are a consequence purely of the symmetry of the system under investigation. The techniques of group theory are deceptively simple, enabling conclusions to be drawn without recourse to detailed mathematics. The underlying theme of symmetry and the elegance of the methods of application combine to produce in applied group theory one of the most beautiful theoretical topics in physics.

Throughout most of this text group theory will be used to exploit the symmetrical configurations of atoms that occur in many molecules and solids. It is, however, worth mentioning briefly some well-known results of other forms of symmetry, which emphasize its significance throughout physics. From a classical viewpoint continuous translational symmetry, that is the absence of a force or the invariance of the energy of a system with respect to some direction in space, leads to the conservation of the linear momentum of the system. Continuous rotational symmetry, the invariance of energy with angle or the absence of a torque about some axis, leads to the conservation of angular momentum about the axis. Also continuous temporal symmetry is equivalent to conservation of energy. Conservation properties form the very root of physics as a discipline; without conservation we would find great difficulty in any attempt to model nature mathematically.

From a quantum-mechanical viewpoint it is necessary to include the ideas of boundary conditions in a discussion of symmetry. Thus, for example, bounded periodic, rather than continuous, translational symmetry leads to the use of discrete **k**-wavevectors to label the states of crystal excitations, and leads to discrete energy eigenvalues. The rotational symmetry of atoms, coupled with the finite and single-valued properties of electronic eigenfunctions, leads to discrete angular momenta and energies and to the use of good quantum numbers to label eigenstates.

Given the extremely powerful results to which continuous symmetries lead it might well be expected that the discrete symmetries with which we are primarily concerned in this text are also to be associated with conservation properties of some sort, and with state labelling. As we shall discover the latter is one of the major results of group theory.

Now the basic problem of quantum mechanics, considered in detail in Chapter 8, is the solution of the Hamiltonian equation in terms of eigenfunctions and energy eigenvalues. The problem can take the following mathematical forms:

$$\mathcal{H}\psi = E\psi; \tag{1.1}$$

$$[a]^{-1}[\mathcal{H}^f - ES^f][a] = [\mathcal{E} - E]; \tag{1.2}$$

$$[\mathcal{H}^f - E_n S^f]\begin{bmatrix} a_{1n} \\ a_{2n} \\ \vdots \end{bmatrix} = 0. \tag{1.3}$$

The details of these equations are not of immediate concern; we need only recognize that the first is a differential equation and that the latter two are *matrix equations*. Solution of the Hamilton equation leads to a set of eigenfunctions ψ_n that describe distributions of the particles in the allowed states of the quantum system, and the energies of these states. In turn a knowledge of the ψ_n allows one to determine how the system would respond to any given external influence (for example an incident beam of electromagnetic radiation, or an applied static field).

Group theory has the following uses in the quantum problem. Firstly it allows one to determine the symmetry properties of the possible states even though the precise form of the distribution functions, ψ_n, might not be known. Secondly, it is often the case that for a given E_n there is more than one distinct eigenfunction; such functions are said to be degenerate. The second use of group theory is in determining such degeneracies of the states of a system. With a knowledge of the state symmetries, further application of group theory allows the form of a response of the system to be predicted. For example one can determine whether or not radiation can excite a molecule from any one particular state to another.

Additionally, if one does wish to solve the matrix equations, which generally requires computational methods, group theory may be used to reduce the dimensions of the matrices one has to deal with, with considerable savings in time.

Because it is difficult to think intuitively about quantum-mechanical eigenfunctions, the techniques of applied group theory will be introduced here by considering instead the coupled-oscillator problem. This problem is solved in classical physics by exactly the same matrix methods as those required for the above quantum problem. Thus if we imagine a set of masses mutually attached by springs, there will be some equilibrium condition—a configuration of the system as a whole—for which all the masses are stationary. The coupled oscillator problem sets out to answer the question: 'If the system is displaced slightly from equilibrium, then what is the subsequent motion of each mass?' Mathematically this problem is manifested by matrix equations which have identical form to equations (1.2) and (1.3) above:

$$[\alpha]^{-1}[K^d - M^d\omega^2][\alpha] = [\Omega - \omega^2 I] \tag{1.4}$$

or

$$[K^d - M^d\omega_i^2]\begin{bmatrix}\alpha_{1i}\\ \alpha_{2i}\\ \vdots\end{bmatrix} = 0. \tag{1.5}$$

In these equations $\boldsymbol{K}^d$ and M^d are known as force and mass matrices and are obtained from a knowledge of the potential and kinetic energies of the system of masses under a set of displacements from their equilibrium positions (Chapter 6). For a system containing N masses, each able to move in any of three perpendicular directions, there are $3N$ solutions for ω_i, and $3N$ associated sets of α_{ji} values. Each solution represents a normal mode of oscillation of the system.

To appreciate the simplicity of the consequences of symmetry consider the model, pictured in Figure 1.1, which is trivial to construct.

I want to concentrate on the motion, along their line of connection, of the two symmetrically placed 100 gramme masses. This system has just two degrees of freedom; the smaller mass is uniquely positioned once the positions of the others are set. Hence any displacement can be defined in terms of two vectors. If mass 1 is displaced by a small amount and mass 2 held at the equilibrium, on releasing the masses, 1 begins to oscillate but its amplitude is damped rapidly and 2 oscillates. The converse occurs and then the motion repeats periodically. However, as far as each mass is concerned, it does not undergo a simple harmonic motion. By direct contrast if both masses are moved to the same side, by the same small amplitude, and released, then they oscillate back and forth, in phase, at some frequency ω_1. Alternatively if they are given equal but opposite initial displacements they subsequently oscillate π out of phase, at a frequency ω_2. These two modes of motion of the system are termed the normal modes; ω_1, ω_2 are the natural frequencies. The more complex motion obtained from any other initial conditions may be modelled by some linear combination of the normal mode displacements.

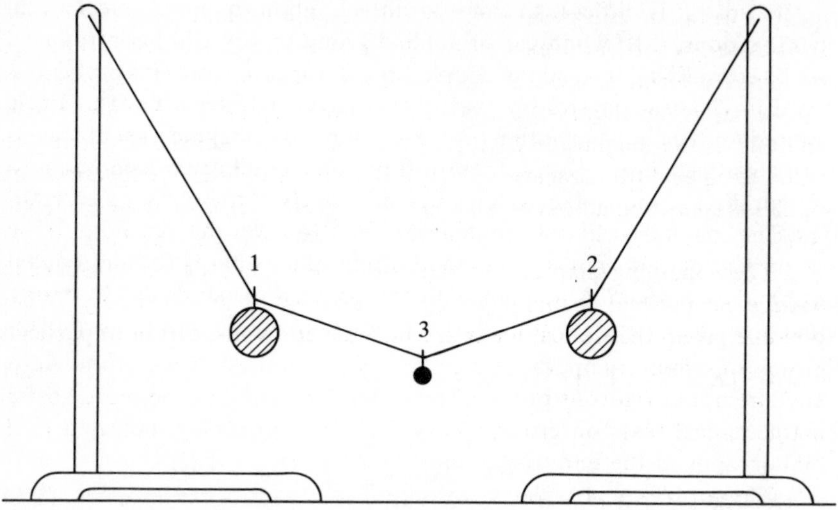

Figure 1.1 Model of a system that demonstrates the symmetry of normal modes. Masses 1 and 2 are 100 grammes each. Mass 3 is far lighter. Effectively the system acts, in the plane of the masses, like a pair of harmonic oscillators coupled together by the connecting line and the small mass. A weak spring can in principle be used instead to couple masses 1 and 2, but the demonstration is easier to set up as indicated.

The most important point to be brought out here though is that both normal modes are highly symmetric—in some way they display the symmetry of the system.

The classical coupled oscillator is a useful macroscopic model for the vibrational motion of the nuclei in a molecule; the nuclei are coupled through the electrostatic forces between all the particles in the molecule. Therefore the first type of problem to which we shall apply group theory will correspond to the analysis of the vibrations of a polyatomic molecule, such as NH_3 (ammonia) or C_6H_6 (benzene) which have high symmetry. In the benzene case it would require the construction and diagonalization of a 36×36 matrix in order to determine the number of distinct natural frequencies, their values and the displacements. Group theory tells us almost immediately that there are only 24 such frequencies and that the solution of eight 2×2 matrices and two 4×4 matrices will give us the remaining information.

Having obtained a set of eigenvectors and frequencies or, in quantum mechanics, a set of eigenfunctions and energy eigenvalues, the next question one asks is how can one excite the vibrations, or how can transitions be excited from one energy level to another. Again group theory can provide the answer. Whether the excitation is to be by the absorption of radiation, or by some scattering process, group theory will tell us if it can be accomplished.

What group theory itself will not tell us, without recourse to further analysis, is the magnitude of any of the parameters discussed above—the value of natural frequencies or energies, or the strengths of excitation—the so-called transition matrix elements. Group theory gives us yes–no selection rules.

The remaining chapters in Part 1 are included in order to introduce the basic concepts and to build up the vocabulary of group theory. Where possible group theoretical ideas are introduced by inspection of particular groups of symmetry operations; certain important results are given without any attempt at rigorous proof. The reader is referred to the more abstract mathematical texts on group theory for these proofs, as indicated in the bibliography at the end of the book.

The problems provided with each chapter serve two purposes. Problems marked with an asterisk should be considered as essential exercises; they are included in order to allow the reader to test his/her understanding of the text material. Other problems are included in order to indicate some of the features of group theory that are not considered to be crucial to progress through the present text.

2

Symmetry operations and the group multiplication table

In this section we are concerned with operations such as rotations and reflections. These operations are defined by considering what they do to an arbitrarily shaped object. Two operations are said to be distinct only if they produce different results when applied to an arbitrary object. (Thus a rotation by π radians is distinct from a rotation by $\pi/2$ radians but indistinct from a rotation by 3π radians.)

We will be dealing with objects that have specific symmetries. That is, there will be a set of distinct operations which, when applied to the object, do not change its appearance, whereas they would change the appearance of an arbitrary object. For a given object the set is known as its *group of symmetry operations*. The following are definitions of most of the operations that will appear in the symmetry groups to be considered in this text. In each case the operation is defined with respect to a fixed reference frame (the observer's reference frame) which should be considered quite separately from the object. The object is introduced into this fixed frame and rotated, reflected etc. accordingly.

(There is an alternative approach to symmetry in which the object is kept stationary and the observer's reference frame is rotated etc. about the object. To avoid confusion this approach will not be pursued.)

2.1 Symmetry operations

Identity operation, E. This operation, conventionally labelled E, appears in all symmetry groups. It indicates the trivial operation—do nothing.

Rotation, C_n. A rotation by angle $2\pi/n$, of a specified sense, about a specified axis. We choose the sense of rotation to be that of a right-handed screw. Looking from above down a vertical axis towards the origin the rotation is anti-clockwise.

Inversion, i. Each point on the object at position (x, y, z) in the reference frame is transformed to $(-x, -y, -z)$. A fixed central point must be defined.

Reflection, σ. Defined with respect to a plane in the reference frame.

Improper rotation, S_n. A rotation C_n, followed by or following a reflection, σ, with respect to the plane normal to the rotation axis. A central point, where the plane intersects the axis, must be defined. One writes $S_n \equiv \sigma C_n$. (The symbol $\equiv$ means 'is equivalent to'.) There is an alternative way of describing an improper rotation, namely a rotation followed by or following inversion, $S_n \equiv iC_m$. It should be emphasized though that the angle of rotation $2\pi/m$ is not necessarily the same as that in the previous definition. For example, an S_6 operation has the possible forms $S_6 \equiv \sigma C_6$ or $S_6 \equiv iC_3$. Inversion and reflection are special cases of the improper rotations; $i \equiv S_2$, $\sigma \equiv S_1$ (see Problem 2.3). We shall meet other forms of operation later, particularly in dealing with the periodicity of crystal lattices.

The *product* AB of two operations A and B is defined as that operation which has the same effect (on an arbitrary object) as obtained by performing operation B *followed by* operation A. (Thus S_n can be described as the product of σ and C_n.)

In order to get the reader thinking about specific symmetry groups two examples will now be presented. The first is a very simple object but has a complicated symmetry group which will not be met again until Chapter 11. The second example, the group of the ammonia molecule, will be used throughout Chapters 1–8 in order to follow through the techniques of group theory for one particular system.

Example Group (a), the Cl_2 molecule

This is a molecule of high symmetry. It is introduced here briefly in order to demonstrate a group with an infinity of symmetry operations—a continuous group. In the next few chapters we shall concentrate on finite groups for reasons that will rapidly become obvious.

In Figure 2.1 the y-axis is defined to be pointing upwards out of the paper. The origin 0 and the axes defined by $0x$, $0y$, $0z$ form a fixed frame of reference and the group of symmetry operations is such that each operation, when applied to the molecule but maintaining the reference frame, appears to leave the molecule unaltered.

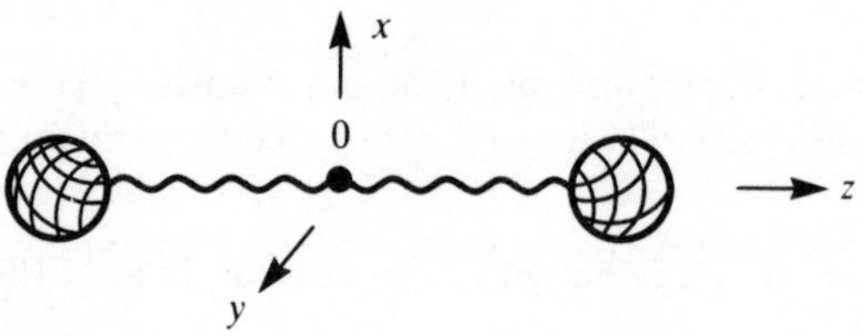

Figure 2.1 Schematic to show the symmetry of the diatomic molecule Cl_2.

Point group. The centre of the molecule, at 0, is clearly unique and must remain unmoved by any symmetry operation. The resulting set of operations is called a point group for the above reason.

The Cl_2 symmetry operations are:

(i) The identity, E.
(ii) Inversion, i.
(iii) Rotation by any angle α from 0 to 2π (exclusive), about the molecular axis $0z$; $C_{2\pi/\alpha}$ or $R(\alpha, z)$.$\alpha=0$ is excluded here as it is the identity operation, already included. $\alpha=2\pi$ is excluded as for any object it has the same effect as E. For the same reason $R(\alpha, z)$ and $R(\alpha+2\pi, z)$ are never included together. (An exception occurs when spin is included in quantum mechanical problems.)
(iv) Any improper rotation about the molecular axis, $S(\alpha, z)$. $S(0, z)$ is the reflection in the x-y plane; $S(\pi, z)$ is the inversion and should not be included again.
(v) Rotation by π about any axis through 0, in the x-y plane, C_2.
(vi) Reflection in any plane containing the molecular axis, σ_v.

These six classes of operation complete the point group symmetries of the Cl_2 group. The group is labelled the point group $D_{\infty h}$; the significance of such labelling will become clear in subsequent sections.

Example Group (b), the NH_3 molecule

The ammonia molecule, NH_3, may be thought of for symmetry purposes as a trio of hydrogen atoms at the corners of an equilateral triangle, and a nitrogen atom at some position on the axis normal to the plane of hydrogen atoms, through the centre of the triangle. In Figure 2.2 we define various useful axes. $0a$, $0b$, $0c$, $0d$ are again to be considered as fixed axes. The symmetry operations for this molecule are:

(i) The identity, E.
(ii) Rotation by $2\pi/3$, in the anti-clockwise direction, about the axis $0d$, looking towards 0 from the nitrogen position, C_3.
(iii) Rotation by $4\pi/3$ about the above axis. We label this C_3^2, recognizing that it corresponds to the product C_3C_3. If there is an operation C_n about some axis then $C_{n/2}$, $C_{n/3}$, . . . , $C_{n/n-1}$ must also be symmetry operations, i.e., rotations by $4\pi/n$, $6\pi/n$, etc. $C_{n/m}$ is labelled C_n^m if n/m is not an integer.
(iv) Reflections with respect to planes 0ad, 0bd, and 0cd; labelled σ_a, σ_b and σ_c respectively for present purposes.

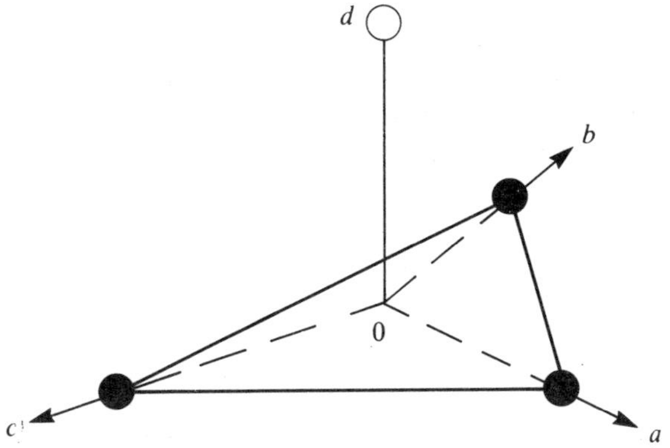

Figure 2.2 Schematic of the NH_3 molecule.

An insight into many of the ideas of group theory, avoiding the abstract mathematics of the theory itself, is obtained by constructing and inspecting the product properties of the above six operations, which constitute the symmetry operations of the group labelled C_{3v}.

2.2 Multiplication table for operations

Table 2.1 shows an almost complete group multiplication table for the C_{3v}

group, obtained in terms of the six operations. There are several comments to be made regarding it:

Table 2.1 Multiplication table for the C_{3v} group operations.

	E	C_3	C_3^2	σ_a	σ_b	σ_c
E	E	C_3	C_3^2	σ_a	σ_b	(1)
C_3	C_3	C_3^2	E	σ_c	σ_a	σ_b
C_3^2	C_3^2	E	C_3	σ_b	σ_c	σ_a
σ_a	σ_a	σ_b	σ_c	E	(3)	C_3^2
σ_b	σ_b	σ_c	σ_a	(4)	E	C_3
σ_c	(2)	σ_a	σ_b	C_3	C_3^2	E

(i) The elements in the table are those operations which have the same effect on an arbitrary object as obtained by successively following the operation on the top row vertically above the element in question, by the operation horizontally alongside at the far left. Thus for example the operation at position (1) is equivalent to σ_c followed by E. That is $(1) \equiv E\sigma_c \equiv \sigma_c$.

(ii) Similarly operation (2) is equivalent to the identity followed by σ_c; $(2) \equiv \sigma_c$. It is a trivial statement to make but in this case it does not matter if we perform σ_c before or after E. If this arbitrariness of order is true for every pair of operations in the group then the group is said to be *Abelian*. We write operations (1) and (2) as $E\sigma_c$ and $\sigma_c E$ respectively, and $E\sigma_c \equiv \sigma_c E \equiv \sigma_c$. Note that the order in which we understand the operations to be performed starts from the right-hand operator, just as it would if we were referring to differential operations in calculus.

(iii) The operation (3) is therefore written, $(3) \equiv \sigma_a\sigma_b$. In order to see what single operation this could be replaced by we use a technique which can be applied for combinations of operations of any type. We need to consider an object of arbitrary sysmmetry. The schematic NH_3 molecule can be converted into such an object by attaching labels 1, 2, 3 to the three hydrogen atoms (Figure 2.3A). Under the operation σ_b atoms 1 and 3 are interchanged, 2 is unaltered; Figure 2.3B is obtained after the operation. We emphasize once again that the axes $0a$, $0b$, $0c$ are fixed. Thus σ_a applied to the object as described in Figure 2.3B interchanges the atoms that are labelled 1 and 2 now, and therefore the combined operation $\sigma_a\sigma_b$ has the effect of changing from

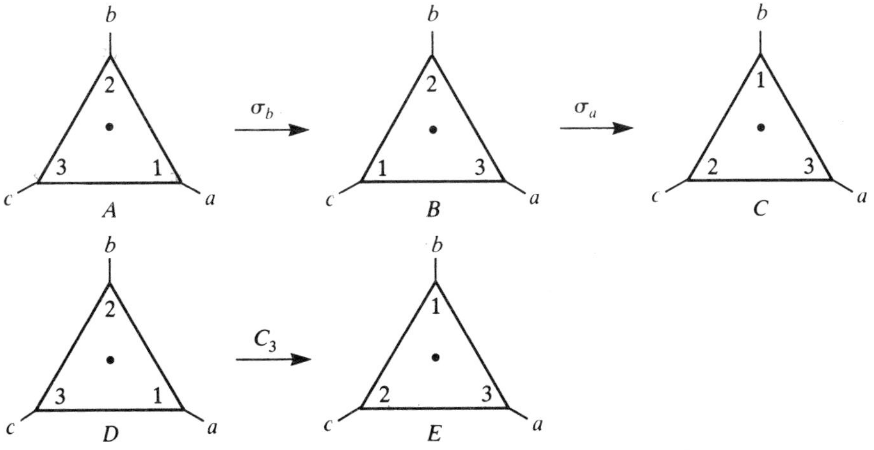

Figure 2.3 Demonstration of the equivalence of $\sigma_a\sigma_b$ and C_3.

A to C. Figures 2.3D and E demonstrate the effect of operation C_3, which is indistinguishable from $\sigma_a\sigma_b$. Thus $(3)\equiv\sigma_a\sigma_b\equiv C_3$.

(iv) ***Commutation.*** Operation (4) considered by the above technique is $(4)\equiv\sigma_b\sigma_a\equiv C_3^2$. Operations (3) and (4) have been dealt with at length, not only to demonstrate the meaning of the 'product' of two operations but also to show that the order of this product may be significant. Not all groups are Abelian, in particular the C_{3v} group is not.

$$\sigma_a\sigma_b\neq\sigma_b\sigma_a$$

We say that operations σ_a and σ_b do not commute, whereas E and σ_c do. In general a pair of operations commute if their product does not depend on the order in which they are applied.

(v) ***Closure.*** Every element in the multiplication table is one of the original six group operations. Obviously if we follow one symmetry operation by another the symmetric object must still look unaltered; the net effect must correspond to some single symmetry operation. This is a property, known as closure, of all groups. If we were to construct a multiplication table and find a product that gives some new (distinct) operation then we should have included this new operation in the group in the first place.

(vi) ***Associative law.*** Note that $\sigma_a\sigma_b$ followed by C_3 produces the same effect as C_3^2. So does σ_b followed by $C_3\sigma_a$. That is:

$$C_3(\sigma_a\sigma_b)\equiv C_3C_3\equiv C_3^2,$$

$$(C_3\sigma_a)\sigma_b\equiv\sigma_c\sigma_b\equiv C_3^2.$$

That the ordering of the pairing process is irrelevant is known as the associative property; it is a general property of symmetry operations.

(vii) ***Rearrangement theorem.*** Each row in the multiplication table contains each element of the group just once; so does each column. It is easy to anticipate this rearrangement theorem for any group of symmetry operations. Given one initial operation producing some specific effect on an arbitrary object each second operation leads to a different net effect, which we have already stated must be contained in the group.

(viii) ***Inverse of a symmetry operation.*** A particular result of the rearrangement theorem is that the identity operation appears just once in each row and once in each column of the table. For a specific operation G_r, there is therefore a unique operation G_s such that $G_rG_s \equiv E$. Then G_s is called the inverse of G_r and is often written as $G_s \equiv G_r^{-1}$. By reference to the table another general result may be noted: if G_s is the inverse of G_r then G_r is the inverse of G_s.

(ix) ***Subgroups.*** The top left-hand corner of the C_{3v} group multiplication table satisfies the closure, associative and rearrangement results discussed above (Table 2.2).

Table 2.2 Multiplication table for E, C_3, C_3^2.

	E	C_3	C_3^2
E	E	C_3	C_3^2
C_3	C_3	C_3^2	E
C_3^2	C_3^2	E	C_3

This set of operations (E, C_3, C_3^2) describes an object of slightly reduced symmetry with respect to the NH_3 molecule. It is called a subgroup of C_{3v}. Note that the operations E, σ_a, σ_b, σ_c, however, do not form a subgroup—closure is not obeyed by this set. There are four other subgroups of C_{3v}: (E, σ_a), (E, σ_b), (E, σ_c) and the trivial case, (E). Figure 2.4 shows objects of symmetry (E, C_3, C_3^2); (E, σ_a); and (E) respectively.

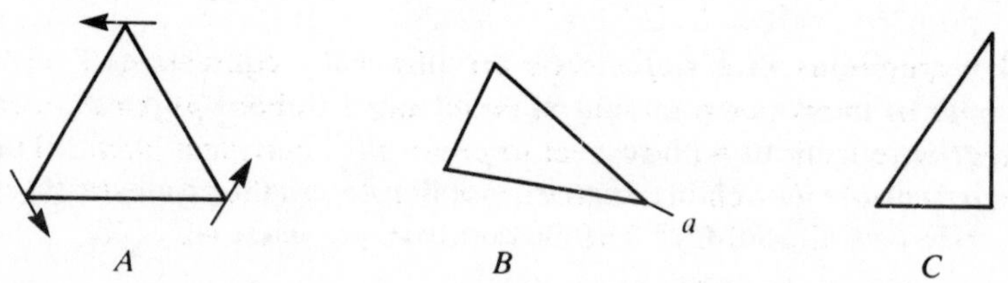

Figure 2.4 Reduction of the C_{3v} group symmetry.

In passing, note that the number of elements of a subgroup is in each case an integer submultiple of the number in the original group.

(x) ***Classes of operation.*** Operations σ_a, σ_b and σ_c are physically entirely equivalent. There is also a high degree of similarity between C_3 and C_3^2; they can be thought of as rotations by the same angle, about the same axis, but in the opposite sense. The six C_{3v} operations are said to comprise three classes: $E, (C_3, C_3^2)$ and $(\sigma_a, \sigma_b, \sigma_c)$. There is a more specific way of determining those operations in the same class as any one given operation. We define the class of the operation G_r to contain all those distinct elements that are obtained by performing the products $G_s^{-1}G_rG_s$ using all operations G_s in the group.

For example, to find the class of σ_a:

$$E^{-1}\sigma_a E \equiv \sigma_a, \qquad C_3^{-1}\sigma_a C_3 \equiv \sigma_c, \qquad (C_3^2)^{-1}\sigma_a C_3^2 \equiv \sigma_b$$

$$\sigma_a^{-1}\sigma_a\sigma_a \equiv \sigma_a, \qquad \sigma_b^{-1}\sigma_a\sigma_b \equiv \sigma_c, \qquad \sigma_c^{-1}\sigma_a\sigma_c \equiv \sigma_b.$$

Thus σ_b and σ_c are in the same class as σ_a.

(xi) ***Similarity transformation of operations.*** $G_s^{-1}G_rG_s$ is known as a similarity transformation performed on G_r. The resultant effect of these three operations on the object is the same as that of one of the operations that are *similar* to G_r.

(xii) ***Conjugate element.*** $G_t \equiv G_s^{-1}G_rG_s$ is said to be the conjugate element to G_r, with respect to G_s. As $G_sG_s^{-1} \equiv E$ another way of expressing conjugates is $G_sG_t \equiv G_rG_s$.

Note (i) that E is always in a class of its own, and (ii) that no element can ever be in two classes. We can prove these results as follows:

(i) $G_s^{-1}EG_s \equiv G_s^{-1}G_s \equiv E$ for any G_s.
(ii) Suppose $G_s^{-1}G_rG_s \equiv G_t$ and $G_s^{-1}G_qG_s \equiv G_t$.

Because $G_sG_s^{-1}G_rG_sG_s^{-1} \equiv G_sG_tG_s^{-1}$, then $G_r \equiv G_sG_tG_s^{-1} \equiv G_q$, and the classes of G_r and G_q must be identical.

The members of each class can usually be ascertained without resort to the use of similarity transformations, using the following rules:

The operations of a single class are physically equivalent. They may be proper or improper rotations by *equal angles*, about *physically equivalent axes*, or reflections with respect to *physically equivalent planes*. Further, if there is no obvious choice for the sense of rotation then a rotational class will contain both C_n and C_n^{-1}. The example of a plane square object is useful for demonstration purposes (Figure 2.5).

Amongst the group symmetry operations are C_2 rotations about the $0x$, $0y$ and $0z$ (perpendicular to the plane) axes, also C_4 and C_4^3 about the $0z$ axis.

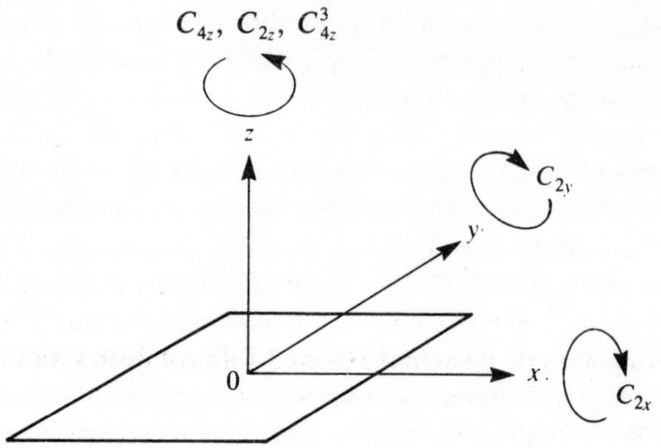

Figure 2.5 Selected symmetry operations of a planar square object.

(Note that we write C_4^2 about the $0z$ axis as a C_2 operation). The $0x$ and $0y$ axes are physically indistinguishable but the $0z$ axis is distinct. The classes of these group operations are: (C_{2x} and C_{2y}), C_{2z}, (C_{4z} and C_{4z}^3).

2.3 Group postulates

In conclusion so far, the properties of a group are:

(i) There is a set of elements (symmetry operations in our case) $G_1, G_2, \ldots$
The set may be infinite or denumerable.
The total number of elements is called the order of the group, g.
(ii) There is a meaning to the product of two elements, $G_r G_s$.
(iii) The product of any two elements is itself contained in the group (closure).
(iv) The associative law holds; $G_r(G_s G_t) \equiv (G_r G_s)G_t$.
(v) The set contains an identity element E such that for any element G_r, $G_r E \equiv E G_r \equiv G_r$.
(vi) For each element G_r there is a unique inverse G_s such that $G_r G_s \equiv G_s G_r \equiv E$.

Such properties are not restricted to symmetry groups: a trivial example of a different type of group is one having as its elements the set of all positive and negative integers (including zero) ... $-2, -1, 0, 1, 2$ ... and having as its 'product' relation the operation addition. Property (iii) is obeyed, $r+s$ is an integer. The associative law holds, $r+(s+t)=(r+s)+t$. There is an identity element, zero: $r+0=0+r=r$. Finally each element has a unique inverse; $r+(-r)=(-r)+r=0$.

There are myriad alternative examples. However, we shall be dealing exclusively either with sets of symmetry operations or with sets of matrices, and the product relations in these cases are successive operation and matrix multiplication respectively. The groups themselves are labelled, in addition to the operations within them. Thus C_{3v} indicates the symmetry operations of an object having a single principal axis of 3-fold rotation symmetry and a set of 3 (necessarily) reflection planes that contain the axis. Other common notations for the groups include: D_n—having an n-fold principal axis and n 2-fold rotation axes perpendicular to the principal one; T—the tetrahedral groups, with three equivalent principal axes; O—the cubic groups, with three principal axes and perpendicular symmetry axes; S_n—having an n-fold improper rotation axis. Others will be met at various stages; they are summarized in Appendix A1(b).

Problems

Note: Problems marked with an asterisk should be considered as essential exercises.

2.1** ***The mathematical diversity of group theory
In order to place the symmetry group results that we are interested in into some perspective, consider the following sets of entities: (i) all real numbers, (ii) all vectors in two dimensions and (iii) all vectors in three dimensions, and the product relations (a) addition, (b) division, (c) scalar product, (d) vector product. State which combinations form groups, naming the identity element in each case. State all those group postulates that are not obeyed in the other cases.

2.2** ***General determination of product relations
A general method for evaluating the product relations required in order to construct group multiplication tables is to consider a general point P on the object at (x,y,z) in the fixed reference frame. Under an operation G_s, P will move to new coordinates (x', y', z') and under G_r (x', y', z') will move to coordinates (x'', y'', z''). The operation that sends P directly to (x'', y'', z'') must therefore be equivalent to G_rG_s.

Use this technique to show that the product of rotations by π about two axes at right angles to each other is a π-rotation about a third axis at right angles to both of the original two.

*2.3 *Improper rotations*

Consider an arbitrary point P on an object, initially at the coordinates (x,y,z) in a fixed frame of reference. Use the definition σC_n given in the text for improper rotations to show that (i) $S_1 \equiv iC_2$, (ii) $S_2 \equiv iE$, (iii) $S_4 \equiv iC_4$. You will need to define the axes and planes carefully. It may also be shown that $S_6 \equiv iC_3$. The point is that all symmetry operations can thereby be divided into pure and improper rotations (counting E as a pure rotation by zero radians), and that the improper rotations can be obtained by employing the products iC_n provided suitable axes are considered. As the C_n operations are relatively easy to find for a given body, if it also has inversion symmetry the improper rotations may be found directly from the products iC_n, for each C_n.

*2.4 *Group multiplication tables*

Obtain the symmetry operations for a non-square planar rectangle, in three-dimensional space. Choose one operation in the group (other than the identity) and find its product relation with each of the group operations. Are the closure relation and the rearrangement theorem satisfied?

Modify the rectangle slightly and obtain the resulting subgroup of symmetry operations.

2.5 *Subgroup order, and cosets*

Suppose that a group E, G_2, G_3, G_4, ... G_g has as one of its subgroups the operations E, G_2, G_3.

Show that G_4E, G_4G_2, G_4G_3 lie outside of this subgroup. Let the above products be equivalent to G_4, G_5 and G_6.

By considering G_7,E, G_7G_2, G_7G_3 prove that the order g of the original group must be an integer multiple of the order of the sub-group, i.e. that any subgroup order must be an integer submultiple of the group order.

The set of elements G_4E, G_4G_2, G_4G_3 is called the *left coset* of the subgroup (with respect to G_4). The *right coset* is EG_4, G_2G_4, G_3G_4.

Show that such cosets cannot contain the identity E and therefore cannot themselves be subgroups.

2.6 *Invariant subgroups, and factor groups*

An invariant subgroup is a subgroup that contains only whole classes of the original group (such as E, C_3, C_3^2 with respect to the C_{3v} group).

Show that the left and right cosets of an invariant subgroup are identical.

A coset may be formed with respect to any of the original group operations, thus the subgroup itself is a coset.

(E, C_3, C_3^2) is an invariant subgroup of C_{3v}; (E, C_3, C_3^2) and $(\sigma_a, \sigma_b, \sigma_c)$ are the only cosets of this subgroup, call them $\mathscr{C}_1$ and $\mathscr{C}_2$. We define a product of cosets as the set of distinct elements obtained from the products of each element of coset–1 with each element of coset–2. With this definition show that $\mathscr{C}_1$ and $\mathscr{C}_2$ form a group. Which is the identity element of this 'group' of cosets?

The result is a general one for invariant subgroups; the group of cosets is known as the *factor group*.

2.7 ***Abelian Groups***

Show that each element of an Abelian group is in a class of its own.

2.8 ***Classes***

The full group of the square planar object (Figure 2.5) in which the two faces are considered to be distinct, contains sixteen symmetry operations. Obtain all sixteen and divide them into classes using the rules described in Chapter 2. Your group should contain S_4 and S_4^3 operations (with respect to the z-axis). Show that these belong to the same class by evaluating $\sigma^{-1} S_4 \sigma$, where the reflection plane σ contains the z-axis (for example the y-z plane).

2.9** ***Cubic symmetry

Describe the 48 symmetry operations of a cube, using the results of Problem 2.3 if necessary. Use the rules of thumb to divide the operations into ten classes.

Note that the groups of operations considered in Problems 2.4 and 2.8 are subgroups of the cubic group. In general one can construct a tree of subgroups as a high order of symmetry is reduced in various ways.

2.10 ***Permutation symmetry***

For a distribution of identical objects any permutation of the objects will leave the system apparently unaltered. It is conventional to write a permutation operation in the style

$$\begin{pmatrix} 1 & 2 & 3 & 4 \\ 3 & 2 & 1 & 4 \end{pmatrix}.$$

This operation takes the object at *position* (1) and moves it to *position* (3) etc. The 24 permutations of four objects are thus described by the 24 permutations of i, j, k, l in

$$\begin{pmatrix} 1 & 2 & 3 & 4 \\ \mathrm{i} & \mathrm{j} & \mathrm{k} & \mathrm{l} \end{pmatrix}.$$

(i) Write down the inverses of

$$\begin{pmatrix}1\,2\,3\,4\\3\,2\,1\,4\end{pmatrix} \quad \text{and} \quad \begin{pmatrix}1\,2\,3\,4\\2\,3\,4\,1\end{pmatrix}.$$

(ii) What is the result of

$$\begin{pmatrix}1\,2\,3\,4\\4\,3\,2\,1\end{pmatrix}\begin{pmatrix}1\,2\,3\,4\\3\,2\,1\,4\end{pmatrix}?$$

(iii) Show that cyclic permutations form a subgroup of the 24 operations.

Note that permutation symmetry plays a significant role in atomic theory, as the electrons in an atom are indistinguishable objects.

2.11 ***The S_3 group?***

Why can there be no group of operations containing only an improper rotation axis of odd order, whereas improper rotations about an even-order axis do form groups?

2.12 ***Point group generators***

All six operations of the C_{3v} point group can be obtained from, for example, the successive application of the two operations C_3 and σ_a. That is:

$$E \equiv \sigma_a^2; \quad C_3; \quad C_3^2 = C_3C_3; \quad \sigma_a; \quad \sigma_b \equiv \sigma_a C_3; \quad \sigma_c \equiv C_3\sigma_a.$$

Other pairs of operations can also be used, but no one operation alone. In contrast the operation C_6 can be used to generate all of the elements of the C_6 group: C_6^6, C_6, C_6^2, C_6^3, C_6^4, C_6^5. C_3,σ_a and C_6 respectively are called *generators* for the two point groups. Obtain a set of generators for the point group considered in Problem 2.4. Show that your set must contain three operations.

3

Matrix representations of groups

The mathematical development of symmetry groups exploits the technique of matrix methods. It is therefore appropriate, at the start of this section, to remind ourselves of the definition of matrix multiplication.

A matrix A is a rectangular array of elements.

$$A = \begin{bmatrix} A_{11} & A_{12} & A_{13} \ldots \\ A_{21} & A_{22} & \\ A_{31} & & \\ \cdot & & \\ \cdot & & \\ \cdot & & \end{bmatrix}.$$

The matrix product $A\ B = C$ is possible providing that the width of A is equal to the height of B. The resulting matrix C is as high as A and wide as B, and has elements,

$$C_{ij} = \sum_k A_{ik} B_{kj}.$$

As a visual method of calculating C_{ij} one takes the ith row of A and the jth column of B, and multiplies the first elements in each, adds the product of the second elements and so on. For example,

$$C_{53}=A_{51}B_{13}+A_{52}B_{23}+\cdots+A_{5n}B_{n3}.$$

$$\underbrace{\begin{bmatrix} & & \\ & & \\ A_{51} & A_{52}\cdots & A_{5n} \end{bmatrix}}_{\text{5th row of } A}\underbrace{\begin{bmatrix} B_{13} \\ B_{23} \\ \cdot \\ \cdot \\ \cdot \\ B_{n3} \end{bmatrix}}_{\text{3rd column of } B}$$

3.1 Matrix representations of operations

Five sets of matrices are given in Table 3.1, with six matrices in each set. Concentrating for the moment on set (1) consider the matrix product dc.

Table 3.1 C_{3v} matrix sets for discussion in Chapter 3. (By a_1 we are to understand the matrix a as presented in set (1), etc.)

(1) $a=\begin{bmatrix}1 & 0\\ 0 & 1\end{bmatrix}$ $b=\begin{bmatrix}-1/2 & -\sqrt{3}/2\\ \sqrt{3}/2 & -1/2\end{bmatrix}$ $c=\begin{bmatrix}-1/2 & \sqrt{3}/2\\ -\sqrt{3}/2 & -1/2\end{bmatrix}$

$d=\begin{bmatrix}-1/2 & -\sqrt{3}/2\\ -\sqrt{3}/2 & 1/2\end{bmatrix}$ $e=\begin{bmatrix}-1/2 & \sqrt{3}/2\\ \sqrt{3}/2 & 1/2\end{bmatrix}$ $f=\begin{bmatrix}1 & 0\\ 0 & -1\end{bmatrix}$

(2) $a=a_1$ $b=b_1$ $c=c_1$

$d=\begin{bmatrix}1/2 & -\sqrt{3}/2\\ -\sqrt{3}/2 & -1/2\end{bmatrix}$ $e=\begin{bmatrix}-1 & 0\\ 0 & 1\end{bmatrix}$ $f=\begin{bmatrix}1/2 & \sqrt{3}/2\\ \sqrt{3}/2 & -1/2\end{bmatrix}$

(3) $a=b=c=d=e=f=[1]$

(4) $a=b=c=[1]$
$d=e=f=[-1]$

(5) $a=\begin{bmatrix} a_1 & & 0\\ & & 0\\ 0 & 0 & a_3\end{bmatrix}=\begin{bmatrix}1 & 0 & 0\\ 0 & 1 & 0\\ 0 & 0 & 1\end{bmatrix}$ etc. to

$f=\begin{bmatrix} f_1 & & 0\\ & & 0\\ 0 & 0 & f_3\end{bmatrix}=\begin{bmatrix}1 & 0 & 0\\ 0 & -1 & 0\\ 0 & 0 & 1\end{bmatrix}$

$$dc = \begin{bmatrix} -1/2 & -\surd 3/2 \\ -\surd 3/2 & 1/2 \end{bmatrix} \begin{bmatrix} -1/2 & \surd 3/2 \\ -\surd 3/2 & -1/2 \end{bmatrix}$$

$$= \begin{bmatrix} (-1/2)(-1/2)+(-\surd 3/2)(-\surd 3/2) & (-1/2)(\surd 3/2)+(-\surd 3/2)(-1/2) \\ (-\surd 3/2)(-1/2)+(1/2)(-\surd 3/2) & (-\surd 3/2)(\surd 3/2)+(1/2)(-1/2) \end{bmatrix}$$

$$= \begin{bmatrix} 1 & 0 \\ 0 & -1 \end{bmatrix} = f$$

Indeed each matrix product can be shown to equal one of the other members of set (1). So we can construct a multiplication table (Table 3.2) just as in the case of symmetry operations.

Table 3.2 Matrix multiplication table for matrix set (1) of Table 3.1.

	a	*b*	*c*	*d*	*e*	*f*
a	*a*	*b*	*c*	*d*	*e*	*f*
b	*b*	*c*	*a*	*f*	*d*	*e*
c	*c*	*a*	*b*	*e*	*f*	*d*
d	*d*	*e*	(*f*)	*a*	*b*	*c*
e	*e*	*f*	*d*	*c*	*a*	*b*
f	*f*	*d*	*e*	*b*	*c*	*a*

Here the bracketed element is the one calculated above, *dc*. This table is formally identical to the symmetry operation Table 2.1; there is a one-to-one correspondence between the operations E to σ_c and the matrices *a* to *f*. Because of this correspondence we may say that the operations and the matrices multiply in the same way. Now look at set (2) of the matrices in Table 3.1: the same multiplication table is obtained. For set (3) the table consists completely of unit one-dimensional matrices; there is then a formal identity with Table 2.1 provided we make the many-to-one correspondence where we replace all of E to σ_c by the matrix [1]. Similarly replacing E, C_3 and C_3^2 by [1] but σ_a, σ_b, σ_c by [−1] we obtain a table formally identical to the set (4) multiplication table. Finally for set (5) there is again a one-to-one correspondence. To confirm this, note that by set (5) we understand each matrix to contain, in its top left-hand corner, the corresponding two-dimensional matrix of set (1), and in its bottom right-hand corner the one-dimensional matrix of set (3). Recognizing the matrix relationship

$$\begin{bmatrix} A & 0 \\ 0 & B \end{bmatrix} \begin{bmatrix} C & 0 \\ 0 & D \end{bmatrix} = \begin{bmatrix} AC & 0 \\ 0 & BD \end{bmatrix},$$

then the correspondence of the matrix set (5) with the group symmetry operation multiplication table follows directly from the fact that sets (1) and (3) give the correspondence.

The results clearly prompt the question: 'what is the physical correspondence between the operations and the matrices?' Suppose we take a pair of perpendicular unit vectors **i** and **j** in the hydrogen plane, originating at the centre of the triangle, as described in Figure 3.1, and ask: 'how do these vectors transform under the group operations?' Note that **i** and **j** are being

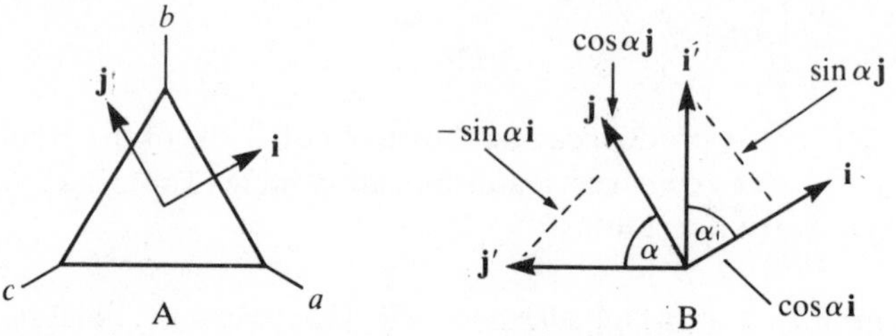

Figure 3.1 The effect of rotations on chosen vectors.

treated as if they were attached to the object rather than being fixed vectors in the external frame of reference. Firstly, under a rotation by angle α, by reference to Figure 3.1(B), **i** is rotated to a new unit vector **i**′ and **j** to **j**′. By considering projection onto the original basis vectors,

$$\mathbf{i}' = \cos\alpha\,\mathbf{i} + \sin\alpha\,\mathbf{j}, \qquad \mathbf{j}' = -\sin\alpha\,\mathbf{i} + \cos\alpha\,\mathbf{j}.$$

In vector notation,

$$[\mathbf{i}',\mathbf{j}'] = [\,\mathbf{i},\mathbf{j}\,]\begin{bmatrix}\cos\alpha & -\sin\alpha \\ \sin\alpha & \cos\alpha\end{bmatrix} = [\,\mathbf{i},\mathbf{j}\,][R(\alpha)].$$

For $\alpha=0$, the identity operation; $\alpha=2\pi/3$, corresponding to a C_3 operation; and for $\alpha=4\pi/3$, this rotational matrix takes the respective forms

$$\begin{bmatrix}1 & 0 \\ 0 & 1\end{bmatrix}, \quad \begin{bmatrix}-1/2 & -\sqrt{3}/2 \\ \sqrt{3}/2 & -1/2\end{bmatrix}, \quad \begin{bmatrix}-1/2 & \sqrt{3}/2 \\ -\sqrt{3}/2 & -1/2\end{bmatrix}.$$

These are just the matrices a, b and c of set (1). By the same token under the σ_c operation **i** remains unchanged and **j** changes direction; that is

$$[\mathbf{i}',\mathbf{j}'] = [\,\mathbf{i},\mathbf{j}\,]\begin{bmatrix}1 & 0 \\ 0 & -1\end{bmatrix}.$$

This matrix is just matrix f of set (1). Finally to understand σ_a and σ_b, using Figures 3.2(A), (B) respectively,

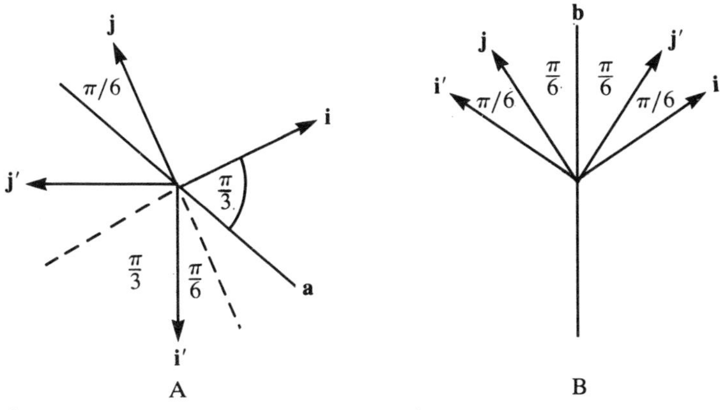

Figure 3.2 The effect of operations σ_a and σ_b on the vectors **i** and **j**.

$$[\mathbf{i}',\mathbf{j}']=[\,\mathbf{i},\mathbf{j}\,]\begin{bmatrix}-\cos\pi/3 & -\cos\pi/6\\ -\cos\pi/6 & \cos\pi/3\end{bmatrix}=[\,\mathbf{i},\mathbf{j}\,]\begin{bmatrix}-1/2 & -\sqrt{3}/2\\ -\sqrt{3}/2 & 1/2\end{bmatrix},$$

$$[\mathbf{i}',\mathbf{j}']=[\,\mathbf{i},\mathbf{j}\,]\begin{bmatrix}-\cos\pi/3 & \cos\pi/6\\ \cos\pi/6 & \cos\pi/3\end{bmatrix}=[\,\mathbf{i},\mathbf{j}\,]\begin{bmatrix}-1/2 & -\sqrt{3}/2\\ \sqrt{3}/2 & 1/2\end{bmatrix}.$$

These are the matrices *d* and *e* respectively. In conclusion then, matrix set (1) is the set of matrices that describe how the particular vectors **i**, **j** transform under the symmetry operations of the C_{3v} group. The matrices are each associated, in order, with E to σ_c. Note that only the matrices corresponding to the reflections relied on our choice of **i** and **j**. The rotation, in a plane, of any **i**, **j** orthogonal pair (having the same sense as above) is described by $R(\alpha)$. This gives us a clue as to the significance of matrix set (2). Matrices *a*, *b* and *c* again correspond to E, C_3 and C_3^2 matrices. To obtain the pair of **i** and **j** that transform like *d*, *e* and *f* we have only to note that we wish matrix *e* to correspond to the σ_b operation in order that the equivalence of the multiplication tables is upheld. Hence we require that **i** is transformed to −**i** and **j** stays the same under the σ_b operation. Figure 3.3 shows the choice of **i** and **j** for which this is true.

For this pair matrices *d* and *f* consistently show how **i**,**j** transform under operations σ_a and σ_c. Having decided that we would like to find unit vectors that transform in accord with our sets of matrices, set (3) is easy to consider. A unit vector **k** pointing towards the nitrogen atom, from the centre of the triangle, is undisturbed by any of the C_{3v} operations, i.e., we could write for each operation,

$$[\mathbf{k}']=[\mathbf{k}]\,[1].$$

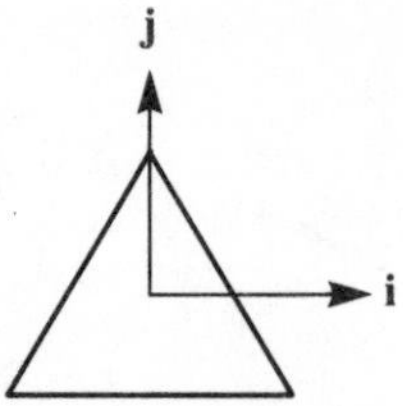

Figure 3.3 Unit vector basis for matrix set (2) of Table 3.1.

A little lateral thinking is required for set (4), which describes how the pseudo-vector $\mathbf{i} \wedge \mathbf{j}$ transforms. This is a pseudo-vector in that it is in the **k**-direction but it does not have a specified sense. Under inversion it is unaffected as ($\mathbf{i} \to -\mathbf{i}, \mathbf{j} \to -\mathbf{j}$) whereas the polar vector **k** changes sign. Either matrix set (1) or set (2) can be used to demonstrate that under the rotational group operations $\mathbf{i} \wedge \mathbf{j}$ is unchanged but that it reverses sign under reflections σ_a, σ_b or σ_c. The particular choice of **i** and **j** is not significant. For example, using set (1) and considering σ_a, the *d* matrix is obtained via

$$(\mathbf{i} \wedge \mathbf{j})' = (-\tfrac{1}{2}\,\mathbf{i} - \tfrac{\sqrt{3}}{2}\,\mathbf{j}) \wedge (-\tfrac{\sqrt{3}}{2}\,\mathbf{i} + \tfrac{1}{2}\,\mathbf{j})$$
$$= (-\tfrac{1}{4}\,\mathbf{i} \wedge \mathbf{j} + \tfrac{3}{4}\,\mathbf{j} \wedge \mathbf{i}) = -(\mathbf{i} \wedge \mathbf{j}) = (\mathbf{i} \wedge \mathbf{j})(-1).$$

Finally the vectors **i** and **j** chosen for set (1), and the vector **k**, transform together in accord with matrix set (5). There are many other matrix sets that could have been written down here, with the same properties as above. One further comment is appropriate in this context. If we had chosen a set of three orthogonal vectors **i**, **j**, **k** pointed arbitrarily with respect to the molecular symmetry directions, then once again we could describe their transformations under the C_{3v} operations by a set of six matrices. This set would again obey the group multiplication table. However, now there would in general be non-zero matrix elements in all positions of these six, three-dimensional matrices.

Matrix representations of a group, Γ

Any *set* of matrices obeying the same multiplication table as the *set* of operations of a symmetry group will therefore obey the group postulates. It is said to be a matrix representation of the group. We will label it Γ.

Matrix representation of a symmetry operation, $\Gamma(G_r)$

Each matrix within the set Γ is associated with a specific symmetry operation which it is said to represent. That matrix associated with the operation G_r is labelled $\Gamma(G_r)$; there are g of them in the set Γ, g being the order of the group. The correspondence of the multiplication tables of the operations and the matrices requires that:

$$\text{if } G_r G_s \equiv G_t, \text{ then } \Gamma(G_r)\Gamma(G_s) = \Gamma(G_t).$$

Basis of a representation

For each of the matrix representations Γ that we considered, there was a set of vectors, which we chose to be unit, orthogonal (orthonormal) vectors, such that the individual matrices $\Gamma(G_r)$ described how the vectors transformed under the operations G_r. Such vectors are said to form a basis for the representation Γ. We will come across different entities, other than vectors, that also form bases for matrix representations.

It was necessary to define how $\Gamma(G_r)$ was to be used; we did this, in effect, by inspection of our matrix sets. More generally, if the operation G_r, defined as in Chapter 2, transforms a set of basis entities $\mathbf{i}, \mathbf{j}, \mathbf{k}, \ldots$ into $\mathbf{i}', \mathbf{j}'$, $\mathbf{k}' \ldots$, then we must treat the basis set as a row vector:

$$[\mathbf{i}', \mathbf{j}', \mathbf{k}', \ldots] = [\mathbf{i}, \mathbf{j}, \mathbf{k}, \ldots]\ \Gamma(G_r).$$

Equivalent representations of a group

Whereas matrix sets (1), (3) and (4) had as their bases, distinct unit vectors, the same could not be said of sets (1) and (2). Indeed our choices of **i** and **j** were just two out of an infinite set obtained by orienting **i** in any direction in the hydrogen plane and then placing **j** perpendicular to it, also in the plane. The different pairs of vectors so obtained are linear combinations of each other. The sets of matrices obtained by these various choices are examples of equivalent matrix representations of the group.

Orthogonal matrices

Having said that different choices of **i**, **j** above are linear combinations of each other it is appropriate to ask whether any or only specific combinations are allowed. We should also like to relate the matrices $\Gamma(G_r)$ and $\Gamma_N(G_r)$ that represent some G_r with respect to two different sets of basis vectors $(\mathbf{i}, \mathbf{j})$ and $(\mathbf{i}_N, \mathbf{j}_N)$.

If we demand that any choice of vector pairs contain two orthogonal unit vectors, then

$$\begin{bmatrix} \mathbf{i} \\ \mathbf{j} \end{bmatrix} \cdot [\,\mathbf{i},\mathbf{j}\,] = \begin{bmatrix} 1 & 0 \\ 0 & 1 \end{bmatrix} = \begin{bmatrix} \mathbf{i}_N \\ \mathbf{j}_N \end{bmatrix} \cdot [\mathbf{i}_N, \mathbf{j}_N].$$

Suppose our two choices are related by a matrix N, describing the linear combination, i.e.

$$[\mathbf{i}_N, \mathbf{j}_N] = [\,\mathbf{i},\mathbf{j}\,]N \quad \text{and} \quad \begin{bmatrix} \mathbf{i}_N \\ \mathbf{j}_N \end{bmatrix} = \tilde{N} \begin{bmatrix} \mathbf{i} \\ \mathbf{j} \end{bmatrix},$$

where $\tilde{N}$ is the transpose of N.

$$\text{Thus} \begin{bmatrix} \mathbf{i}_N \\ \mathbf{j}_N \end{bmatrix} \cdot [\mathbf{i}_N, \mathbf{j}_N] = \tilde{N} \begin{bmatrix} \mathbf{i} \\ \mathbf{j} \end{bmatrix} \cdot [\,\mathbf{i},\mathbf{j}\,]N = \tilde{N} \begin{bmatrix} 1 & 0 \\ 0 & 1 \end{bmatrix} N,$$

$$\text{so} \quad \tilde{N}N = \begin{bmatrix} 1 & 0 \\ 0 & 1 \end{bmatrix} \quad \text{or } \tilde{N} = N^{-1}.$$

This argument can be applied to any number of mutually orthogonal real basis vectors. In general, a matrix N that transforms one basis set into an equivalent set must be an orthogonal matrix, meaning that it has the property, $\tilde{N} = N^{-1}$.

3.2 Similarity transformation of matrices

Having performed a transformation N on the basis vectors, the effect on the matrix representation of the operation G_r can be found by reference to Figure 3.4.

Suppose that Figure 3.4(A) demonstrates the effect of N in changing the choice of bases.

$$[\mathbf{i}_N, \mathbf{j}_N] = [\,\mathbf{i},\mathbf{j}\,]N. \tag{3.1}$$

Figure 3.4(B) demonstrates the effect of some symmetry operation G_r on the original vectors. The matrix representation of this operation $\Gamma(G_r)$ is by definition such that

$$[\mathbf{i}', \mathbf{j}'] = [\,\mathbf{i},\mathbf{j}\,]\Gamma(G_r). \tag{3.2}$$

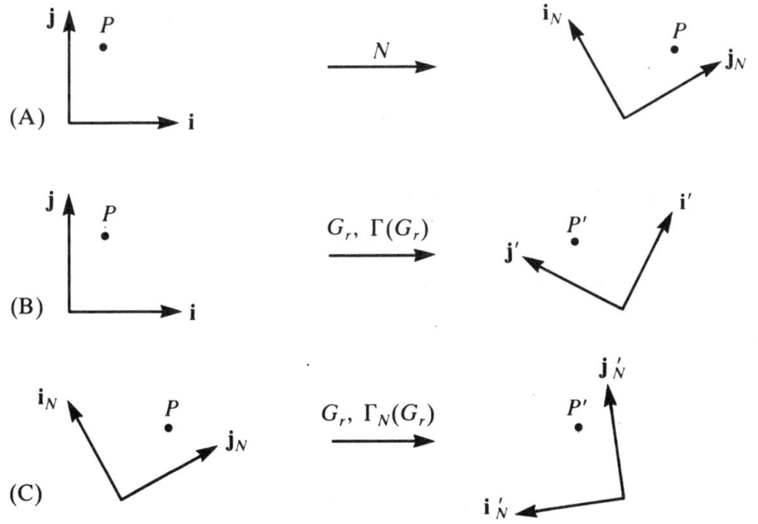

Figure 3.4 The effect of operations on axes and coordinates.

Similarly for the second choice of bases, Figure 3.4(C), we define $\Gamma_N(G_r)$ such that

$$[\mathbf{i}'_N, \mathbf{j}'_N] = [\mathbf{i}_N, \mathbf{j}_N]\ \Gamma_N(G_r). \tag{3.3}$$

In addition, consider a point P which moves with our symmetric body under the group operations. P is the same point on the body, regardless of which choice of basis vectors we made. Hence the vector from the origin to P is undisturbed by the choice of bases, i.e., if (x,y) are the coordinates of P, then

$$[\,\mathbf{i},\mathbf{j}\,]\begin{bmatrix} x \\ y \end{bmatrix} = [\mathbf{i}_N,\mathbf{j}_N]\begin{bmatrix} x_N \\ y_N \end{bmatrix}. \tag{3.4}$$

From (3.1) and (3.4), we obtain

$$[\,\mathbf{i},\mathbf{j}\,]\begin{bmatrix} x \\ y \end{bmatrix} = [\,\mathbf{i},\mathbf{j}\,]N\begin{bmatrix} x_N \\ y_N \end{bmatrix}, \qquad \text{so} \qquad \begin{bmatrix} x \\ y \end{bmatrix} = N\begin{bmatrix} x_N \\ y_N \end{bmatrix}.$$

Also after the operation G_r, P has moved to P' such that the vector OP' is independent of choice of basis:

$$[\mathbf{i}',\mathbf{j}']\begin{bmatrix} x' \\ y' \end{bmatrix} = [\mathbf{i}'_N,\mathbf{j}'_N]\begin{bmatrix} x'_N \\ y'_N \end{bmatrix}. \tag{3.5}$$

But as the basis vectors rotate with the point, under the operation G_r the coordinates of the point cannot alter:

$$\begin{bmatrix} x' \\ y' \end{bmatrix} = \begin{bmatrix} x \\ y \end{bmatrix}; \quad \begin{bmatrix} x'_N \\ y'_N \end{bmatrix} = \begin{bmatrix} x_N \\ y_N \end{bmatrix}. \tag{3.6}$$

Hence starting from (3.5), on the left-hand side:

$$[\mathbf{i}',\mathbf{j}'] \begin{bmatrix} x' \\ y' \end{bmatrix} = [\mathbf{i}, \mathbf{j}]\Gamma(G_r) \begin{bmatrix} x \\ y \end{bmatrix} = [\mathbf{i}, \mathbf{j}]\Gamma(G_r)N \begin{bmatrix} x_N \\ y_N \end{bmatrix}.$$

On the right-hand side,

$$[\mathbf{i}'_N,\mathbf{j}'_N] \begin{bmatrix} x'_N \\ y'_N \end{bmatrix} = [\mathbf{i}_N,\mathbf{j}_N]\Gamma_N(G_r) \begin{bmatrix} x_N \\ y_N \end{bmatrix} = [\,\mathbf{i},\mathbf{j}\,]N\Gamma_N(G_r) \begin{bmatrix} x_N \\ y_N \end{bmatrix}.$$

Comparing these two sides,

$$\Gamma(G_r)N = N\Gamma_N(G_r),$$

which is more commonly written

$$\Gamma_N(G_r) = N^{-1}\Gamma(G_r)N.$$

This formula enables us to obtain the second matrix representation in terms of the first. The combination of matrix multiplications is known as a similarity transformation: the original basis vectors have been transformed by N into physically similar ones. Γ_N and Γ are equivalent representations of the group.

Note here that the group operations we have considered leave both the magnitude of any vector and the angle between any two vectors undisturbed. Hence they will definitely transform any orthogonal set into a second orthogonal set—the $\Gamma(G_r)$ are orthogonal matrices.

The important conclusion here though is that the matrix representations for each operation in two equivalent representations of a group are related by a similarity transformation, $\Gamma_N \equiv N^{-1}\Gamma N$.

3.3 Reducible and irreducible representations of groups

Reducible representations

In the example given, we have noted that an arbitrary trio of vectors $(\mathbf{i},\mathbf{j},\mathbf{k})$ would lead to a three-dimensional matrix representation with few, if any, zero elements in any of the six matrices (one for each operation). However, there must be some transformation to different vectors $(\mathbf{i}_N, \mathbf{j}_N, \mathbf{k}_N)$, in particular to the set we chose in order to construct matrix set (5) of Table 3.1, which will transform the original six matrices to a set of six *block diagonal* matrices. (Matrix set (5) consisted of a set of non-zero blocks down the diagonal, and zeros elsewhere.) The original set of matrices form what is called a reducible representation of the group.

Representation in reduced form

Once the block diagonalization has been accomplished, by some similarity transformation, each matrix in the set being in identical blocked form; and given that no simpler blocking can be achieved by a different transformation, then the representation of the group is said to be in reduced form.

Matrix set (5) is of such form; there is no transformation that can reduce the 2×2 blocks of all the matrices into two 1×1 blocks. From a physical point of view any vector in the hydrogen plane must transform into a linear combination of itself and a second perpendicular vector, at least under the rotational C_3 and C_3^2 operations.

Irreducible representations of a group

Those matrix representations, such as sets (1) and (4), that cannot be reduced and are not in block diagonalized form (in that the matrices in them do not *all* have zero elements in the corners away from the diagonal) are said to be irreducible matrix representations of the group. The individual sets of blocks that appear in the reduced form of larger representations (matrix set (5)) are just these irreducible representations.

Inequivalent irreducible matrix representations of a group

Two irreducible representations that cannot be obtained from each other by applying a similarity transformation are said to be inequivalent to each other. Matrix sets (1), (3) and (4) are inequivalent irreducible matrix representations of the C_{3v} group. Herein we shall use the term *irrep.* as an abbreviation for such representations.

The first irrep. statement

The concept of inequivalent irreducible representations is extremely important in group theory. The reason is that whilst there may be an infinite number of matrix representations of a group, given that equivalent representations may exist:

> *the number of irreps of any group is equal to the number of classes in the group.*

For finite groups, with small numbers of distinct classes of operation, this means that we have only to deal with a few irreps. This important result is given here without proof.

The second irrep. statement

If l_j is the dimension of the jth irrep. (i.e. the number of basis entities for Γ^j) then, summing over all irreps,

$$\sum_j l_j^2 = g.$$

Again we give this result without proof; a justification for it, based on inspection of group properties, is considered in Problem 5.2.

For example, in the C_{3v} group there are six symmetry operations ($g=6$) and the only way to satisfy the two irrep. statements, with integer l_j values, is:

$$\sum_j l_j^2 = 1^2 + 1^2 + 2^2 = 6.$$

There are three irreps (three classes), two being one-dimensional matrix sets and one being two-dimensional. Thus our sets (3), (4), and (1) form a complete collection of inequivalent irreducible matrix representations of the C_{3v} group. They are given the labels shown in Table 3.3, where by A_1 for example we are to understand a set of six matrices $A_1(E)$ to $A_1(\sigma_c)$.

Table 3.3 The C_{3v} irreps and notations.

Matrix set	(3)	(4)	(1)
Group theory notation, Γ^j	A_1	A_2	E
Dimension, l_j	1	1	2
Basis vector(s)	**k**	$\mathbf{i} \wedge \mathbf{j}$	**i** and **j**

We use a slightly different form of letter E here in order to be quite definite in distinguishing this label for an irrep. from the label for the identity symmetry operation E.

An understanding of the distinctions between the various types of matrix representation is essential in the following chapters. For this reason a second example is now given.

Consider a pyramid-shaped object with a non-square rectangular base as pictured in Figure 3.5(A). The symmetry operations for the pyramid are as follows: E, C_2 about the vertical axis, σ_a and σ_b defined in Figure 3.5(B).

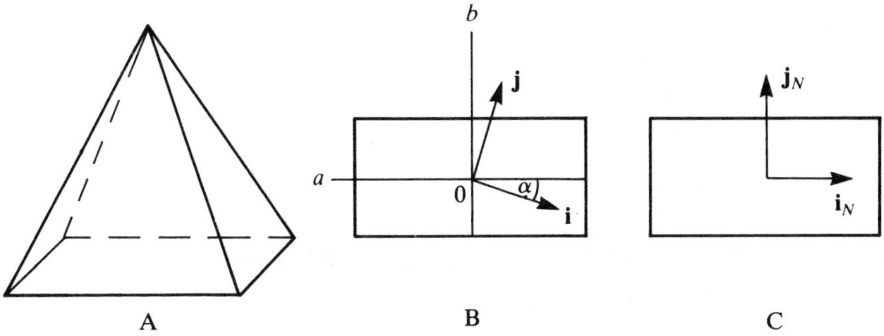

Figure 3.5 Symmetry operations for a pyramid with a non-square rectangular base.

Suppose that orthogonal unit vectors **i** and **j** are defined as shown, with **i** at an arbitrary angle α to $0a$. The matrix representations of the group operations, with **i** and **j** as basis are:

$$\Gamma(E)=\begin{bmatrix}1 & 0\\ 0 & 1\end{bmatrix},\quad \Gamma(C_2)=\begin{bmatrix}-1 & 0\\ 0 & -1\end{bmatrix},\quad \Gamma(\sigma_a)=\begin{bmatrix}\cos 2\alpha & \sin 2\alpha\\ \sin 2\alpha & -\cos 2\alpha\end{bmatrix},$$
$$\Gamma(\sigma_b)=\begin{bmatrix}-\cos 2\alpha & -\sin 2\alpha\\ -\sin 2\alpha & \cos 2\alpha\end{bmatrix}.$$

In contrast, for the vectors $\mathbf{i}_N$ and $\mathbf{j}_N$ of Figure 3.5(C) the representations of the four operations are:

$$\Gamma_N(E)=\begin{bmatrix}1 & 0\\ 0 & 1\end{bmatrix},\quad \Gamma_N(C_2)=\begin{bmatrix}-1 & 0\\ 0 & -1\end{bmatrix},\quad \Gamma_N(\sigma_a)=\begin{bmatrix}1 & 0\\ 0 & -1\end{bmatrix},$$
$$\Gamma_N(\sigma_b)=\begin{bmatrix}-1 & 0\\ 0 & -1\end{bmatrix}.$$

For the Γ set, some but *not all* of the matrices are completely diagonal. By changing the basis to $\mathbf{i}_N$, $\mathbf{j}_N$, however, complete diagonalization of all four matrices is achieved.

(i) The matrix set Γ is therefore *reducible*.
(ii) No further diagonalization can be achieved so that the set Γ_N is in *reduced form*.

(iii) Each matrix in Γ_N is in the form

$$\begin{bmatrix} \Gamma_1(G_r) & 0 \\ 0 & \Gamma_2(G_r) \end{bmatrix}.$$

The set Γ_1 for example is an *irreducible matrix representation of the group*:

$$\Gamma_1(E)=[1], \quad \Gamma_1(C_2)=[-1], \quad \Gamma_1(\sigma_a)=[1], \quad \Gamma_1(\sigma_b)=[-1].$$

(iv) The set Γ_2 can not be obtained from Γ_1 by any similarity transformation.

$$\Gamma_2(E)=[1], \quad \Gamma_2(C_2)=[-1], \quad \Gamma_2(\sigma_a)=[-1], \quad \Gamma_2(\sigma_b)=[1].$$

Γ_1 and Γ_2 are therefore *inequivalent irreducible matrix representations* (irreps) of the group.

There are in fact four irreps for the group of this pyramid, because there are four physically distinct symmetry operations and thus four classes. All of the irreps are one-dimensional in this example. In the C_{3v} group (NH_3) example we encountered a two-dimensional irrep. Irreps of higher dimension (matrix representations that cannot be brought into simpler block diagonalized form) will be encountered for the cubic and rotational groups in later chapters.

Problems

3.1** ***Matrix multiplication

Verify that the block pattern is preserved when multiplying similarly blocked matrices, using the following example

$$\begin{bmatrix} 2 & 2 & & & & \\ 1 & 4 & & & & \\ & & 6 & & & \\ & & & 1 & 3 & 5 \\ & & & 1 & 1 & 1 \\ & & & 1 & 2 & 1 \end{bmatrix}\begin{bmatrix} 2 & 1 & & & & \\ 3 & 2 & & & & \\ & & 2 & & & \\ & & & 4 & 1 & 2 \\ & & & 1 & 3 & 1 \\ & & & 2 & 1 & 1 \end{bmatrix}$$

3.2 Verify that matrices obey the associative law.

3.3 Prove that irreps of abelian groups are one-dimensional.

***3.4** A group consists of two symmetry elements, E and σ.

(i) Starting with a molecule of NH_3, show how a reduction in its symmetry, by the replacement of one or more atoms by a different isotope, can bring it into the above symmetry group, called C_{1h}.

(ii) Find a three-dimensional matrix representation for the group such that the two matrices are not both diagonal. Indicate the basis vectors you use.

(iii) Is your representation in reduced form? If not, how could it be put into such form?

(iv) How many inequivalent irreducible matrix representations of the group are there?

(v) Describe some possible basis vectors for irreducible representations.

3.5 ***The regular representation***

Show that the following four matrices represent a group, by constructing the group multiplication table and demonstrating that the group postulates are obeyed.

$$a=\begin{bmatrix}1&0&0&0\\0&1&0&0\\0&0&1&0\\0&0&0&1\end{bmatrix}\qquad b=\begin{bmatrix}0&1&0&0\\1&0&0&0\\0&0&0&1\\0&0&1&0\end{bmatrix}$$

$$c=\begin{bmatrix}0&0&1&0\\0&0&0&1\\1&0&0&0\\0&1&0&0\end{bmatrix}\qquad d=\begin{bmatrix}0&0&0&1\\0&0&1&0\\0&1&0&0\\1&0&0&0\end{bmatrix}$$

Compare the positions of a to d in the table with the positions of the unit elements in the matrices themselves. This is an example of a *regular representation* of a group. In general a regular representation of a symmetry group is constructed by reconstructing the original multiplication table so as to force the identity onto the major diagonal, as shown below:

$$\begin{array}{c|ccccc}
 & E & G_r & G_s & G_t & \cdots \\
\hline
E & E & & & & \\
G_r^{-1} & & E & & & \\
G_s^{-1} & & & E & & \\
\vdots & & & & \ddots &
\end{array}$$

The matrix representation of a particular operation in the group, $\Gamma(G_r)$, has units at positions corresponding to the presence of G_r in the table and zeros elsewhere.

3.6 Is the matrix representation of Problem 3.5 in reducible, reduced or irreducible form? How many classes are there in the group?

Hint: consider the two irrep. statements and the fact that every group must have a one-dimensional irrep.

***3.7** Obtain the multiplication table for the planar object pictured in Figure 3.6.

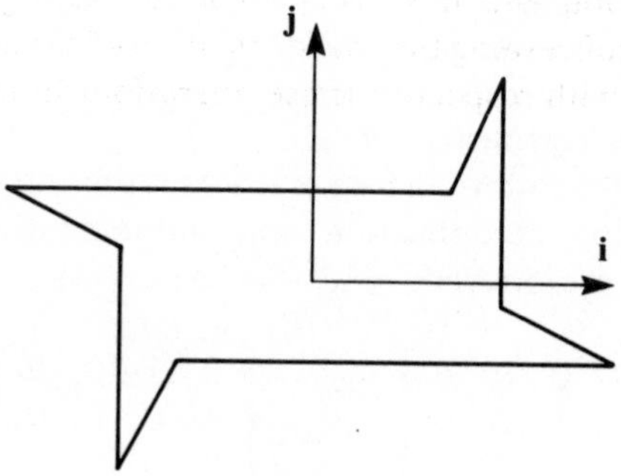

Figure 3.6

Consider the polar vectors **i**, **j**, **k** (pointing out of the page) and the axial vectors $\mathbf{i}\wedge\mathbf{j}$, $\mathbf{j}\wedge\mathbf{k}$ and $\mathbf{k}\wedge\mathbf{i}$. Obtain a set of six-dimensional matrices that describe the transformation properties of these vectors under the group operations. Which form is this matrix representation in? Select the irreducible matrix representations of the group.

3.8 ***Isomorphism***

Obtain the multiplication table for the group of water molecule symmetry operations (Fig. 3.7).

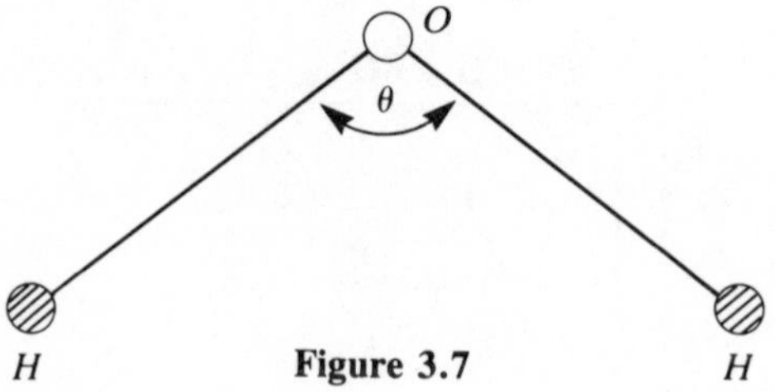

Figure 3.7

The symmetry group of Problem 3.5 is called C_{2h}, that of H_2O is C_{2v}. Both of these and the D_2 group (with elements E, C_{2x}, C_{2y}, C_{2z}) have the same multiplication table (if the group elements are written in appropriate orders). Such groups are said to be isomorphic. A group with this particular multiplication table is called the four-group.

Is the C_4 group with elements E, C_{4z}, C_{2z}, C_{4z}^3 isomorphic with C_{2h}, C_{2v} and D_2?

3.9 Consider the vectors **i** and **j**, centred at the point group origin, which form the basis of the *E* irrep. of the C_{3v} group. Reduce the symmetry and state the subgroup you obtain. Do **i** and **j** still form the basis of a representation and if so is it an irrep? Can you make a general statement concerning the transformation of a symmetry adapted basis of a group with respect to those operations in one of its subgroups?

4

Character tables

4.1 Group characters

Almost all the information we need in group theory is contained in a complete set of irreps for the group in question. However, if there are irreps of two or higher dimensionality in the set then there is a certain arbitrariness about which of the equivalent irreps to use. It would therefore be appropriate if we could find some property of irreps which defines each one uniquely but which is independent of the choice amongst equivalent irreps. We would then be in a position to employ this property to characterize the irreps, and the group.

The fact that the matrices in equivalent irreps are related by a similarity transformation gives us the property we are after. The traces (sum over the diagonal elements) of two matrices related in this way are equal; the proof is:

$$\Gamma_N(G_r)=N^{-1}\Gamma(G_r)N.$$

Hence,

$$\text{trace }[\Gamma_N(G_r)]=\sum_{ijk} N_{ij}^{-1}\Gamma(G_r)_{jk}\,N_{ki}$$

$$=\sum_{jk}\Gamma(G_r)_{jk}\,\{\sum_i N_{ki}N_{ij}^{-1}\}$$

(as the order in which we write the matrix *elements* is irrelevant)

$$=\sum_{jk}\Gamma(G_r)_{jk}\,\delta_{kj}$$

$$=\text{trace }[\Gamma(G_r)].$$

In group theory the trace of an irrep. of a particular operation is known as its *character*, $\chi^j(G_r)$;

$$\chi^j(G_r)=\sum_{k=1}^{l_j}\Gamma_{kk}^j(G_r).$$

So we have demonstrated that the characters of an irrep. of a group (one character for each G_r) are independent of the choice of bases. Furthermore, because the operations in a particular class are related by a similarity transformation,

$$G_t = G_s^{-1} G_r G_s \qquad \text{or} \qquad \Gamma(G_t) = \Gamma^{-1}(G_s)\,\Gamma(G_r)\,\Gamma(G_s),$$

then the characters for any two operations in the same class are equal. (Note that the *superscript j* refers to the jth irrep. Apart from l_j a *subscript j* refers above to a matrix element.)

Group character tables

We can now construct a table in which we note each class of a given group, each irrep., and the sets of characters for each irrep. From the first irrep. statement the table will be square. It is conventional also to include, in the notation for each class, the number g_r of operations in the class.

For the C_{3v} group then we have the *character table* shown as Table 4.1.†

Table 4.1 C_{3v} character table; values of $\chi^j(G_r)$.

$\Gamma^j \downarrow$	E	$2C_3$	3σ $\leftarrow G_r$
A_1	1	1	1
A_2	1	1	-1
E	2	(-1)	0

The bracketed element, for example, corresponds to $\chi^j(C_3)$, where $\Gamma^j = E$; it has been found in our case by taking matrix set (1) of Table 3.1 and summing over the diagonal elements of the particular matrix b that represented the operation C_3. Note that the trace of matrix c is also -1, as it should be because c represents C_3^2, which is in the same class as C_3. Similarly matrices d, e and f all have the same trace, 0. Obviously we could alternatively have used matrix set (2) in order to find this third row of characters in the table. We do indeed obtain the same results, as we must.

Just as the group multiplication table was a 'magic square' with many interesting properties, so also is its character table. The following general comments are usefully made, by inspection of the C_{3v} table (Table 4.1).

(i) The first column in the table contains the dimensions of the irreps, l_j. This is to be expected as each basis entity is unaltered by the identity operation; $\Gamma^j(E)$ is the unit diagonal matrix of dimension l_j, with trace therefore equal to l_j.

†Please note the use of the symbol *E* to label the two-dimensional irrep. This is used to avoid ambiguity with the use of E for the identity operation.

(ii) The first row of the characters in the table consists exclusively of ones. Every character table must contain such a row, which exhibits the fact that there is always at least one basis entity that transforms into itself under all the symmetry operations. In our example the vector **k** formed such a basis. Quite generally a spherically symmetric basis *function* centred at the point group centre will always form a basis for this, the *symmetric irrep*.

(iii) The table is square, as required; the number of irreps is the same as the number of classes.

(iv) ***First orthogonality theorem for characters.*** If we consider the characters $\chi^j(G_r)$ of a given irrep. Γ^j, to be the elements of a vector of dimension g, then the vectors so formed for each of the irreps are all of magnitude $g^{1/2}$ and are orthogonal to each other.

$$[\chi^j(G_1), \chi^j(G_2) \ldots \chi^j(G_g)] \begin{bmatrix} \chi^i(G_1) \\ \chi^i(G_2) \\ \vdots \\ \chi^i(G_g) \end{bmatrix} = g\delta_{ij}.$$

This property is called the first orthogonality theorem for characters, or the irreducibility criterion. It can be used, with $i=j$, to check whether a given representation is irreducible. For C_{3v} we have, as examples of this theorem,

$$[1,1,1,1,1,1] \begin{bmatrix} 2 \\ -1 \\ -1 \\ 0 \\ 0 \\ 0 \end{bmatrix} = [1,1,1,-1,-1,-1] \begin{bmatrix} 2 \\ -1 \\ -1 \\ 0 \\ 0 \\ 0 \end{bmatrix} = 0,$$

$$[2,-1,-1,0,0,0] \begin{bmatrix} 2 \\ -1 \\ -1 \\ 0 \\ 0 \\ 0 \end{bmatrix} = 6 = g.$$

In general, recognizing that we will come across complex characters the theorem is expressed as:

$$\sum_{r=1}^{g} \chi^j(G_r)^* \chi^i(G_r) = g\delta_{ij}.$$

(v) ***Second orthogonality theorem for characters.*** A second set of vectors can be constructed from the columns of the character table, with elements

$\chi^j(G_r)$ where G_r is fixed and j varied. Again these are seen to be orthogonal:

$$[\chi^1(G_r), \chi^2(G_r) \ldots] \begin{bmatrix} \chi^1(G_s) \\ \chi^2(G_s) \\ \vdots \end{bmatrix} = \frac{g}{g_r} \Delta_{sr}.$$

Δ_{sr} is unity if G_s and G_r are in the same class and zero otherwise. At this point it is also useful to note that the number of elements in any class (g_r indicates the order of the class containing G_r) is an integer submultiple of the order of the group; g/g_r is an integer. For C_{3v}, examples of the theorem are:

$$[1,1,-1] \begin{bmatrix} 1 \\ -1 \\ 0 \end{bmatrix} = [1,1,2] \begin{bmatrix} 1 \\ -1 \\ 0 \end{bmatrix} = 0,$$

$$[1,-1,0] \begin{bmatrix} 1 \\ -1 \\ 0 \end{bmatrix} = 2 = \frac{g}{g_\sigma}.$$

In general this second orthogonality theorem for characters is given by

$$\sum_j \chi^j(G_r)^* \, \chi^j(G_s) = \frac{g}{g_r} \Delta_{sr}.$$

A particular result of this theorem is the second irrep. statement: setting $G_s = G_r = E$, $g_r = 1$ and $\chi^j(G_r) = l_j$, we obtain $\Sigma_j \, l_j^2 = g$.

For almost all groups likely to be encountered in physics, the character table is available in one or another set of tabulations. However, it is both satisfying to know how they can be constructed and useful to know how to construct them if necessary. The above set of statements and theorems provides the rules for this construction.

In practice one does need to know, however, how to obtain the character for some representation, usually a reducible one, which has some specified basis. This can be considered to be the second major task of any group theory problem in physics, the first task having been to obtain the appropriate set of symmetry operations for the system being studied. Now, because only diagonal elements of any $\Gamma(G_r)$ appear in its character, consider a specific basis entity and ask, under a typical operation G_r in some class, 'How does this entity transform into itself: does it go into itself, minus itself, into one of the other basis entities, or into some linear combination of entities that contains a fraction f times the original basis of interest?' For these four cases the diagonal of $\Gamma(G_r)$ contains the value $+1$, -1, 0 or f respectively. Repeat this consideration for the same G_r but for each basis in the set and you will build up the trace of $\Gamma(G_r)$. The point is that you never need to worry about the fraction of other bases that appear on transforming the one of interest; non-diagonal elements of $\Gamma(G_r)$ are irrelevant in so far as the character table is concerned.

Obviously this procedure is all the more simple if one can choose as the original basis set entities that transform simply under the group operations.

Appendix A2 contains the character tables that will be used in this text. These include tables for the 32 crystal point groups, the full rotation and axial groups plus a few additional groups. A selection of the character tables are also reproduced below, partly for use in problems and in the following few chapters. Examples of systems with these symmetries are also given.

4.2 Selected character tables

This section consists of selected character tables and example systems. Details of notation, and a full set of tables, may be found in Appendices A1 and A2 respectively. The figures show simple, example systems of the associated symmetries.

Table 4.2

C_1	E
A	1

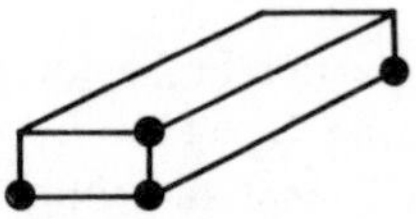

Table 4.3

C_2	E	C_2
A	1	1
B	1	−1

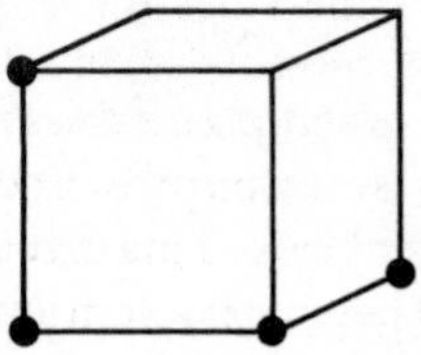

Table 4.4

C_{2v}	E	C_2	σ_v	σ'_v
A_1	1	1	1	1
A_2	1	1	−1	−1
B_1	1	−1	1	−1
B_2	1	−1	−1	1

Table 4.5

C_{3v}	E	$2C_3$	$3\sigma_v$
A_1	1	1	1
A_2	1	1	−1
E	2	−1	0

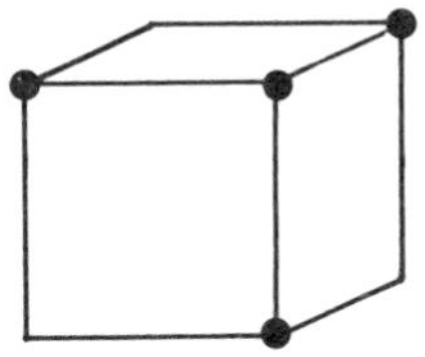

Table 4.6

C_{4v}	E	$2C_4$	C_2	$2\sigma_v$	$2\sigma_d$
A_1	1	1	1	1	1
A_2	1	1	1	−1	−1
B_1	1	−1	1	1	−1
B_2	1	−1	1	−1	1
E	2	0	−2	0	0

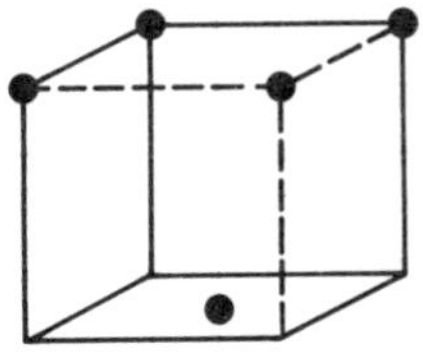

Table 4.7

C_{1h}	E	σ_h
A'	1	1
A''	1	−1

Table 4.8

C_{2h}	E	C_2	i	σ_h
A_g	1	1	1	1
B_g	1	−1	1	−1
A_u	1	1	−1	−1
B_u	1	−1	−1	1

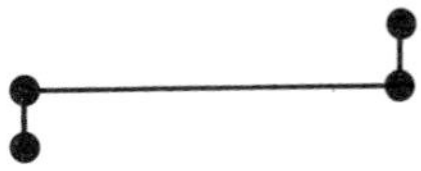

Table 4.9

D_2	E	C_{2z}	C_{2y}	C_{2x}
A_1	1	1	1	1
B_1	1	1	−1	−1
B_2	1	−1	1	−1
B_3	1	−1	−1	1

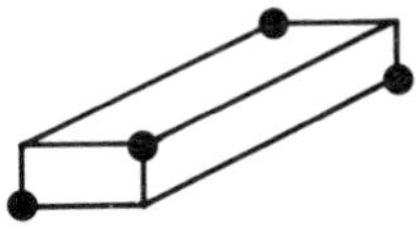

Table 4.10

D_3	E	$2C_3$	$3C'_2$
A_1	1	1	1
A_2	1	1	−1
E	2	−1	0

Table 4.11

D_4	E	$2C_4$	C_2	$2C'_2$	$2C''_2$
A_1	1	1	1	1	1
A_2	1	1	1	−1	−1
B_1	1	−1	1	1	−1
B_2	1	−1	1	−1	1
E	1	0	−2	0	0

Table 4.12

D_{2d}	E	$2S_4$	C_2	$2C'_2$	$2\sigma_d$
A_1	1	1	1	1	1
A_2	1	1	1	−1	−1
B_1	1	−1	1	1	−1
B_2	1	−1	1	−1	1
E	2	0	−2	0	0

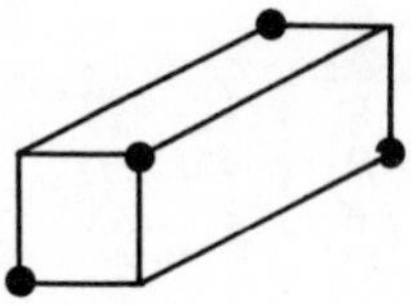

Table 4.13

D_{2h}	E	C_{2z}	C_{2y}	C_{2x}	i	σ_{xy}	σ_{xz}	σ_{yz}
A_g	1	1	1	1	1	1	1	1
B_{1g}	1	1	−1	−1	1	1	−1	−1
B_{2g}	1	−1	1	−1	1	−1	1	−1
B_{3g}	1	−1	−1	1	1	−1	−1	1
A_u	1	1	1	1	−1	−1	−1	−1
B_{1u}	1	1	−1	−1	−1	−1	1	1
B_{2u}	1	−1	1	−1	−1	1	−1	1
B_{3u}	1	−1	−1	1	−1	1	1	−1

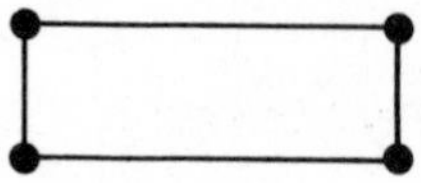

Table 4.14

D_{3h}	E	$2C_3$	$3C'_2$	σ_h	$2S_3$	$3\sigma_v$
A'_1	1	1	1	1	1	1
A'_2	1	1	−1	1	1	−1
E'	2	−1	0	2	−1	0
A''_1	1	1	1	−1	−1	−1
A''_2	1	1	−1	−1	−1	1
E''	2	−1	0	−2	1	0

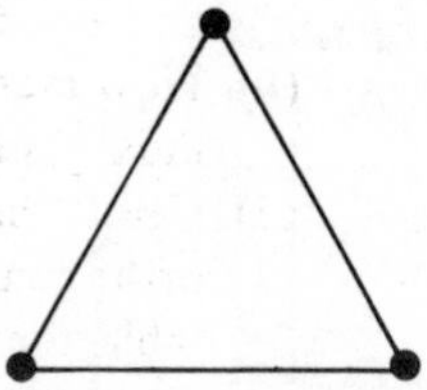

Table 4.15

S_2	E	i
A_g	1	1
A_u	1	−1

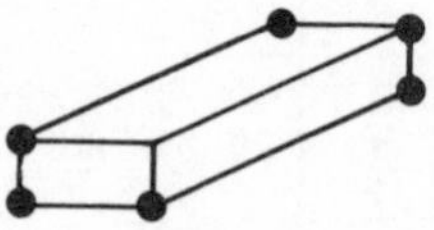

Table 4.16

T_d	E	$8C_3$	$6S_4$	$3C_2$	$6\sigma_d$
A_1	1	1	1	1	1
A_2	1	1	−1	1	−1
E	2	−1	0	2	0
T_1	3	0	1	−1	−1
T_2	3	0	−1	−1	1

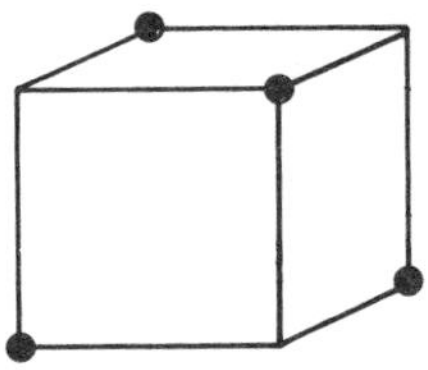

Table 4.17

O	E	$8C_3$	$6C_4$	$3C_2$	$6C'_2$
A_1	1	1	1	1	1
A_2	1	1	−1	1	−1
E	2	−1	0	2	0
T_1	3	0	1	−1	−1
T_2	3	0	−1	−1	1

Problems

4.1 Write down the character table for the C_{1h} symmetry group, using the results obtained in Problem 3.4.

***4.2** A molecule consists of four identical atoms at the corners of a non-square rectangle, and a fifth atom displaced normally above the centre of the rectangle.

(i) Write down the operations of the symmetry group of the molecule and divide them into classes.

(ii) How many inequivalent irreducible representations of the group must there be, and what are their dimensions?

(iii) Given that the characters are all real integers, use the character orthogonality relations in order to construct the entire character table for the group.

4.3 ***Complex characters***

The C_4 cyclic group consists of the physically distinct operations E, C_4, C_2 and C_4^3.

Show that $\chi^j(C_2)=[\chi^j(C_4)]^2$ for any irrep. of the group. Hence show that the characters cannot all be real integers in this example.

Use the fact that $E\equiv C_4^4$ in order to find the full (complex) character table.

*4.4 The point group of the methane molecule is T_d. Look up this group in the character tables and demonstrate that the character orthogonality relations are obeyed, by selecting a few example relations.

4.5 Use the first orthogonality relation for characters in order to obtain an expression for

$$\sum_{r=1}^{g} \chi^i(G_r).$$

Confirm by inspection that various character tables obey this relation.

4.6 ***Isomorphic groups***
Look up the character tables selected in Table 4.2. Write down all groups that are isomorphic with each other. C_{2h} is isomorphic with C_{2v} even though the characters are not identical in their tables as presented. Reorder the irreps and/or the group operations so as to make the tables identical.

For the three isomorphic groups of order 2 find the irreps which have as their bases the unit vectors **i**, **j** and **k**. (In so doing you will see why the groups are distinct even though they are isomorphic.)

5

Bases for representations

5.1 Basis vectors

We have noted that the set of orthogonal vectors, **i**, **j**, **k** (appropriately oriented) have transformation properties that are completely described by matrix representation set (5) of Table 3.1; **i**, **j**, **k**, are said to form a *basis* for this particular matrix set.

Similarly any arbitrarily oriented set, $\mathbf{i}_N$, $\mathbf{j}_N$, $\mathbf{k}_N$ forms a basis for some reducible representation of the group; one that is in general not in the block-diagonal form of set (5).

Symmetry-adapted basis vectors

The fact that set (5) is in reduced (block-diagonal) form tells us that of the three basis vectors the pair **i** and **j** transform into linear combinations of each other under every group operation. Also **k** transforms into a constant times itself (the constant being +1). Clearly **i**,**j** and **k** display the symmetry of the group in some manner. More especially the transformation properties of **i**,**j** are given by the irreducible matrix representation E (set (1)) that forms the top left-hand corner of set (5); and the transformation of **k** is given by the irreducible representation A_1 (set (3)).

$$G_r[\mathbf{i},\mathbf{j}] = [\mathbf{i},\mathbf{j}]\, E(G_r).$$

$$G_r[\mathbf{k}] = [\mathbf{k}]\, A_1(G_r).$$

Quite generally, any vectors or sets of vectors whose transformation properties are given by the *irreducible* representations are known as *symmetry-adapted basis vectors*. They may be said to transform 'in accord with' the irreducible representation for which they form a basis. The similarity transform (N) that converts some arbitrary matrix representation to reduced form is just that which transforms the original basis vectors into the symmetry-adapted vectors.

We will come across other bases for representations than unit vectors centred at the point group origin. For example, in dealing with the vibrations of atoms in molecules and solids we shall employ basis displacements, and in quantum-mechanical problems electronic wave functions form bases. Whatever entity (generalized vector) we are dealing with, though, we shall require that the effect of a symmetry operation upon the entity is well defined, and that it sends the entity into a linear combination of similar entities, namely the other bases of the same representation. We will also require a definition of a general scalar product so that we can exploit the results of orthonormality.

Basis displacements

The first problem we are going to tackle involves the vibrations of the hydrogen atoms of NH_3 in the plane of the hydrogen triangle. Figure 5.1 shows schematically the six *basis displacements* we will use in this problem. In principle we should consider twelve displacements. In addition to the six considered there are three hydrogen displacements normal to the plane of the paper, and three mutually orthogonal nitrogen displacements. The resulting problem is considered to be too complicated for demonstration purposes; the six hydrogen displacements pictured in Figure 5.1 do however form an ideal demonstration model. These six displacements (one for each degree of freedom in the hydrogen plane) allow any planar displacement of the molecule to be described. The full NH_3 problem is considered in Problem 5.6. Having mastered the planar motion model the reader should find that extension to the full problem is straightforward.

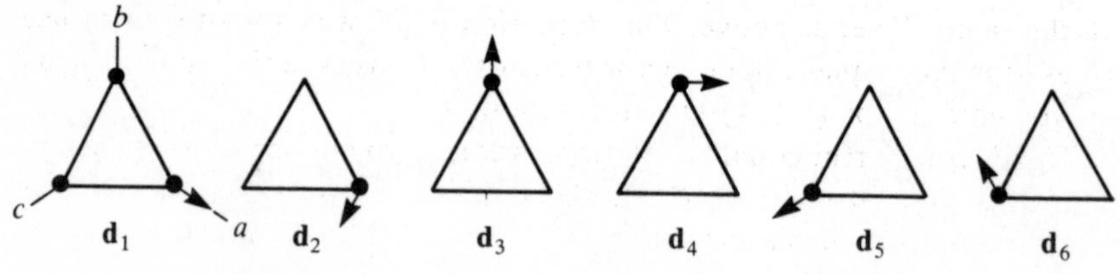

Figure 5.1 Definition of six basis displacements of NH_3. By the displacement $\mathbf{d}_1$ we understand that the atom that is in the σ_a plane of a fixed external reference frame is displaced unit distance radially outwards from the centre of the triangle. $\mathbf{d}_2 \rightarrow \mathbf{d}_6$ are defined similarly, in directions chosen to be radial and tangential with respect to the central point.

An arbitrary displacement of the configuration of all three atoms may be written:

$$\mathbf{disp} = \sum_{i=1}^{6} r_i\,\mathbf{d}_i.$$

Zero displacement is understood to be the equilibrium configuration of the atoms.

In order to define what we are to understand by the effect of a symmetry operation on a displacement, concentrate on the atom displaced along $\mathbf{d}_1$ by distance r_1. Under the C_3 operation this atom rotates anticlockwise to the top of the triangle. The displaced configuration ($r_1'\ \mathbf{d}_1'$) therefore looks like a displacement by r_1 in the $\mathbf{d}_3$ direction:

$$C_3[r_1\mathbf{d}_1] = r_1'\mathbf{d}_1' = r_1\mathbf{d}_3 \quad \text{or} \quad C_3[\mathbf{d}_1] = \mathbf{d}_1' = \mathbf{d}_3\,.$$

This gives us one column in the matrix representation for C_3.

$$\Gamma^d(C_3) = \begin{bmatrix} 0 & . & . & . \\ 0 & . & . & . \\ 1 & . & . & . \\ 0 & . & . & . \\ 0 & . & . & . \\ 0 & . & . & . \end{bmatrix},$$

where in general

$$G_r[\mathbf{d}_1 \ldots \mathbf{d}_6] = [\mathbf{d}_1' \ldots \mathbf{d}_6'] = [\mathbf{d}_1 \ldots \mathbf{d}_6]\ \Gamma^d[G_r].$$

With this definition the matrices $\Gamma^d(G_r)$ obey the same multiplication table as the group symmetry operations. (There was an alternative definition we could have used. We could have asked which **d** transformed *into* $\mathbf{d}_1$ under C_3; for such a definition the **d**'s would have transformed as column vectors

$$G_r \begin{bmatrix} \mathbf{d}_1 \\ \vdots \\ \mathbf{d}_6 \end{bmatrix} = \Gamma^d(G_r) \begin{bmatrix} \mathbf{d}_1 \\ \vdots \\ \mathbf{d}_6 \end{bmatrix},$$

with the same Γ^d-set as above. The definition made was a matter of choice such as to make displacement vector transformations take the same form as previous unit vector (**i**, **j**, **k**) transformations.)

To generate orthogonality relations we need to choose those displacements of the same atom to be in orthogonal directions ($\mathbf{d}_1.\mathbf{d}_2 = 0$); and, on the understanding that displacements of different atoms correspond to a completely different set of coordinates, to define any such pairs of displacements to be orthogonal, regardless of direction ($\mathbf{d}_1.\mathbf{d}_3 = 0$ by definition). Thus

$$(\Sigma_i\, r_i\,\mathbf{d}_i)\ .\ (\Sigma_j\, r_j\,\mathbf{d}_j) = \Sigma_i\, r_i\, r_j\, \delta_{ji}.$$

The equivalent definitions for basis functions are presented at the beginning of Chapter 9.

5.2 Reduction of a representation

The six basis displacements $\mathbf{d}_1 \ldots \mathbf{d}_6$ form a basis for a six-dimensional representation of the C_{3v} group. There is a set of six-dimensional matrices which show how $\mathbf{d}_1 \ldots \mathbf{d}_6$ transform under the six symmetry operations of the group. The set of matrices obeys the group multiplication table. We will use the notation Γ^d for this representation of the group and $\Gamma^d(E)$ to $\Gamma^d(\sigma_c)$ for the individual representations of each operation. From the definition of the effect of a group operation on $\mathbf{d}_i$ we have:

$$E\,[\mathbf{d}_1,\mathbf{d}_3,\mathbf{d}_3,\mathbf{d}_4,\mathbf{d}_5,\mathbf{d}_6]=[\mathbf{d}_1,\mathbf{d}_2,\mathbf{d}_3,\mathbf{d}_4,\mathbf{d}_5,\mathbf{d}_6]\ ,$$
$$C_3[\mathbf{d}_1,\mathbf{d}_3,\mathbf{d}_3,\mathbf{d}_4,\mathbf{d}_5,\mathbf{d}_6]=[\mathbf{d}_3,\mathbf{d}_4,\mathbf{d}_5,\mathbf{d}_6,\mathbf{d}_1,\mathbf{d}_2]\ , \quad \text{etc.}$$

Thus

$$\Gamma^d(E)=\begin{bmatrix} 1 & & & & & \\ & 1 & & 0 & & \\ & & 1 & & & \\ & & & 1 & & \\ & 0 & & & 1 & \\ & & & & & 1 \end{bmatrix}, \qquad \Gamma^d(C_3)=\begin{bmatrix} 0&0&0&0&1&0\\ 0&0&0&0&0&1\\ 1&0&0&0&0&0\\ 0&1&0&0&0&0\\ 0&0&1&0&0&0\\ 0&0&0&1&0&0 \end{bmatrix}, \qquad \text{etc.}$$

This Γ^d representation must be reducible. Firstly it was established in Chapter 4 that the irreducible matrix representations of the C_{3v} group are either one-dimensional (A_1 or A_2) or two-dimensional (E). Hence Γ^d cannot be an irrep. Secondly Γ^d is not in the same block diagonalized form, with one-dimensional and two-dimensional irreps down the diagonal, for each group operation, and so Γ^d is not in reduced form—it is *reducible*.

The next major task of group theory is to find the reduced (block diagonal) form of Γ^d. At first sight this requires us to find a transformation N that achieves the block-diagonalization,

$$N^{-1}\,\Gamma^d(G_r)\,N=\Gamma^D(G_r),$$

where $\Gamma^D\,(G_r)$ takes the same block diagonal form for each G_r. However, N is not a unique matrix, and in fact the blocked form of Γ^D can be determined without our first finding an N. N is not unique for two reasons. Firstly there is no uniquely specified order for the blocks in $\Gamma^D\,(G_r)$, providing the blocked form is the same for each G_r. Secondly, if one or more of the irreps appearing as the diagonal blocks is of dimension 2 or more, then these blocks are not uniquely specified, as there are many equivalent irreps in these cases.

The various choices of N correspond to different choices of *symmetry-adapted basis displacements*, these being sets of displacements that transform in accord with the irreps. For the six-dimensional problem described above there are six symmetry-adapted displacements, for which we introduce the notation $\mathbf{D}_1$, $\mathbf{D}_2 \ldots$, $\mathbf{D}_6$. This orthonormal set of displacements is related to $\mathbf{d}_1 \ldots, \mathbf{d}_6$ through a transformation N:

$$[\mathbf{D}_1, \mathbf{D}_2 \ldots, \mathbf{D}_6] = [\mathbf{d}_1, \mathbf{d}_2 \ldots, \mathbf{d}_6]\, N,$$

with $\mathbf{D}_i \,.\, \mathbf{D}_j = \delta_{ij}$.

The reduced form of the matrix representation can now be written:

$$\Gamma^D = N^{-1}\,\Gamma^d N = a_1 A_1 \oplus a_2 A_2 \oplus a_3\, E.$$

The equation and the symbol $\oplus$ means that, for each operation G_r, the reduced representation contains, as blocks on the diagonal, $A_1(G_r)$ a_1-times, $A_2(G_r)$ a_2-times and $E(G_r)$ a_3-times. If this result is to hold then, bearing in mind that a similarity transformation maintains the character of Γ, we must have the character relationship,

$$\chi^D(G_r) = \chi^d(G_r) = a_1 \chi^1(G_r) + a_2\, \chi^2(G_r) + a_3\, \chi^3(G_r).$$

$\chi^j(G_r)$ refers to the character for G_r, in the jth irrep. As there will be one such relation for each class of operation (for any two operations in the same class the relations are identical), and since the number of classes (n_c) equals the number of irreps, then we have n_c equations in the n_c variables a_j. The a_j are independent of G_r as the block-diagonalization must have the same form for each G_r. We expect then a unique solution for the a_j, also if they are to be meaningful they must all be positive integers (or zero), as they tell us the number of times a matrix appears in Γ^D. We need therefore only construct the $\chi^d(G_r)$ in order to obtain the form of the matrix reduction. To do so we exploit the statements made in Chapter 4.

(i) Under the operation E all displacements stay the same, so $\Gamma^d(E)$ must be the six-dimensional unit matrix. We have already noted this. Hence $\chi^d(E) = 6$.

(ii) Under C_3 all atoms change position, therefore each diagonal term of $\Gamma^d(C_3)$ must be zero, so that $\chi^d(C_3) = 0$. Again we have already discovered this in constructing $\Gamma^d(C_3)$.

(iii) Having found $\chi^d(C_3)$ for one member of the class of C_3 we do not need to consider other members, $\chi^d(C_3^2) = \chi^d(C_3)$.

(iv) Under a typical member of the class of σ, σ_a say, only one atom remains unmoved; only $\mathbf{d}_1$ and $\mathbf{d}_2$ transformations can contribute to $\chi^d(\sigma_a)$. Under σ_a, $\mathbf{d}_1$ is unaltered and therefore contributes $+1$ to $\chi^d(\sigma_a)$; $\mathbf{d}_2$ transforms to $-\mathbf{d}_2$ and contributes -1. In total then, $\chi^d(\sigma_a) = (0 + 1 - 1) = 0$.

It is useful in summary to present the character table for the irreps along with the $\chi^d(G_r)$.

Table 5.1 Characters for the reducible representation Γ^d of the C_{3v} group.

	E	$2C_3$	3σ	
A_1	1	1	1	
A_2	1	1	-1	$\chi^j(G_r)$
E	2	-1	0	
Γ^d	6	0	0	$\chi^d(G_r)$

We therefore require $\Gamma^D = a_1 A_1 \oplus a_2 A_2 \oplus a_3 E$ such that

$$\chi^D(E) = \chi^d(E) = 6 = a_1.1 + a_2.1 + a_3.2 ,$$
$$\chi^D(C_3) = \chi^d(C_3) = 0 = a_1.1 + a_2.1 + a_3.-1 ,$$
$$\chi^D(\sigma) = \chi^d(\sigma) = 0 = a_1.1 + a_2.-1 + a_3.0 .$$

The solution is $a_1 = 1$, $a_2 = 1$, $a_3 = 2$: that is,

$$\Gamma^D = A_1 \oplus A_2 \oplus 2E.$$

For each G_r,

$$\Gamma^D(G_r) = \begin{bmatrix} A_1(G_r) & & & 0 \\ & A_2(G_r) & & \\ & & E(G_r) & \\ 0 & & & E(G_r) \end{bmatrix}.$$

In future we shall also use the notation, $\Gamma^d \Rightarrow A_1 \oplus A_2 \oplus 2E$, meaning that on reduction (i.e. change to symmetry-adapted bases) Γ^d takes the block diagonal form $[A_1 \oplus A_2 \oplus 2E]$. As required, all of the a_j coefficients are positive integers (none happens to be zero in this example). If one does not obtain integers or zeroes, then either the characters $\chi^d(G_r)$ have been calculated incorrectly, or the solution of the simultaneous equations for a_j is in error. Group theory provides its own check in this respect. A second simple check that the reduction is reasonable is to confirm that the total dimension of Γ^D is identical to the initial number of displacements. This is in effect putting into words the relation,

$$\chi^D(E) = \chi^d(E) = a_1 l_1 + a_2 l_2 + a_3 l_3,$$

where l_j is the dimension of the jth irrep.

In many circumstances it is simple to derive the a_j by inspection of the character table. If this proves difficult one can turn to the simultaneous equations. However there proves to be a relation, derivable from orthogonality expressions (see Problem 5.1) that facilitates the calculation. This is the most important relation so far presented from the point of view of the day-to-day application of group theory. It is given here without proof.

The reduction formula

A reducible matrix representation Γ with characters $\chi(G_r)$ may be reduced to the block diagonal form $\Sigma_j a_j\Gamma^j$, where the irreps Γ^j have characters $\chi^j(G_r)$. The coefficients are,

$$\boxed{a_j=\frac{1}{g}\sum_{r=1}^{g}\chi^j(G_r)^*\ \chi(G_r).}$$

The * indicates the complex conjugate, complex characters will be met later.

For the C_{3v} example, $g=6$;

$$\begin{aligned}
&j=1, \quad \Gamma^j=A_1, \quad a_1=1/6\ (1.6+0)=1,\\
&j=2, \quad \Gamma^j=A_2, \quad a_2=1/6\ (1.6+0)=1,\\
&j=3, \quad \Gamma^j=E, \quad a_3=1/6\ (2.6+0)=2.
\end{aligned}$$

$$\Gamma^D=A_1\oplus A_2\oplus 2E.$$

5.3 Symmetry-adapted displacements

The above form of Γ^D tells us that six symmetry-adapted displacements can be constructed such that they will transform in accord with:

$$G_r[\mathbf{D}_1,\mathbf{D}_2]=[\mathbf{D}_1,\mathbf{D}_2\ldots]\begin{bmatrix} A_1(G_r) & & & 0\\ & A_2(G_r) & & \\ & & E(G_r) & \\ 0 & & & E^N(G_r)\end{bmatrix}.$$

This form can be decomposed to give,

$$\begin{aligned}
&G_r[\mathbf{D}_1]=[\mathbf{D}_1]\ A_1(G_r),\\
&G_r[\mathbf{D}_2]=[\mathbf{D}_2]\ A_2(G_r),\\
&G_r[\mathbf{D}_3,\mathbf{D}_4]=[\mathbf{D}_3,\mathbf{D}_4]\ E(G_r),\\
&G_r[\mathbf{D}_5,\mathbf{D}_6]=[\mathbf{D}_5,\mathbf{D}_6]\ E^N(G_r).
\end{aligned}$$

The following comments are appropriate:

(i) ***Belonging.*** $\mathbf{D}_1$ is some linear combination of $\mathbf{d}_1$ to $\mathbf{d}_6$, such that under the group operations it transforms in accord with A_1, namely into itself under each G_r. We say that $\mathbf{D}_1$ *belongs* to the irrep. A_1, for which it forms a basis.

(ii) Similarly $\mathbf{D}_2$ is a combination of the $\mathbf{d}$'s, which belongs to A_2. It transforms into itself under E, C_3, C_3^2 but minus itself under $\sigma_a,\sigma_b,\sigma_c$.

(iii) ***Partner vectors.*** $\mathbf{D}_3,\mathbf{D}_4$ transform into linear combinations of the pair.

They are called *partner vectors* and belong to different rows of the irrep. E.

(iv) For the second appearance that E has in the reduced representation an additional superscript N has been added. This is to indicate that the precise forms of $E(G_r)$ and $E^N(G_r)$ may differ; that is, they may be two different but equivalent irreducible representations.

(v) E^N can be obtained in form identical to E. There is a particular choice of $\mathbf{D}_5$ and $\mathbf{D}_6$ by comparison to the chosen $\mathbf{D}_3$ and $\mathbf{D}_4$, for which $E^N(G_r)=E(G_r)$ for all G_r. If this is achieved, then $\mathbf{D}_5$ is said to belong to the *same* row of E as $\mathbf{D}_3$, and $\mathbf{D}_6$ to the same row as $\mathbf{D}_4$. This is an important statement to have made, from the applications viewpoint to be considered in the following sections. $\mathbf{D}_5$ and $\mathbf{D}_6$ are of course partner vectors (they transform into each other) regardless of the form of E^N.

(vi) Not only is there a certain arbitrariness on the construction of $\mathbf{D}_5$, $\mathbf{D}_6$ by comparison to $\mathbf{D}_3$,$\mathbf{D}_4$, but even if we forced E^N and E to be equal, any of the equivalent forms of E could be used. They lead to different sets of vectors $\mathbf{D}_3 \rightarrow \mathbf{D}_6$.

(vii) It is often possible to construct several of the **D**'s by inspection. In the present case, from statement (i) above, we require a linear combination of the **d**'s that is invariant under all the symmetry operations; $\mathbf{D}_1$ needs to display the full symmetry of the triangle. The normalized vector $(\mathbf{d}_1+\mathbf{d}_3+\mathbf{d}_5)/\sqrt{3}$, shown below, has this property. Similarly $(\mathbf{d}_2+\mathbf{d}_4+\mathbf{d}_6)/\sqrt{3}$ demonstrates the properties required of $\mathbf{D}_2$. And by reference to the fact that we know that **i** and **j** unit vectors form a basis for E, there is a likelihood that the final two displacements schemed will do so too.

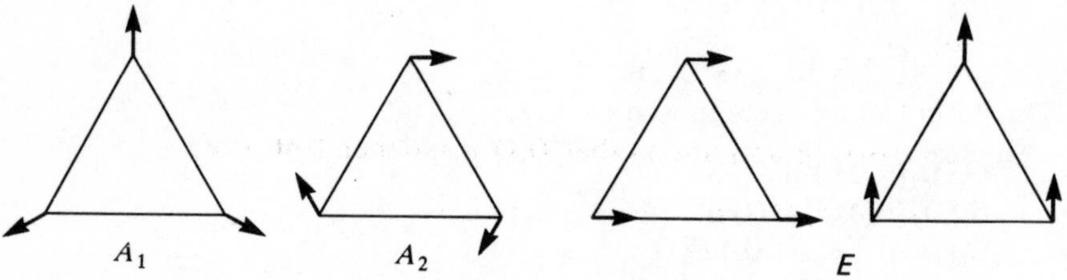

Figure 5.2 Likely symmetry-adapted basis displacements for the C_{3v} example.

Matrix formula for symmetry-adapted bases

As with the derivation of the a_j, there are formulae that may be used to obtain the **D**'s; we do not have to rely on inspection. Again we give the result without proof, although some justification is given in Problem 5.2:

$$\mathbf{D}^j_k(\mathbf{Q}) = c^j_{kl}\ (\mathbf{Q}) \sum_{r=1}^{g} \Gamma^j(G_r)^*_{kl}\ G_r\mathbf{Q}.$$

(i) $\mathbf{Q}$ is any chosen basis displacement. In practice one chooses a displacement that transforms simply, as the summation requires one to find the transformations of $\mathbf{Q}$ under every group operation.

(ii) $c^j_{kl}(\mathbf{Q})$ is a normalization constant that depends on the irreducible representation Γ^j for which a basis displacement is to be found. The constant also depends on the choices of $\mathbf{Q}$ and of the matrix element indices k,l.

(iii) Unless it turns out that the sum cancels to zero, which it often does, the formula generates one of the symmetry-adapted displacements belonging to the kth row of Γ^j. Different $\mathbf{Q}$ vectors produce different displacements belonging to the same row.

(iv) With this formula the $\mathbf{D}$'s so obtained are orthogonal to each other. On the right-hand side the c's are not predetermined, only the summation part can be calculated. The c's are to be calculated by normalizing the displacements. One thereby obtains sets of orthonormal symmetry-adapted displacements.

Character formula for symmetry-adapted displacements

Although we happen to have available for the C_{3v} group the full matrices of the irreps, in general only the characters of the irreps are known. The above equation is then not suitable for obtaining the $\mathbf{D}$'s. One can however obtain a suitable set of $\mathbf{D}$'s by using an adaptation of the $\mathbf{D}^j(\mathbf{Q})$ formula.

Note that:

$$\frac{\mathbf{D}^j_k(\mathbf{Q})}{c^j_{kk}(\mathbf{Q})} = \sum_r \Gamma^j(G_r)^*_{kk}\ G_r\mathbf{Q}.$$

Summing over k and noting that $G_r\mathbf{Q}$ is independent of k,

$$\sum_k D^j_k(\mathbf{Q})/c^j_{kk}\ (\mathbf{Q}) = \sum_r \left[\sum_k \Gamma^j(G_r)^*_{kk}\right] G_r\mathbf{Q},$$

which we rewrite as

$$\boxed{\mathbf{D}^j(\mathbf{Q}) \propto \sum_{r=1}^{g} \chi^j(G_r)^*\ G_r\mathbf{Q}.}$$

(i) This formula is much easier to use than the matrix formula; it ranks alongside the a_j formula in importance regarding the application of group theory to the problems considered in this text.

(ii) The resulting $\mathbf{D}$'s, although they transform in accord with a particular irrep., are no longer mutually orthogonal. Neither can they be said to belong to a particular row of the Γ^j. By this statement we mean that if Γ^j appears more than once, the $\mathbf{D}^j$'s that are generated can still be grouped into partner vectors (by inspection), but these groups will form bases for different (equivalent) irreducible representations.

The C_{3v} example will now be used to demonstrate (i) the differences in the results of the two expressions for $\mathbf{D}$ and (ii) the simplest technique for using these expressions.

5.4 Technique for constructing symmetry-adapted basis displacements

(a) Matrix formula

In using either the matrix formula or the character formula for the $\mathbf{D}^j$ one needs to select some $\mathbf{Q}$ displacements and obtain expressions for all $G_r\mathbf{Q}$, i.e., for the effect of every group operation on the chosen $\mathbf{Q}$. To this end it is useful to construct a table of the $G_r\mathbf{Q}$. This could prove extremely troublesome unless care is taken.

Firstly, choose for possible $\mathbf{Q}$ vectors only simple displacements that transform into trivial combinations of each other. Secondly, construct the table only for physically distinct $\mathbf{Q}$ vectors; not for every possible mutually orthogonal $\mathbf{Q}$. The resulting table will give sufficient information for use in the matrix formula and can easily be adapted for use in the character formula.

For the six planar displacements of the hydrogen atoms in NH_3 we would choose $\mathbf{d}_1$ and $\mathbf{d}_2$ as two possible $\mathbf{Q}$ vectors. $\mathbf{d}_3$ is physically indistinct from $\mathbf{d}_1$, and we do not bother to tabulate for it, etc. Note that we defined the $\mathbf{d}$ vectors in such a way that they did transform very simply. Had we chosen a set such as indicated in Figure 5.3, then more complicated transformations would have arisen. That is, whilst each of $\mathbf{d}_1$ to $\mathbf{d}_6$ transforms into one other specified $\mathbf{d}$ under any group operation, $\mathbf{e}_1$ to $\mathbf{e}_6$ transform into linear combinations of two $\mathbf{e}$ vectors under some of the operations, C_3 for example.

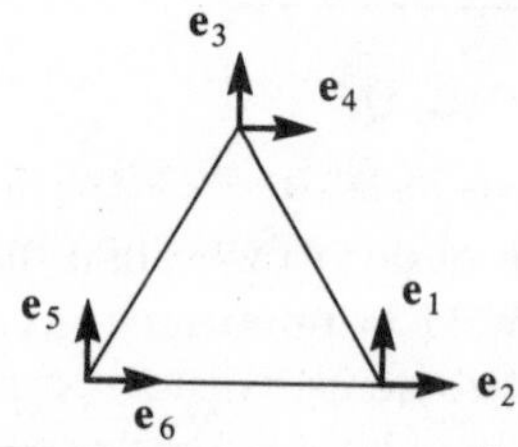

Figure 5.3 Alternative basis vectors for the C_{3v} problem.

Table 5.2 $G_r\mathbf{Q}$ values for physically distinct displacements of the C_{3v} example.

		E	C_3	C_3^2	σ_a	σ_b	σ_c	
$\mathbf{Q}$	$\mathbf{d}_1$	1	3	5	1	5	3	$G_r\mathbf{Q}$
	$\mathbf{d}_2$	2	4	6	−2	(−6)	−4	

Table 5.2 is the $G_r\mathbf{Q}$ table required for **d** vectors. With regard to the table note the following points.

(i) Each group operation must be included, not just the classes.
(ii) For simplicity of notation only the **d**-subscripts need be written in the table. Thus, for example, the bracketed element corresponds to the fact that under the operation σ_b the basis displacement $\mathbf{d}_1$ reflects to the position of $-\mathbf{d}_6$.

Table 5.3 Irrep. matrices for the C_{3v} group, taken from Table 3.1.

	$\Gamma^j(E)$	$\Gamma^j(C_3)$	$\Gamma^j(C_3^2)$	$\Gamma^j(\sigma_a)$	$\Gamma^j(\sigma_b)$	$\Gamma^j(\sigma_c)$
$j=1;\ \Gamma^j=A_1$	1	1	1	1	1	1
$j=2;\ \Gamma^j=A_2$	1	1	1	−1	−1	−1
$j=3;\ \Gamma^j=E$	$\begin{bmatrix}1 & 0\\ 0 & 1\end{bmatrix}$	$\begin{bmatrix}-\frac{1}{2} & -\frac{\sqrt{3}}{2}\\ \frac{\sqrt{3}}{2} & -\frac{1}{2}\end{bmatrix}$	$\begin{bmatrix}-\frac{1}{2} & -\frac{\sqrt{3}}{2}\\ -\frac{\sqrt{3}}{2} & -\frac{1}{2}\end{bmatrix}$	$\begin{bmatrix}-\frac{1}{2} & -\frac{\sqrt{3}}{2}\\ -\frac{\sqrt{3}}{2} & -\frac{1}{2}\end{bmatrix}$	$\begin{bmatrix}-\frac{1}{2} & \frac{\sqrt{3}}{2}\\ -\frac{\sqrt{3}}{2} & -\frac{1}{2}\end{bmatrix}$	$\begin{bmatrix}1 & 0\\ 0 & -1\end{bmatrix}$

The matrix sets for inequivalent C_{3v} irreducible representations are reproduced in Table 5.3. Combining these with the $G_r\mathbf{Q}$-table we finally obtain the following.

For $j=1,\ \Gamma^j=A_1,\ k=l=1$:

$$\mathbf{D}_1^1(\mathbf{d}_1) \propto [1.\mathbf{d}_1 + 1.\mathbf{d}_3 + 1.\mathbf{d}_5 + 1.\mathbf{d}_1 + (1.\mathbf{d}_5) + 1.\mathbf{d}_3];$$

therefore

$$\mathbf{D}_1^1(\mathbf{d}_1) = (\mathbf{d}_1 + \mathbf{d}_3 + \mathbf{d}_5)/\sqrt{3}.$$

$$\mathbf{D}_1^1(\mathbf{d}_2) \propto [1.\mathbf{d}_2 + 1.\mathbf{d}_4 + 1.\mathbf{d}_6 + 1.(-\mathbf{d}_2) + 1.(-\mathbf{d}_6) + 1.(-\mathbf{d}_4)];$$

therefore

$$\mathbf{D}_1^1(\mathbf{d}_2) = 0.$$

(i) The bracketed element in $\mathbf{D}_1^1(\mathbf{d}_1)$ is $\Gamma^1(\sigma_b)^*_{11}\ \sigma_b\mathbf{d}_1$.
(ii) For this one-dimensional representation obviously we are restricted to $k=l=1$.
(iii) The constant factor outside of the summation was ignored until the form of $\mathbf{D}^j(\mathbf{Q})$ had been obtained, it was then calculated by normalization.

(iv) Very often $\mathbf{D}^j(\mathbf{Q})$ turns out to be zero, as in the case of $\mathbf{D}^1_1(\mathbf{d}_2)$. However it is very easy to anticipate, by inspecting $G_r\mathbf{Q}$ and $\Gamma^j(G_r)$, when this will happen, thereby avoiding the necessity of writing down the summation. With experience it becomes possible to ignore immediately those $\mathbf{Q}$ vectors that will lead to null results.

(v) We knew that A_1 only appeared once in Γ^D. Hence there was only one $\mathbf{D}$ belonging to A_1. Thus having found a non-zero result for $\mathbf{D}^1_1(\mathbf{d})$ we need not have gone on to consider $\mathbf{D}^1_1(\mathbf{d}_2)$ at all.

(vi) Had we chosen other $\mathbf{Q}$ vectors, then $\mathbf{d}_3$, $\mathbf{d}_5$ or any combination of $\mathbf{d}_1$, $\mathbf{d}_3$, and $\mathbf{d}_5$ would all have given

$$\mathbf{D}^1_1(\mathbf{Q}) = (\mathbf{d}_1 + \mathbf{d}_3 + \mathbf{d}_5)/\sqrt{3};$$

$\mathbf{d}_2, \mathbf{d}_4, \mathbf{d}_6$ combinations would all have given a null result.

With these comments in mind we can now proceed to find the remaining five $\mathbf{D}$ expressions.

For $j=2$, $\Gamma^j = A_2$, $k=l=1$:

$$\mathbf{D}^2_1(\mathbf{d}_2) \propto [1.\mathbf{d}_2 + 1.\mathbf{d}_4 + 1.\mathbf{d}_6 + (-1).(-\mathbf{d}_2) + (-1).(-\mathbf{d}_6) + (-1).(-\mathbf{d}_4)],$$

therefore

$$\mathbf{D}^2_1(\mathbf{d}_2) = (\mathbf{d}_2 + \mathbf{d}_4 + \mathbf{d}_6)/\sqrt{3}$$

For $j=3$, $\Gamma^j = E$

and $k=l=1$: $\mathbf{D}^3_1(\mathbf{d}_1) \propto [1.\mathbf{d}_1 + -\tfrac{1}{2}\mathbf{d}_3 + -\tfrac{1}{2}\mathbf{d}_5 + -\tfrac{1}{2}\mathbf{d}_1 + -\tfrac{1}{2}\mathbf{d}_5 + 1.\mathbf{d}_3]$,

$$\mathbf{D}^3_1(\mathbf{d}_1) = (\mathbf{d}_1 + \mathbf{d}_3 - 2\mathbf{d}_5)/\sqrt{6}.$$

For $k=l=1$: $\mathbf{D}^3_1(\mathbf{d}_2) \propto [1.\mathbf{d}_2 + -\tfrac{1}{2}\mathbf{d}_4 + -\tfrac{1}{2}\mathbf{d}_6 + -\tfrac{1}{2}\mathbf{d}_2 + -\tfrac{1}{2}(-\mathbf{d}_6) + 1.(-\mathbf{d}_4)]$,

$$\mathbf{D}^3_1(\mathbf{d}_2) = (\mathbf{d}_2 - \mathbf{d}_4)/\sqrt{2}.$$

For $k=l=2$: $\mathbf{D}^3_2(\mathbf{d}_1) = (\mathbf{d}_3 - \mathbf{d}_1)/\sqrt{2}$.

For $k=l=2$: $\mathbf{D}^3_2(\mathbf{d}_2) = (\mathbf{d}_2 + \mathbf{d}_4 - 2\mathbf{d}_6)/\sqrt{6}$.

For $k=1$, $l=2$: $\mathbf{D}^3_1(\mathbf{d}_1) = (\mathbf{d}_1 + \mathbf{d}_3 - 2\mathbf{d}_5)/\sqrt{6}$.

In connection with the displacements belonging to E:

(i) The formula generates four mutually orthogonal displacements, labelled here $\mathbf{D}^3_1(\mathbf{d}_1)$, $\mathbf{D}^3_1(\mathbf{d}_2)$, $\mathbf{D}^3_2(\mathbf{d}_1)$, $\mathbf{D}^3_2(\mathbf{d}_2)$.

(ii) Having obtained the above four displacements we had no need to go further. Using other matrix elements or other $\mathbf{d}$ vectors for $\mathbf{Q}$ will give the same results or null results. One example for $k=1$, $l=2$ is included above to demonstrate that it leads to exactly the result obtained using $k=l=1$. Thus $\mathbf{D}^j$ is independent of the choice of l.

(iii) The four displacements do fall into two pairs:
$\mathbf{D}^3_1(\mathbf{d}_1)$ and $\mathbf{D}^3_2(\mathbf{d}_1)$ are partner vectors, as are $\mathbf{D}^3_1(\mathbf{d}_2)$ and $\mathbf{D}^3_2(\mathbf{d}_2)$.

(iv) The manner in which $\mathbf{D}_1^3(\mathbf{d}_1)$ and $\mathbf{D}_2^3(\mathbf{d}_1)$ transform is identical to the manner in which $\mathbf{D}_1^3(\mathbf{d}_2)$ and $\mathbf{D}_2^3(\mathbf{d}_2)$ transform. Hence $\mathbf{D}_1^3(\mathbf{d}_1)$ and $\mathbf{D}_1^3(\mathbf{d}_2)$ can both be said to belong to the same row of E—the $k=1$ row, etc.

(v) There was a choice in overall sign in formulating each of the normalized **D** vectors. The signs chosen here were such that comment (iv) followed. As we shall discover when we come to use **D** vectors to physical ends this is not a point that need worry us.

(vi) Because the irrep. *E is two-dimensional* there is an arbitrariness in the choice amongst the equivalent irreducible representations that we use for the Γ^j. For each different choice we would obtain different **D** vectors. For example, were we to have used matrix set (2), then we would find

$$\mathbf{D}_1^3(\mathbf{d}_1) \propto (\mathbf{d}_5 - \mathbf{d}_1), \quad \mathbf{D}_2^3(\mathbf{d}_1) \propto (\mathbf{d}_5 + \mathbf{d}_1 - 2\mathbf{d}_3), \quad \text{etc.}$$

That is, we would generate linear combinations of the partner vectors found above. This is consistent with our understanding of equivalent representations.

(vii) Choosing $\mathbf{Q} = \mathbf{d}_3$ one obtains:

$$\mathbf{D}_1^3(\mathbf{d}_3) = (\mathbf{d}_3 + \mathbf{d}_5 - 2\mathbf{d}_1)/\sqrt{6}, \quad \mathbf{D}_2^3(\mathbf{d}_3) = (\mathbf{d}_5 - \mathbf{d}_3)/\sqrt{2}.$$

These are again linear combinations of $\mathbf{D}_1^3(\mathbf{d}_1)$ and $\mathbf{D}_2^3(\mathbf{d}_1)$. They are examples of specific combinations that transform in exactly the same way as $\mathbf{D}_1^3(\mathbf{d}_1)$ and $\mathbf{D}_2^3(\mathbf{d}_1)$; that is, they belong to the same rows of the same irreps.

No extra information is obtained by running through further **d** vectors once the four orthogonal **D** vectors have been found.

(viii) Because the irrep. *E occurs twice* in Γ^D there is a further arbitrariness in the **D** vectors we find. Suppose we had considered $\mathbf{D}_1^3(a_1\mathbf{d}_1 + a_2\mathbf{d}_2)$. This would have generated a symmetry adapted displacement of the form $(b_1\mathbf{D}_1^3(\mathbf{d}_1) + b_2\mathbf{D}_1^3(\mathbf{d}_2))$. The same combination of second-row vectors would have been generated. This simply demonstrates the general result that if we take some linear combination of vectors belonging to row 1 of some irrep., and the same combination of vectors belonging to row 2, etc., then the resulting combinations will transform in exactly the same way as the original vectors. They also belong to rows 1, 2, . . . of the irrep. Formally this is expressed by the following argument:

$$G_r[\mathbf{D}_1^j(1), \mathbf{D}_2^j(1) \ldots] = [\mathbf{D}_1^j(1), \mathbf{D}_2^j(1), \ldots]\, \Gamma^j(G_r),$$

$$G_r[\mathbf{D}_1^j(2), \mathbf{D}_2^j(2) \ldots] = [\mathbf{D}_1^j(2), \mathbf{D}_2^j(2), \ldots]\, \Gamma^j(G_r), \text{ etc.,}$$

then

$$G_r[\sum_n \mathbf{D}_1^j(n)\,\gamma_n, \sum_n \mathbf{D}_2^j(n)\,\gamma_n, \ldots] = [\sum_n \mathbf{D}_1^j(n)\,\gamma_n, \sum_n \mathbf{D}_2^j(n)\,\gamma_n, \ldots]\, \Gamma^j(G_r).$$

This result will prove to be very important when we come to study the normal modes of vibration of molecules.

The scheme of symmetry-adapted displacements obtained above is shown in Figure 5.4. As we anticipated we find that the A_1 displacement is an expansion of the triangle and the A_2 displacement is a rotation. The E

A_1 symmetry. $\mathbf{D}_1=\mathbf{D}_1^1(\mathbf{d}_1)=(\mathbf{d}_1+\mathbf{d}_3+\mathbf{d}_5)/\sqrt{3}$.

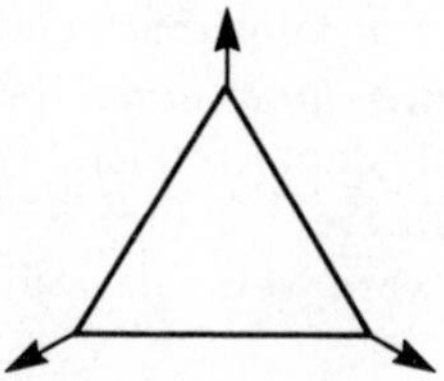

A_2 symmetry. $\mathbf{D}_2=\mathbf{D}_1^2(\mathbf{d}_2)=(\mathbf{d}_2+\mathbf{d}_4+\mathbf{d}_6)/\sqrt{3}$.

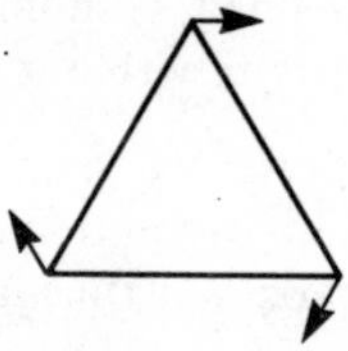

E symmetry. $\mathbf{D}_3=\mathbf{D}_1^3(\mathbf{d}_1)=(\mathbf{d}_1+\mathbf{d}_3-2\mathbf{d}_5)/\sqrt{6}$
$\mathbf{D}_4=\mathbf{D}_2^3(\mathbf{d}_1)=(\mathbf{d}_3-\mathbf{d}_1)/\sqrt{2}$ } Partner vectors

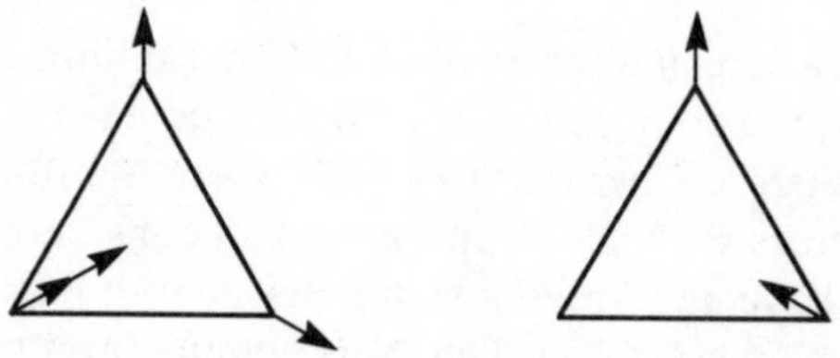

E symmetry. $\mathbf{D}_5=\mathbf{D}_1^3(\mathbf{d}_2)=(\mathbf{d}_2-\mathbf{d}_4)/\sqrt{2}$
$\mathbf{D}_6=\mathbf{D}_2^3(\mathbf{d}_2)=(\mathbf{d}_2+\mathbf{d}_4-2\mathbf{d}_6)/\sqrt{6}$ } Partner vectors

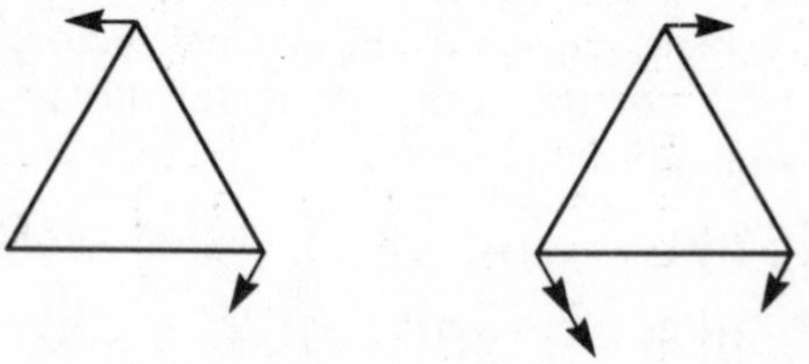

Figure 5.4 Typical set of symmetry adapted displacements for the C_{3v} problem.

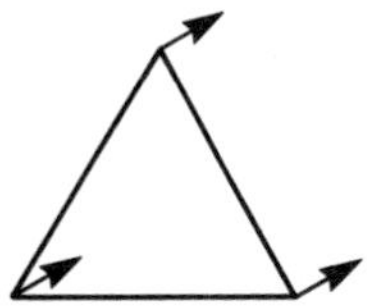

Figure 5.5 The linear combination of $\mathbf{D}_3$ and $\mathbf{D}_5$ above, that corresponds to a translation.

displacements, however, do not look as anticipated. However, we have noted in (vii) above that any linear combination of **D** vectors belonging to the same row of the same irrep. is also a basis (belonging to the same row). Looking at $\mathbf{D}_3$ and $\mathbf{D}_5$ it is immediately obvious that they both have the same reflection symmetry with respect to the σ_c plane, and that $\mathbf{D}_4$ and $\mathbf{D}_6$ are antisymmetric with respect to this same plane. Furthermore it is obvious that a linear combination, $(\mathbf{D}_3-\mathbf{D}_5)/\sqrt{2}$ to be precise, describes a displacement of each atom in the direction associated with the σ_c plane (Figure 5.5). The σ_c plane comes in here in a preferred manner because the irreducible representation matrix set (1) which we used to obtain the **D** vector was itself obtained with **i** and **j** vectors parallel and normal to the σ_c plane. In order to generate the **D** vectors given in Figure 5.1, we would have had to choose matrix set (2), which was defined with σ_b as the preferred plane, and use for **Q** a specific combination such as $(2\mathbf{d}_1-\mathbf{d}_2)/\sqrt{5}$; this is a displacement in the 'horizontal' direction.

(b) Technique for constructing D vectors: character formula

As this is the method one tends to use in practice we will present here all the tables to be used in the construction of **D** vectors, and the procedures.

$$\mathbf{D}^j \propto \sum_r \chi^j(G_r)^* \, G_r\mathbf{Q}.$$

Table 5.4 The tabulation required to assist in constructing the symmetry adapted displacements in the C_{3v} example.

		E	C_3	C_3^2	σ_a	σ_b	σ_c	
Q	$\mathbf{d}_1$	1	3	5	1	5	3	$G_r\mathbf{Q}$
	$\mathbf{d}_2$	2	4	6	−2	−6	−4	
Γ^j	A_1	1	1		1			$\chi^j(G_r)$
	A_2	1	1		−1			
	E	2	−1		0			

For $j=1$, $\Gamma^j=A_1$: $\mathbf{D}^1(\mathbf{d}_1)=(\mathbf{d}_1+\mathbf{d}_3+\mathbf{d}_5)/\sqrt{3}=\mathbf{D}_1$.

For $j=2$, $\Gamma^j=A_2$: $\mathbf{D}^2(\mathbf{d}_2)=(\mathbf{d}_2+\mathbf{d}_4+\mathbf{d}_6)/\sqrt{3}=\mathbf{D}_2$.

(For the one-dimensional representations the procedure is identical to that for the matrix formula for $\mathbf{D}^j_{k\cdot}$.)

For $j=3$, $\Gamma^j=E$:

$$\mathbf{D}^3(\mathbf{d}_1)\propto 2.\mathbf{d}_1+-1.(\mathbf{d}_3+\mathbf{d}_5)+0.(\mathbf{d}_1+\mathbf{d}_5+\mathbf{d}_3)=(2\mathbf{d}_1-\mathbf{d}_3-\mathbf{d}_5)/\sqrt{6}.$$

$$\mathbf{D}^3(\mathbf{d}_2)=(2\mathbf{d}_2-\mathbf{d}_4-\mathbf{d}_6)/\sqrt{6}.$$

At this point we run out of physically distinct $\mathbf{Q}$ values to place in the formula, but we know that there are four $\mathbf{D}$ vectors belonging to E. Now we turn to other $\mathbf{d}$ vectors that are physically similar to $\mathbf{d}_1$ and $\mathbf{d}_2$. We could go through the same procedure as above but it is simpler to note that, because of the physical equivalences, $G_r\mathbf{d}_3$ is obtained from $G_r\mathbf{d}_1$ by the cyclic permutation $\mathbf{d}_1\rightarrow\mathbf{d}_3\rightarrow\mathbf{d}_5$, etc. Hence the $\mathbf{D}^3$ may be formed by such permutation:

$$\mathbf{D}^3(\mathbf{d}_3)=(2\mathbf{d}_3-\mathbf{d}_5-\mathbf{d}_1)/\sqrt{6},$$

$$\mathbf{D}^3(\mathbf{d}_4)=(2\mathbf{d}_4-\mathbf{d}_6-\mathbf{d}_2)/\sqrt{6}.$$

(i) As we have already stated, the $\mathbf{D}$ vectors obtained from the character formula do not belong to specific rows of the irreps. Furthermore the partner vectors, those that transform into linear combinations of each other $\mathbf{D}^3(\mathbf{d}_1)$ and $\mathbf{D}^3(\mathbf{d}_3)$ above for example, are not orthogonal.

(ii) It is possible, if necessary, to construct orthogonal $\mathbf{D}$ vectors using the Gram–Schmidt orthogonalization procedure. For example construct $\mathbf{D}^3=(a\mathbf{D}^3(\mathbf{d}_1)+b\mathbf{D}^3(\mathbf{d}_3))$ such that $\mathbf{D}^3.\ \mathbf{D}^3(\mathbf{d}_1)=0$. These latter two vectors now form one choice of orthogonal pair. To construct another pair, from $\mathbf{D}^3(\mathbf{d}_2)$ and $\mathbf{D}^3(\mathbf{d}_4)$, that transform exactly as the first pair (belong to the same rows of E) is not such a simple procedure, although it must in principle be possible.

(iii) If we went on to generate $\mathbf{D}^3(\mathbf{d}_5)$ and $\mathbf{D}^3(\mathbf{d}_6)$ we would get no new information. These are symmetry-adapted displacements but they are just linear combinations of those already generated:

$$\mathbf{D}^3(\mathbf{d}_5)=(2\mathbf{d}_5-\mathbf{d}_1-\mathbf{d}_3)/\sqrt{6}=-(\mathbf{D}^3(\mathbf{d}_1)+\mathbf{D}^3(\mathbf{d}_2))/\sqrt{2},\text{ etc.}$$

5.5 Summary so far

The following brief summary is included to remind readers new to group theory of the major conclusions of Chapters 2 to 5. Problem 5.7, the benzene problem, will give readers the opportunity to test all those manipulative skills of group theory that have been covered so far.

We have established that there is a wide assortment of sets of matrices that represent symmetry groups, and that under some similarity transforms all members of a given set can be simultaneously block-diagonalized. The reduction process (block-diagonalization) is achieved in effect by finding specific linear combinations of the bases of the original matrix representation such that these combinations exhibit the symmetry of the group in some manner. In particular these symmetry-adapted bases transform simply amongst themselves in small groups. Each group forms the basis for one of the irreducible representations that form the blocks in the reduced matrix representation.

We have introduced two crucial formulae: one tells us the form of the block-diagonalization of the reduced representation (the a_j coefficients); the other tells us, at least for the basis displacement case, the form of the symmetry-adapted bases (the $\mathbf{D}_k^j$ vectors).

$$\mathbf{D} = \mathbf{d}N$$

$$\Gamma^{D} = N^{-1}\,\Gamma^{d}\,N = \sum_{j} a_j\,\Gamma^{j}.$$

(We have introduced the abbreviated notation $\mathbf{D}$ here, to indicate the row vector $[\mathbf{D}_1, \mathbf{D}_2 \ldots]$ etc.

The useful formulae to remember are

$$a_j = \frac{1}{g}\sum_{r=1}^{g} \chi^j(G_r)^*\chi(G_r),$$

$$\mathbf{D}^j \propto \sum_{r=1}^{g} \chi^j(G_r)^*\;G_r\mathbf{Q}$$

Clearly it is now appropriate to formulate the particular physical problems whose solution is to be facilitated by use of the above results. The first to be tackled, in part two of this text, is a classical representation of all the problems of interest; it is the normal-mode problem.

Problems

***5.1** The matrix sets (1), (3) and (4) of Table 3.1 form a complete set of *inequivalent irreducible matrix representations* (irreps) for the C_{3v} group.

(i) Form a six-dimensional vector whose components are the (1,1) elements of the matrix set (1).

Form five more such vectors by taking each set of (i, j) elements of the three matrix sets and demonstrate that the vectors so formed are orthogonal to each other. Write down the magnitudes of the six vectors. The result is a particular example of the *general orthogonality theorem*.

$$\sum_r \Gamma^{j'}_{k'l'}(G_r)^* \, \Gamma^j_{kl}(G_r) = \frac{g}{l_j}\delta_{jj'}\delta_{kk'}\delta_{ll'} \qquad (5.1)$$

(ii) Check that you understand the terms in this question, and verify the factor g/l_j using the above results.

(iii) Put $k=l$, $k'=l'$ and sum over k, k' to show that the characters of the irreps obey the orthogonality law:

$$\sum_r \chi^{j'}(G_r)^* \, \chi^j(G_r) = g\ \delta_{jj'}. \qquad (5.2)$$

(The results (a) $\sum_j l_j^2 = g$, and (b) number of irreps = number of classes, can be derived from the orthogonality relations (5.1) and (5.2) respectively.)

(iv) Suppose that some reducible representation can be written as

$$\Gamma(G_r) = \sum_j a_j\, \Gamma^j(G_r),$$

where the Γ^j are irreps.

Then $\chi(G_r) = \Sigma_j\, a_j\, \chi^j(G_r)$,

where $\chi(G_r)$ is the character of $\Gamma(G_r)$.

Multiply by $\chi^j(G_r)^*$ and sum over r to show that the coefficient a_j can be obtained by the relation

$$a_j = \frac{1}{g}\sum_r \chi^j(G_r)^* \chi(G_r).$$

5.2 If $\mathbf{D}^j_k$ and $\mathbf{D}^j_l$ are *symmetry-adapted basis displacements* (sabd) which belong to the jth irreducible representation of a group (Γ^j) we say that they are the partner vectors, belonging to the kth and lth rows of Γ^j, and that they must obey the relation

$$G_r \mathbf{D}^j_l = \sum_{i=1}^{lj} \Gamma^j(G_r)_{kl}\, \mathbf{D}^j_k.$$

(i) Make sure you understand the above equation.

Hence $\sum_r \Gamma^{j'}(G_r)^*_{k'l'}\, G_r \mathbf{D}^j_l = \sum_r \sum_k \Gamma^{j'*}_{k'l'}\, \Gamma^j_{kl}\, \mathbf{D}^j_k$.

(ii) Use the orthogonality theorem (1) of Problem 5.1 to show that

$$\sum_r \Gamma^{j'}(G_r)^*_{k'l'}\, G_r \mathbf{D}^j_l = 0 \text{ if } j \neq j',\ l \neq l';$$
$$\propto \mathbf{D}^j_{k'} \text{ otherwise.}$$

(iii) Hence show that for a general displacement $\mathbf{Q}$

$$\sum_r \Gamma^j(G_r)^*_{kl}\, G_r\mathbf{Q} = 0 \text{ or } \mathbf{D}^j_{k}.$$

That is, the sabd obtained using Γ^j in the equation is the sabd belonging to the kth row of the jth irrep.

$\Sigma_r\, \Gamma^j(G_r)^*_{kl}\, G_r$ is an *operator* which projects out of $\mathbf{Q}$ the component which is an sabd of the above form. It is called a *projection operator*.

(iv) Using $\Sigma\, \chi^j\, (G_r)^* G_r\mathbf{Q}$ to find the sabds; will the vector found belong to a particular irrep. or row of an irrep?

5.3 Define unit polar vectors and axial vectors, originating at the point group origin, in such a way that these vectors form bases for irreducible representations of the following groups. (Use the character tables of section 4.2).

(i) The D_{2h} group of Problem 2.4, and the reduced group you considered in that problem.
(ii) The O_h group of cubic operations, Problem 2.9.
(iii) The C_{2h} group, Problem 3.7, and the isomorphic D_2 group.

***5.4** A molecule consists of 6 atoms which lie on the face centres of a cube.

(i) In three dimensions, how many displacements do you need in order to define a molecular configuration?
(ii) Considering only proper rotations (and the identity operation), find the group operations for the molecule and divide them into classes.
(iii) Using the atomic displacements as a basis for a representation (Γ), find the characters for the rotational operation matrix representations.
(iv) One of the irreducible matrix representations of this group has as its basis vectors the set **i**, **j**, **k** centred at the cube centre. What are the characters of this irrep. (which is labelled T_1)?
(v) How many times does T_1 appear in Γ?
(vi) How would you find symmetry-adapted basis displacements belonging to T_1? Don't go quite as far as getting the sabds.
(vii) How many such displacements should such a method give, for T_1? Are any of them partner vectors of each other?
(viii) You should be able to write down three of the sabds belonging to T_1 without recourse to the formulae—can you?

5.5 Obtain the reduced form of the regular representation introduced in Problem 3.5. Extend your result to cover the regular representation of any group.

5.6 The molecule allene C_3H_4 has three carbon atoms on the major axis of a square-based cuboid and hydrogen atoms at the corners as shown in Figure 5.6.

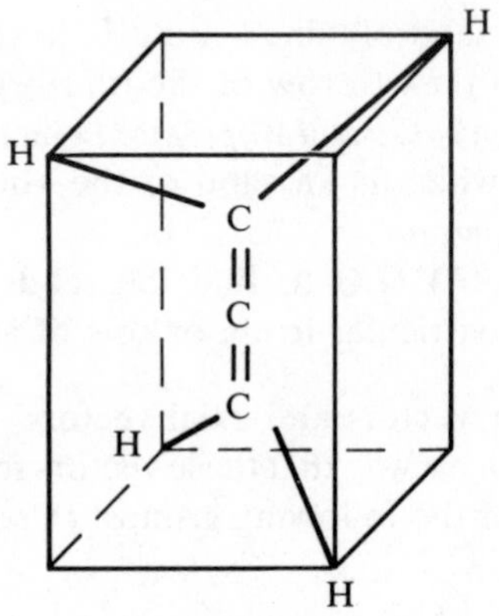

Figure 5.6

The displacements of the four hydrogen atoms, in directions radial with respect to the centre of the cuboid, form the basis for a representation of the allene symmetry group.

Determine the symmetry group. Find the characters for the above representation and reduce it. Find the corresponding symmetry-adapted basis displacements.

What is the matrix N that transforms the original basis set into the normalized, symmetry-adapted set?

Your symmetry group should contain the operation S_4. Obtain $\Gamma^d(S_4)$ and $\Gamma^D(S_4)$ for this operation and confirm that $\Gamma^D(S_4) = N^{-1}\Gamma^d(S_4)N$. Two of your symmetry-adapted displacements should belong to the two rows of the E irrep. Describe two other partner vector displacements of the molecule, that belong to E. Do the two displacements that you have described belong to the same rows of E as the first pair? If not, then redefine them so that they do.

***5.7** Consider all 12 orthogonal displacements of the atoms of NH_3.

(i) Obtain the characters for the 12-dimensional matrix representations of the group.
(ii) Reduce this representation in terms of A_1, A_2, and E.
(iii) Find linear combinations of your original 12 displacements that form bases for the reduced representation, stating to which irrep. each belongs.
(iv) What is the most general displacement you can write down that belongs to A_1?

5.8 ***The benzene problem—symmetry operations and matrix representations***
This and the following problem are included in order to test the manipulative skills learnt so far. As we introduce further concepts in group theory we shall return to benzene as an ongoing example (Chapters 8, 12, 15, 16).

(i) Describe the 24 *symmetry operations* for the benzene molecule C_6H_6 depicted in Figure 5.7. Divide the operations into *classes* by inspection.

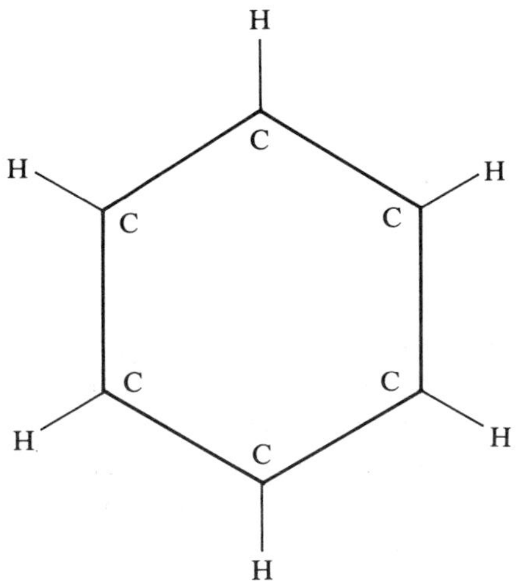

Figure 5.7

(ii) Pick any three non-trivial products of operations and demonstrate that *closure* is obeyed for your group of operations.

(iii) Show that the operations C_6 and C_6^5 are in the same class by using the *similarity transform* technique. Apply the transformation on C_6 using three further selected operations, to demonstrate that no other operations are in this class.

(iv) Modify the symmetry of the molecule by changing one or more hydrogen atoms to deuterium. Demonstrate that the modified symmetry group is a *subgroup* of the original one.

(v) For the operations you considered in part (ii), obtain matrices that represent the operations, having as a *basis* the unit vectors **i**, **j** in the plane of the molecule originating from the centre of the molecule.

(vi) Demonstrate that if $G_rG_s \equiv G_t$, then $\Gamma(G_r)\Gamma(G_s) = \Gamma(G_t)$ for the above *matrix representations*.

5.9 *The benzene problem—displacement vectors*

Use as a basis for a matrix representation of the benzene group, the 12 orthonormal displacements of the carbon atoms defined below.

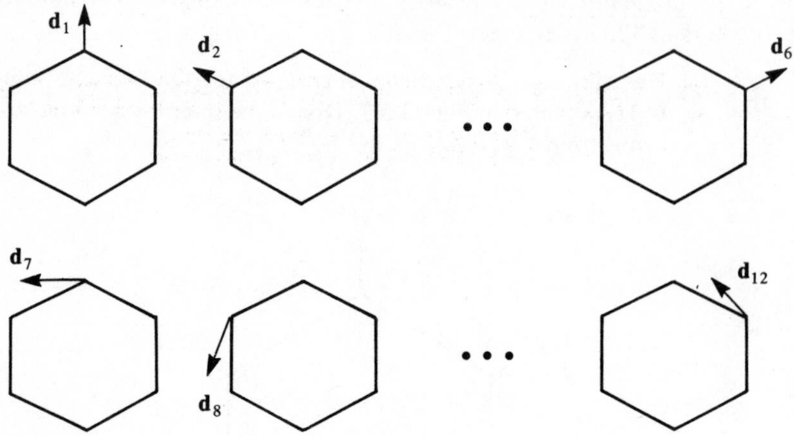

Figure 5.8

(i) What are the characters of the *reducible representation*, $\Gamma_c(12)$, with the above basis?

(ii) If the representation were put into *reduced form* what would the matrices look like in terms of the *irreducible representations* of the benzene group?

(iii) Repeat (i) and (ii) for the six carbon displacements normal to the plane of the molecule ($\mathbf{d}_{13} \rightarrow \mathbf{d}_{18}$) that form a basis for $\Gamma_c(6)$.

(iv) State why there are 36 orthonormal basis displacements for the entire molecule and why one could write $\Gamma(36)$ as six six-dimensional blocks down the diagonal, even without using character tables. From (ii) and (iii) reduce the $\Gamma(36)$ matrix.

(v) Draw up a table showing how $\mathbf{d}_1$, $\mathbf{d}_7$ and $\mathbf{d}_{13}$ transform under the group operations. Add the relevant characters, in the style of Table 5.3.

(vi) From your table find the *symmetry-adapted displacements* for the one-dimensional irreps in $\Gamma_c(12)$ and $\Gamma_c(6)$.

(vii) Use the table to find some of the bases for the two-dimensional irreps and use cyclic permutation to obtain the remaining sabds.

(viii) Show that an sabd of a two-dimensional irrep, found using $\mathbf{d}_4$, is just a linear combination of those you have already obtained, and state why this must be so.

(ix) Sketch your sabds. Do they display any symmetries?

(x) Comment on sabds that involve the hydrogen displacements.

PART TWO

APPLICATION OF SYMMETRY TO NORMAL MODES

6

The normal-mode problem

6.1 Displacement notations

The major difficulty experienced in learning group theory for the first time is often the confusion created by the abundance of new terms and apparently similar notations. In the previous sections it has been necessary to define carefully the various types of representation, etc.

For displacement problems we have also had to introduce two types of displacement vectors: (i) a set of orthonormal basis displacements for which we use the *lower case* **d** notation, selected such as to have fairly simple transformation properties but otherwise arbitrary; and (ii) a set of symmetry-adapted displacements, *capital* **D**. The two sets are related by the similarity transformation orthogonal matrix N. We spell this out yet again because we are about to introduce a third set of displacements, the normal-mode eigenvectors, for which we are going to use the notation *script* $\mathfrak{D}$. These orthonormal vectors will be related to the basis displacements **d** by a matrix α, and to the symmetry-adapted displacements **D** by a matrix β. Thus we have,

$$\mathbf{D}=\mathbf{d}N,$$
$$\mathfrak{D}=\mathbf{d}\alpha,$$
$$\mathfrak{D}=\mathbf{D}\beta.$$

We also note, from the equality $\mathfrak{D}=\mathbf{D}\beta=\mathbf{d}N\beta$, that

$$\alpha=N\beta.$$

In order to introduce these parameters in a problem with as simple a physical interpretation as possible, we shall solve the normal-mode problem for a case with only two degrees of freedom.

6.2 One-dimensional motion of the diatomic molecule

Consider the system, shown in Figure 6.1, of two equal masses m (atoms), held in equilibrium by a restoring force (bond) that is manifested by a Hooke's law spring of force constant k.

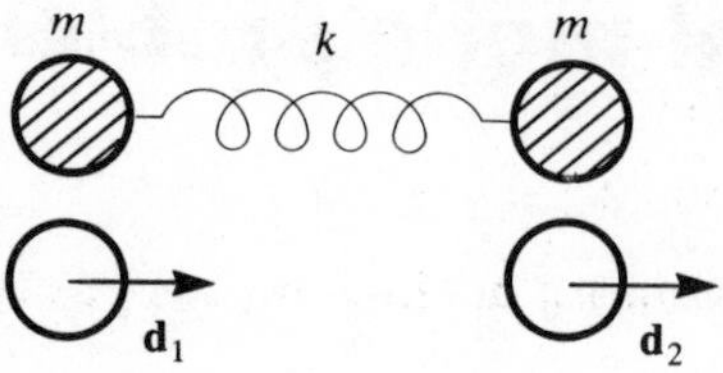

Figure 6.1 Displacement vectors for a diatomic molecule.

$\mathbf{d}_1$ and $\mathbf{d}_2$ are unit displacement vectors for the two masses, in the direction along the molecular axis. Hence some arbitrary initial displacement of the molecule (at time $t=0$) is

$$\mathbf{disp}(0)=r_1(0)\ \mathbf{d}_1+r_2(0)\ \mathbf{d}_2.$$

The problem to be solved is 'What is the time dependence of the motion of the two atoms?'

We will make one restriction: the initial condition will be considered to be simple displacement of the atoms from equilibrium to new positions $r_i(0)$ away, with the initial velocities restricted to be zero, $\dot{r}_i(0)=0$.

The Hooke's law equations of motion we wish to solve are:

$$\begin{aligned} m\ddot{r}_1(t)&=k[r_2(t)-r_1(t)],\\ m\ddot{r}_2(t)&=k[r_1(t)-r_2(t)], \end{aligned} \tag{6.1}$$

The coupling of these linear differential equations, manifested by the presence of $r_2(t)$ in the first equation, etc., prevents a trivial solution. The method used is one of trial solutions. Consider some displacement in which both atoms oscillate at the same frequency, ω_i, but with different amplitudes of oscillation (the notation becomes clear below). This trial solution is

$$\mathbf{disp}_i(t)=\alpha_{1_i}\cos(\omega_i t)\ \mathbf{d}_1+\alpha_{2_i}\cos(\omega_i t)\ \mathbf{d}_2.$$

Substituting into equation (6.1) above, the coefficients α_{j_i} must obey

$$\begin{aligned} -m\omega_i^2\ \alpha_{1_i}&=-k\alpha_{1_i}+k\alpha_{2_i},\\ -m\omega_i^2\ \alpha_{2_i}&=+k\alpha_{1_i}-k\alpha_{2_i}. \end{aligned} \tag{6.2}$$

There are two solutions to this pair of simultaneous equations, corresponding to the fact that our system has two degrees of freedom. The solutions are obtained from

$$\alpha_{2_i}/\alpha_{1_i}=(k-m\omega_i^2)/k=k/(k-m\omega_i^2).$$

Solution $i=1$; $\omega_1=0$, $\alpha_{21}/\alpha_{11}=1$, $\mathbf{disp}_1(t)=\alpha_{11}(\mathbf{d}_1+\mathbf{d}_2)$.

Solution $i=2$, $\omega_2=\sqrt{(2k/m)}$, $\alpha_{22}/\alpha_{12}=-1$,
$\mathbf{disp}_2(t)=\alpha_{12}(\mathbf{d}_1-\mathbf{d}_2)\cos[\sqrt{(2k/m)}t]$.

As the equations (6.1) and (6.2) are linear the premultiplying factors α_{11}, α_{12} can take any chosen values in these solutions, and any linear combination of the two solutions is also a solution. However, there is just one solution that obeys the initial conditions, namely:

$$\begin{aligned}\mathbf{disp}\,(t)&=\{\alpha_{11}(\mathbf{d}_1+\mathbf{d}_2)+\alpha_{12}(\mathbf{d}_1-\mathbf{d}_2)\cos[\sqrt{(2k/m)}t]\}\\&=\{\alpha_{11}+\alpha_{12}\cos[\sqrt{(2k/m)}t]\}\,\mathbf{d}_1+\{\alpha_{11}-\alpha_{12}\cos[\sqrt{(2k/m)}t]\}\,\mathbf{d}_2,\end{aligned}$$

such that $\mathbf{disp}\,(0)=r_1(0)\,\mathbf{d}_1+r_2(0)\,\mathbf{d}_2$.

That is, $\alpha_{11}+\alpha_{12}=r_1(0)$ and $\alpha_{11}-\alpha_{12}=r_2(0)$.

Normal modes of oscillation

On analysing these results we note that there are two initial conditions that lead to very simple subsequent motions. Namely if we had established either $\alpha_{12}=0$ or $\alpha_{11}=0$ respectively, i.e., $r_2(0)=\pm r_1(0)$, then the subsequent motions would have been exactly of the form of the trial solution—harmonic motion where both atoms move at the same frequency. These are the normal modes of oscillation of the molecule.

Quite generally we define a normal mode of a system as a mode of motion in which all components of the system oscillate in phase, at the same frequency. The amplitude of the oscillations generally differs for different components of the system and may in fact be zero for some components. By 'in phase' we are to understand that each component reaches its maximum deviation from equilibrium at the same time.

Eigenvectors and eigenfrequencies

The normal modes are characterized firstly by *orthogonal eigenvectors*

$$\mathfrak{D}_1=(\mathbf{d}_1+\mathbf{d}_2)/\sqrt{2};\ \mathfrak{D}_2=(\mathbf{d}_1-\mathbf{d}_2)/\sqrt{2},$$

which are obtained by normalizing the solutions $i=1$, 2; i.e. by setting $\alpha_{11}=1/\sqrt{2}$, $\alpha_{12}=1/\sqrt{2}$ etc. The second characteristic of the normal-mode solution is the frequency at which each atom oscillates—*the eigenfrequency*.

$$\omega_1=0;\ \omega_2=\sqrt{(2k/m)}.$$

If a normal mode is set up initially, $\mathbf{disp}(0) = \mathcal{R}_i(0)\ \mathfrak{D}_i$, then the subsequent motion is simply $\mathbf{disp}(t) = \mathcal{R}_i(0) \cos(\omega_i t)\mathfrak{D}_i$. Note the use of a *script* $\mathcal{R}$ here to indicate the amplitude of the displacement, in terms of the vector $\mathfrak{D}$. $\mathfrak{D}_i$ is an eigenvector of eigenfrequency ω_i.

Displacement coordinates

There are then two possible descriptions for the motion of the diatomic system. The first, in terms of the basis displacements $\mathbf{d}_i$, involves coordinates $r_i(t)$ having rather complicated time-dependences:

$$\mathbf{disp}(t) = \sum_i r_i(t)\ \mathbf{d}_i.$$

The second, in terms of normal-mode eigenvectors $\mathfrak{D}_i$ and their coordinates $\mathcal{R}_i(t)$, is

$$\mathbf{disp}(t) = \sum_i \mathcal{R}_i(t)\ \mathfrak{D}_i = \sum_i \mathcal{R}_i(0) \cos(\omega_i t)\ \mathfrak{D}_i.$$

It is useful to comment that the normal-mode solutions do appear to have a certain symmetry. The $\omega_1 = 0$ mode is a simple translation of the molecule by an amount $\mathcal{R}_1(0)$. Had we not restricted the initial configuration to be a stationary one ($\dot{r}_i(0) = 0$) then this mode would correspond to the constant-velocity linear translation of the entire system, with the spring (bond) length remaining at its equilibrium value. The $\omega_2 = \sqrt{(2k/m)}$ mode is initiated by symmetric (equal but opposite) displacements of the two atoms (Figure 6.2).

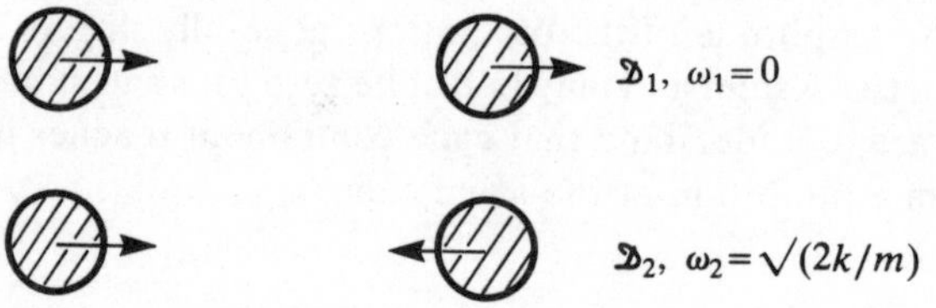

Figure 6.2 The normal modes of the diatomic system.

These are almost identical to the modes of the system described in Chapter 1. The $\omega_1 = 0$ mode in that case had a non-zero frequency because of the restrictions upon translation of the system. Nevertheless we did find in our demonstration that the mode for which equal displacements in the same direction were given to the two masses was the lower-frequency mode.

The α matrix

We have in effect established a transformation α above that takes us from the basis displacements **d** to the mode eigenvectors $\mathfrak{D}$. For our particular example,

$$\mathfrak{D} = \mathbf{d}\ \alpha, \qquad \text{where } \alpha = \frac{1}{\sqrt{2}}\begin{bmatrix} 1 & 1 \\ 1 & -1 \end{bmatrix}.$$

Force and mass matrices

The equations of motion (6.2) take on a matrix form:

$$\begin{bmatrix} k-\omega_i^2 m & -k \\ -k & k-\omega_i^2 m \end{bmatrix}\begin{bmatrix} \alpha_{1i} \\ \alpha_{2i} \end{bmatrix} = \begin{bmatrix} 0 \\ 0 \end{bmatrix}. \tag{6.3}$$

We will now demonstrate that the elements in the 2×2 matrix have a specific physical significance. Suppose we consider the kinetic and potential energies of the system at some stage of its motion. Force and mass matrices will be defined, with respect to the basis displacements **d**, by the following system of equations. Under an arbitrary system displacement,

$$\mathbf{disp}(t) = r_1(t)\mathbf{d}_1 + r_2(t)\ \mathbf{d}_2,$$

the *force matrix* K^d is defined such that the system has a potential energy

$$\text{PE} = \tfrac{1}{2}[r_1,\ r_2]\ K^d\begin{bmatrix} r_1 \\ r_2 \end{bmatrix};$$

and the *mass matrix* M^d is such that the kinetic energy is:

$$\text{KE} = \tfrac{1}{2}[\dot{r}_1,\ \dot{r}_2]\ M^d\begin{bmatrix} \dot{r}_1 \\ \dot{r}_2 \end{bmatrix}.$$

The definition of K^d gives

$$\text{PE} = \tfrac{1}{2}[K_{11}^d\, r_1^2 + K_{22}^d\, r_2^2 + (K_{12}^d + K_{21}^d) r_1 r_2].$$

Note that because there is no physical distinction between terms in $r_1 r_2$ and $r_2 r_1$, there is an arbitrariness in the separate values of K_{12}^d and K_{21}^d, providing they sum to the correct value. It is conventional to choose all such pairings to be equal:

$$K_{12}^d = K_{21}^d.$$

Thus in the present example:

$$\text{PE} = \tfrac{1}{2}k|r_2 - r_1|^2, \qquad K^d = \begin{bmatrix} k & -k \\ -k & k \end{bmatrix};$$

$$\text{KE} = \tfrac{1}{2}m(\dot{r}_1^2 + \dot{r}_2^2), \qquad M^d = \begin{bmatrix} m & 0 \\ 0 & m \end{bmatrix}.$$

The normal-mode equations

With the above definitions we note that the normal-mode matrix equations of motion, (6.3), become

$$[K^d - \omega_i^2 M^d] \begin{bmatrix} \alpha_{1i} \\ \alpha_{2i} \end{bmatrix} = \begin{bmatrix} 0 \\ 0 \end{bmatrix} \tag{6.4}$$

Now this result could be obtained from the slightly more general expression,

$$\alpha^{-1}[K^d - \omega^2 M^d]\, \alpha = [\Omega - \omega^2 I], \tag{6.5}$$

where I is the unit diagonal matrix and Ω is a diagonal matrix of elements

$$\Omega_{ii} = \omega_i^2.$$

To demonstrate the equivalence of the two expressions, pre-multiply by α on both sides and consider the ji matrix elements

$$\sum_k (K^d - \omega^2 M^d)_{jk}\, \alpha_{ki} = \sum_k \alpha_{jk}\, (\Omega - \omega^2 I)_{ki} = \alpha_{ji}\, (\Omega - \omega^2 I)_{ii}.$$

If ω is set equal to the normal-mode frequency ω_i, then the right-hand side is zero regardless of the value of j. Equating all ji elements on the left to zero gives the required result:

$$[K^d - \omega_i^2 M^d] \begin{bmatrix} \alpha_{1i} \\ \alpha_{2i} \end{bmatrix} = \begin{bmatrix} 0 \\ 0 \end{bmatrix}.$$

It is useful to note, in view of the next subsection, that we could multiply the right-hand side of equation (6.5) by any diagonal matrix and still regain result (6.4). The matrix α is therefore not yet uniquely defined; this is because normalization has not been built into equation (6.4).

It is informative to ask what equivalent expressions we should obtain if we started with a knowledge of the mode eigenvectors $\mathfrak{D}$. $K^{\mathfrak{D}}$ and $M^{\mathfrak{D}}$ may be defined by direct analogy to K^d and M^d.

For a general displacement, $\mathbf{disp}(t) = \Sigma_i R_i(t)\mathfrak{D}_i$ the potential and kinetic energies are obtained by noting that

$$\sum_i R_i(t)\mathfrak{D}_i = \frac{1}{\sqrt{2}}\{R_1(t) + R_2(t)\}\mathbf{d}_1 + \frac{1}{\sqrt{2}}\{R_1(t) - R_2(t)\}\mathbf{d}_2.$$

$$\mathrm{PE} = \tfrac{1}{2}k\left[\frac{2R_2}{\sqrt{2}}\right]^2; \text{ so } K^{\mathfrak{D}} = \begin{bmatrix} 0 & 0 \\ 0 & 2k \end{bmatrix}.$$

$$\mathrm{KE} = \tfrac{1}{2}m\left[\frac{\dot{R}_1 + \dot{R}_2}{\sqrt{2}}\right]^2 + \tfrac{1}{2}m\left[\frac{\dot{R}_1 - \dot{R}_2}{\sqrt{2}}\right]^2; \quad \text{so } M^{\mathfrak{D}} = \begin{bmatrix} m & 0 \\ 0 & m \end{bmatrix}.$$

Thus $[K^{\mathfrak{D}} - \omega^2 M^{\mathfrak{D}}] = \begin{bmatrix} -\omega^2 m & 0 \\ 0 & 2k - \omega^2 m \end{bmatrix} = m \begin{bmatrix} \omega_1^2 - \omega^2 & 0 \\ 0 & \omega_2^2 - \omega^2 \end{bmatrix}.$

The point to make is that we have demonstrated simply that the force and mass matrices are diagonal with respect to the normal-mode eigenvectors.

Furthermore, we know that $\mathcal{D} = \mathbf{d}\alpha$, so that the two arbitrary displacements

$$\sum_i r_i \mathbf{d}_i \text{ and } \sum_i \mathcal{R}_i \mathcal{D}_i$$

are equal providing

$$\mathbf{d}r = (\mathbf{d}\alpha)\mathcal{R} \text{ or } r = \alpha\mathcal{R},\ \tilde{r} = \tilde{\mathcal{R}}\alpha^{-1}.$$

(r is the column vector with elements r_1 and r_2 etc., and we have employed the orthogonality of α). The two descriptions of the displacement cannot lead to different energies. Thus for example,

$$\tfrac{1}{2}\tilde{\mathcal{R}}K^{\mathcal{D}}\mathcal{R} = \tfrac{1}{2}\tilde{r}K^{d}r = \tfrac{1}{2}\tilde{\mathcal{R}}\alpha^{-1}K^{d}\alpha\mathcal{R},$$

so that we expect

$$\alpha^{-1}[K^{d} - \omega^2 M^{d}]\alpha = [K^{\mathcal{D}} - \omega^2 M^{\mathcal{D}}]. \tag{6.6}$$

In summary the matrix α we are looking for is one which diagonalizes the force and mass matrices and is an orthogonal matrix. The normal-mode eigenfrequencies appear to be obtained by setting the resulting diagonal elements equal to zero;

$$\omega_i^2 = K_{ii}^{\mathcal{D}}/M_{ii}^{\mathcal{D}}.$$

Normal-mode coordinates

We will justify the above results for the general case, in Section 6.3. Before we do so though the status of equation (6.5) needs some comment. The form of (6.5) is somewhat strange as, in order to be correct dimensionally, the matrix α needs be of dimension $(\text{mass})^{-1/2}$. To put equation (6.6) into the form of (6.5), suppose we defined a set of 'vectors' $\mathcal{Q}$ such that

$$\mathcal{Q}_i = \mathcal{D}_i(M_{ii}^{\mathcal{D}})^{-1/2}.$$

A displacement equal to $\sum_i \mathcal{R}_i(t)\,\mathcal{D}_i$ can now be written as $\sum_i \mathcal{q}_i(t)\,\mathbf{Q}_i$, where $\mathcal{q}_i(t) = \mathcal{R}_i(t)\ (M_{ii}^{\mathcal{D}})^{1/2}$, and the 'force and mass matrices' defined through

$$\text{PE} = \tfrac{1}{2}[\mathcal{q}_1 \quad \mathcal{q}_2]\ K^{\mathcal{Q}} \begin{bmatrix} \mathcal{q}_1 \\ \mathcal{q}_2 \end{bmatrix}$$

etc., are:

$$K^{\mathcal{Q}} = \begin{bmatrix} K_{11}^{\mathcal{D}}/M_{11}^{\mathcal{D}} & 0 \\ 0 & K_{22}^{\mathcal{D}}/M_{22}^{\mathcal{D}} \end{bmatrix}, \qquad M^{\mathcal{Q}} = \begin{bmatrix} 1 & 0 \\ 0 & 1 \end{bmatrix} = I,$$

Therefore

$$[K^{\mathcal{Z}} - \omega^2 M^{\mathcal{Z}}] = [\Omega - \omega^2 I],$$

where Ω is diagonal with elements $\omega_i^2 = K_{ii}^{\mathfrak{D}}/M_{ii}^{\mathfrak{D}}$.

Equation (6.5) is of form $\alpha^{-1}[K^d - \omega^2 M^d]\alpha = [K^{\mathcal{Z}} - \omega^2 M^{\mathcal{Z}}]$. The $\mathcal{q}_i(t)$ are conventionally called the normal-mode coordinates. From these expressions we can see their relationship to the normal-mode eigenvectors. As far as the vectors $\mathcal{Z}_i$ are concerned, they are not only of strange dimension for vectors but also are not normalizable in the conventional sense, although they are easily related to the $\mathfrak{D}_i$.

In the remainder of the text we shall concentrate on the physically clear eigenvectors $\mathfrak{D}_i$. The α we require is defined in equation (6.6) and, from the note we made following the analysis of (6.5) we can see that the following two equations are compatible for the solution for an orthogonal matrix α:

$$\alpha^{-1}[K^d - \omega^2 M^d]\alpha = [K^{\mathfrak{D}} - \omega^2 M^{\mathfrak{D}}],$$

$$(K^d - \omega_i^2 M^d) \begin{bmatrix} \alpha_{1i} \\ \alpha_{2i} \end{bmatrix} = \begin{bmatrix} 0 \\ 0 \end{bmatrix},$$

with $[\mathfrak{D}_1, \mathfrak{D}_2] = [\mathbf{d}_1, \mathbf{d}_2]\alpha$.

We now give the generalization for many atoms, anticipating that the matrices K^d and M^d can be found through the definitions from the potential and kinetic energies.

6.3 Lagrangian solution for normal modes

The Lagrangian for a system of N particles, $\mathcal{L} = \text{KE} - \text{PE}$, obeys the set of equations,

$$\frac{\mathrm{d}}{\mathrm{d}t}\frac{\partial \mathcal{L}}{\partial \dot{q}_i} - \frac{\partial \mathcal{L}}{\partial q_i} = 0,$$

where the q_i here refer to *any* set of $3N$ coordinates for the particles. Three sets of coordinates r_i, $\mathcal{R}_i$, $\mathcal{q}_i$ have been used in Section 6.2; each will be considered.

(a) Displacement coordinates

For a general displacement $\Sigma r_i \mathbf{d}_i$ we can take the r_i to form sets of three amplitudes, each set corresponding to the displacements of a specific atom in orthogonal directions. Thus

$$\mathrm{KE}=\tfrac{1}{2}\sum_{i=1}^{3N} m_i\dot{r}_i^2,$$

where m_i is the mass of atom i for $i=1, 2, 3$, etc. The point to be made is that M^d is automatically diagonal in this representation: $\mathrm{KE}=\frac{1}{2}\sum_i M_{ii}^d\dot{r}_i^2$. To express the potential energy in terms of the r_i we expand the potential with respect to that in the equilibrium configuration, for which all $r_i=0$ by definition.

$$\mathrm{PE}=V_{eq}+\sum_i\left[\frac{\partial V}{\partial r_i}\right]_{eq} r_i+\tfrac{1}{2}\sum_i\sum_j\left[\frac{\partial^2 V}{\partial r_i\partial r_j}\right]_{eq} r_i r_j+\cdots.$$

We can choose V_{eq} to be the zero of potential energy and, as the potential is a minimum for stable equilibrium, $(\partial V/\partial r_i)_{eq}=0$ for all r_i. Finally if we limit ourselves to *small displacements* from equilibrium, then terms in $r_i r_j r_k$ etc. can be ignored. The resulting, harmonic approximation corresponds to considering a Hooke's law restoring force in the diatomic example.

Hence we have $\mathrm{PE}=\frac{1}{2}\sum_i\sum_j K_{ij}^d r_i r_j$, where the elements of the force matrix are, by definition, just the curvatures of V at equilibrium. For any given system of atoms of known masses, interacting via assumed bonds of known force constants, then the elements of M^d and K^d can be found simply using

$$\mathrm{KE}=\tfrac{1}{2}\sum_{\text{atoms}} m_{\text{atom}}\,(\text{atom velocity})^2,$$

$$\mathrm{PE}=\tfrac{1}{2}\sum_{\text{bonds}} k_{\text{bond}}\,|\text{bond extension}|^2$$

(ignoring bond-bending energies).

For the present purposes we have

$$\mathcal{L}=\tfrac{1}{2}\sum_i M_{ii}^d\dot{r}_i^2-\tfrac{1}{2}\sum_i\sum_j K_{ij}^d r_i r_j,$$

$$\frac{\mathrm{d}}{\mathrm{d}t}\frac{\partial\mathcal{L}}{\partial\dot{r}_i}-\frac{\partial\mathcal{L}}{\partial r_i}=M_{ii}^d\ddot{r}_i+\sum_j K_{ij}^d r_j=0.$$

This is a set of coupled equations of motion, corresponding to the diatomic equations (6.2). The complication in solution is brought about by the coupling terms r_j which describe basically the effects of the displacement of one atom on the motion of another.

(b) Normal-mode eigenvectors

If we were able to find displacement vectors $(\mathfrak{D}_i)$ such that both the kinetic and potential energies were diagonalized:

$$\mathrm{KE}=\tfrac{1}{2}\sum_i M_{ii}^{\mathfrak{D}}\dot{R}_i^2, \qquad \mathrm{PE}=\tfrac{1}{2}\sum_i K_{ii}^{\mathfrak{D}}R_i^2,$$

for a general displacement $\sum_i R_i(t)\mathfrak{D}_i$, then the Lagrangian equations of motion would decouple. The solution is

$$M_{ii}^{\mathfrak{D}}\ddot{R}_i+K_{ii}^{\mathfrak{D}}R_i=0;$$

that is, harmonic oscillation just as in the diatomic case.

$$R_i(t)=R_i(0)\cos(\omega_i t+\phi); \qquad \omega_i^2=K_{ii}^{\mathfrak{D}}/M_{ii}^{\mathfrak{D}}.$$

The significance of the diagonalization procedure that the matrix α brings about is thus that the equations of motion decouple thereby.

(c) Normal-mode coordinates

For completeness we note that if a set of $\mathcal{Q}$ 'vectors' can be found such that

$$\mathrm{KE}=\tfrac{1}{2}\sum_i \dot{q}_i^2 \quad \text{and} \quad \mathrm{PE}=\tfrac{1}{2}\sum_i \Omega_{ii}q_i^2,$$

then again the equations of motion are decoupled:

$$\ddot{q}_i+\Omega_{ii}q_i=0.$$

The oscillatory solutions are

$$q_i(t)=q_i(0)\cos(\Omega_{ii}^{1/2}t+\phi).$$

(Additional phase factors have been included in the two normal-mode solutions in order to accommodate all initial conditions.)

6.4 Technique of normal-mode solution

The results above establish the physical problem. How do we solve the matrix diagonalization problem?

$$\alpha^{-1}[K^d-\omega^2M^d]\alpha=[K^{\mathfrak{D}}-\omega^2M^{\mathfrak{D}}].$$

We do not yet know any of the terms on the right-hand side of this equation, nor do we know the matrix elements of α. There are various computer techniques for the solution of this eigenvalue problem, which are applicable given that we can construct K^d and M^d. In effect they must correspond to the following procedure.

(a) *The frequency eigenvalues*

As the right-hand side is diagonal, then it has a zero determinant if any of its diagonal elements is zero, i.e., if $\omega^2=\omega_i^2=K_{ii}^{\mathcal{D}}/M_{ii}^{\mathcal{D}}$, for any of the $3N$ i-values. The eigenfrequencies are thus the $3N$ roots of the determinant of the left-hand side also. But as the determinant of a matrix product is just the product of the determinants, we have

$$\begin{aligned}&|\alpha^{-1}[K^d-\omega^2M^d]\alpha|\\ &=|\alpha^{-1}||K^d-\omega^2M^d||\alpha|=|\alpha^{-1}||\alpha||K^d-\omega^2M^d|\\ &=|\alpha^{-1}\alpha||K^d-\omega^2M^d|=|K^d-\omega^2M^d|.\end{aligned}$$

Thus the eigenfrequencies are the $3N$ roots of the known determinant $|K^d-\omega^2M^d|$, and are relatively easy to calculate.

(b) *The eigenvectors*

Having obtained the eigenfrequencies they can be used one at a time in the vector equation of type (6.4),

$$[K^d-\omega_i^2M^d]\begin{bmatrix}\alpha_{1i}\\ \alpha_{2i}\\ \cdot\\ \cdot\end{bmatrix}=\begin{bmatrix}0\\ 0\\ \cdot\\ \cdot\end{bmatrix},$$

and this set of $3N$ simultaneous equations solved to find the $3N$ values of α_{ji} for $j=1$ to $3N$. There is one such problem for each of the eigenfrequencies.

Obviously for a large molecule the full solution to the normal-mode problem using this technique can be very effort-consuming. The point of group theory is that the $3N\times3N$ matrix equations can be broken up into considerably smaller problems. In Chapter 7 we show how this is achieved.

6.5 Translations and rotations (centre-of-mass)

For a free molecule containing N atoms one requires $3N$ coordinates in order to specify the configuration of the entire molecule. Therefore choosing some equilibrium configuration as an origin it was necessary to define $3N$ orthonormal displacement vectors. The amplitudes of displacements along these vector directions then defined any possible displacement, **disp**.

As far as the group-theoretical analysis is concerned it is almost always necessary to introduce all $3N$ displacements in order to obtain a matrix representation for the molecular group. However, the normal-mode problem must produce certain solutions that could have been predicted directly; these are the translation and rotation modes.

If every atom of a molecule is displaced from equilibrium by the same amount in the same direction the molecule will simply translate to a new equilibrium position and stay there. We could therefore rewrite the displacement as

$$\mathbf{disp}(t) = \Sigma r_i(0)\mathbf{d}_i = R_i(0)\cos(\omega_i t)\mathfrak{D}_i,$$

where $\mathfrak{D}_i$ is the normalized translation displacement, $R_i(0)$ the amplitude of the displacement, and the eigenfrequency ω_i is zero.

For a free molecule there must be three orthogonal *translation* eigenvectors with corresponding eigenfrequencies equal to zero. There must also be three normal-mode eigenvectors that describe the *rotations* of the whole molecule. Again no changes of the relative positions of the atoms occur, so the potential energy of the system is independent of the amplitude of the rotation, just as it is independent of the amplitude of a translation. A rotated molecule will remain in its new equilibrium position. The three rotational eigenvectors will therefore also have zero eigenfrequencies.

In general there are then $(3N-6)$ remaining *vibrational* normal modes of finite frequencies. The case of a linear molecule is the single exception. Rotation about the molecular axis cannot be described in terms of the displacement vectors of the molecule and plays no role in normal-mode theory; there are only two rotational modes of such molecules and therefore $(3N-5)$ vibrations. Throughout this text we will talk of the $3N$ *degrees of freedom* of an N-atom molecule, and divide these into translational, rotational and vibrational degrees of freedom, as given in Table 6.1. (Note that the convention of using N for the number of atoms in the molecule has been adhered to. Confusion with the transformation matrix N should not arise.)

Table 6.1 Degrees of freedom of a free molecule. Parentheses refer to linear molecules.

Degrees of freedom	*Number*	*Eigenfrequency*
Translation	3	0
Rotation	3, (2)	0
Vibration	$3N-6$, $(3N-5)$	non-zero

In a crystal lattice the atoms in each unit cell can be considered to form pseudo-molecules for many purposes. However, both the translational and rotational motion is now restricted; there will be six zero-frequency modes for the entire crystal rather than for each unit cell.

Rather than start with displacements of the atoms with respect to equilibrium it must in principle be possible to define a molecular configuration by any $3N$ orthogonal linear combinations of the atomic coordinates. The amplitudes $R_i(t)$ are just such a set. Alternatively one could use the three coordinates of the molecular centre of mass and $(3N-3)$ relative coordinates. Given some initial equilibrium configuration the configuration following a translation is defined uniquely by the three orthogonal displacements of the centre-of-mass and only by these three displacements. The remaining $(3N-3)$ normal modes must therefore be describable by the $(3N-3)$ relative coordinates. But these are the rotational and vibrational modes. Consequently the centre-of-mass must remain stationary in the latter modes. Having selected out translational modes then the rotational eigenvectors must describe rotations about axes that pass through the centre-of-mass of the system, which might not be at the molecular point-group origin. Further the centre-of-mass must remain fixed in all vibrational modes.

6.6 Degenerate normal modes

The translations and rotations are the simplest examples of degenerate normal modes, that is modes which have identical eigenfrequencies. Degenerate vibrational modes of finite frequency also occur. A unique mode at some frequency is termed non-degenerate or singly degenerate; two modes of the same frequency are doubly degenerate; the translations and rotations of a free molecule are six-fold degenerate.

Degeneracies may come about precisely due to the symmetry of the system, in which case they will be predicted by group theory, as will be shown in Chapter 7. Degeneracy may also originate from the precise form of the energy of the system rather than its symmetry. This is termed *accidental degeneracy*.

An example of *symmetry-induced degeneracy* is the equality of the translational eigenfrequencies of the homonuclear diatomic molecule for orthogonal translations transverse to the nuclear axis. If the molecule were not free but placed in some potential field, these zero-frequency modes would become finite-frequency vibrations. However, providing the field environment maintained the original symmetry of the system ($C_{\infty v}$) the modes would remain degenerate. In this simple case the degeneracy is expected from a physical point of view as there is nothing physically distinct about the two orthogonal directions. In contrast the equality of the eigenfrequency for the translation along the normal axis (longitudinal) with that in the transverse direction is an accidental degeneracy; it arises because the molecule is free in this example. Again in a non-uniform potential field of $C_{\infty v}$ symmetry this mode could become a vibrational mode. However, it will in general have a different frequency to the transverse modes as it is physically distinct from them.

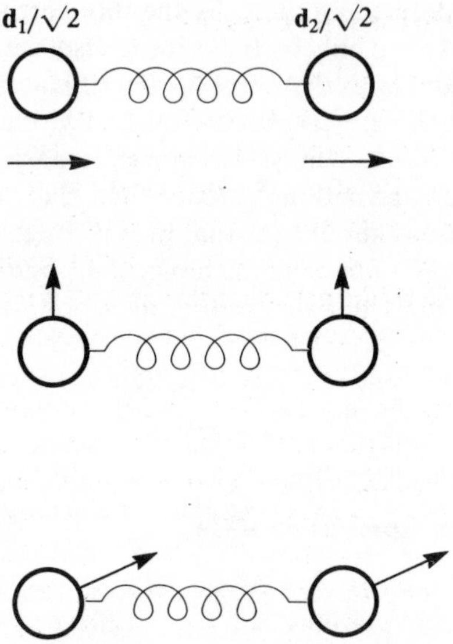

Figure 6.3 The translational normal modes of a diatomic molecule.

Degenerate normal modes are seldom uniquely defined. Clearly if two modes of eigenvectors $\mathfrak{D}_1$ and $\mathfrak{D}_2$ share the same frequency ω, then if an initial molecular displacement is any linear combination of these modes, the subsequent motion will be harmonic, with frequency ω. That is, if

$$\mathbf{disp}(0) = \mathcal{R}_1(0)\mathfrak{D}_1 + \mathcal{R}_2(0)\mathfrak{D}_2,$$

then

$$\mathbf{disp}\ (t) = \mathcal{R}_1(0)\ \cos\omega t\ \mathfrak{D}_1 + \mathcal{R}_2(0)\ \cos\omega t\ \mathfrak{D}_2.$$

Hence the normalized displacement $[\mathcal{R}_1(0)\mathfrak{D}_1 + \mathcal{R}_2(0)\mathfrak{D}_2]/[\mathcal{R}_1^2(0) + \mathcal{R}_2^2(0)]^{1/2}$ could also be used as an eigenvector of frequency ω. Any orthonormal pair of linear combinations of $\mathfrak{D}_1$ and $\mathfrak{D}_2$ could be used to describe the modes. Similar arguments apply to modes of higher degeneracy.

Thus, for the free diatomic motion one could in principle choose any set of axes to define the translation directions and the rotations. Indeed any linear combination of translation and rotation is in principle an eigenvector of zero eigenfrequency. Even in the presence of a restricting potential field those vibrations that originate from the transverse translations of the free molecule are not uniquely defined if there is no preferred pair of orthogonal directions normal to the molecular axis.

6.7 Lattice vibrations in crystals

The number of degrees of freedom for the atoms in a crystal is of course extremely large. Typical crystal atomic densities are of the order of $10^{22}\,cm^{-3}$, so that for a sample of volume $1\,cm^3$ one is dealing with $3N \simeq 3\times 10^{22}$ degrees of freedom. Nevertheless the general considerations of Section 6.3 still hold. There are $3N$ normal modes of motion for the atomic lattice. The task of solving for this huge number of modes is alleviated considerably by the periodic nature of the crystal. It is not appropriate in this chapter to detail how this is achieved, but the following discussion is included in order to introduce the reader to the topic.

All crystals may be divided into cells, containing a few atoms (or perhaps only one), such that the crystal as a whole can be constructed by stacking many identical cells together in a periodic manner. The individual cells will each have a symmetry which is mirrored in the macroscopic symmetry of the entire crystal.

For a normal mode the atoms in a specific cell have some particular motions. In an adjacent cell the same motions will occur but may be delayed slightly in time. Moving through a number of cells (in the same direction) one eventually obtains a repeat of the initial motion. If the repeat distance is l, then all cells separated by l have identical motion (their atoms move *in phase*). These lattice normal modes are known as phonons; the repeat distance can be considered to be the wavelength of the phonon and $2\pi/l$ to be the phonon wave vector. This discussion may be extended to three dimensions, leading to one possible wave vector ($\mathbf{k}$) for every cell in the lattice. This corresponds to possible repeat distances of one cell, two cells, three cells, etc. The case for which the repeat distance is only one cell, that is for which each cell has identical motion, is particularly important. Rather than consider this to be of a phonon of wavelength equal to the cell size it is conventional to consider it as a mode of infinite repeat distance, or a phonon of infinite wavelength, and thus of wave vector $\mathbf{k}=0$.

Let us now suppose that the crystal, of N atoms, consists of $\mathcal{N}$ cells each containing n atoms. There are $\mathcal{N}$ possible wave vectors and for each of these there are $3n$ different modes of motion of atoms in any specific cell. In particular there are $3n$ modes for $\mathbf{k}=0$. Of these, three modes must be the translations of the entire crystal, as every cell moves in phase under a translation. In phonon language these modes are called the $\mathbf{k}=0$ acoustic phonon modes. (Sound propagation corresponds to the generation of acoustic phonon modes at **k** very close to zero.) The remaining $3n-3$ modes are known as the $\mathbf{k}=0$ optic phonons, because they are observed in the optical experiments of infrared absorption and Raman spectroscopy. (To be strictly precise modes of **k** extremely close to zero are observed in optical experiments.)

Even for $\mathbf{k}=0$ phonons a full treatment for the eigenfrequencies and eigenvectors is far more complicated than for molecular modes, because of the coupling between atoms in one cell and those in adjacent cells. Nevertheless we will find that group-theoretical ideas can be applied in a rather similar manner, from which we can deduce many of the properties of the phonons without knowing their eigensolutions in detail.

Problems

6.1 ***Diagonalization of block diagonal matrices***

A matrix $(K^d-\omega^2M^d)$, based on three displacements $\mathbf{d}_1$, $\mathbf{d}_2$, $\mathbf{d}_3$, has the form:

$$\begin{bmatrix} A & 0 & 0 \\ 0 & B & C \\ 0 & C & D \end{bmatrix}$$

(i) Show that the eigenfrequencies are obtained by considering independently, the blocks

$$[A] \qquad \text{and} \qquad \begin{bmatrix} B & C \\ C & D \end{bmatrix}.$$

(ii) Show that one of the eigenvectors is $\mathbf{d}_1$ and that the others are combinations of only $\mathbf{d}_2$ and $\mathbf{d}_3$.

This question is a simple example of the important result that the blocks can be considered independently when diagonalizing block diagonal matrices. This point will be used repeatedly in the following chapters.

6.2 ***The normal modes of a diatomic molecule***

Define six orthonormal displacement vectors for the homonuclear diatomic molecule.

Write down the potential and kinetic energies in terms of amplitudes of your displacements. Hence generate a 6×6, $[K^d-\omega^2M^d]$ matrix.

Comment on the solutions of this normal-mode problem for (a) the eigenfrequencies and (b) the eigenvectors. What modification do you obtain for a heteronuclear diatomic molecule?

Evaluate the position of the centre-of-mass of any vibrational normal modes you have obtained.

6.3 *The normal modes of CO_2

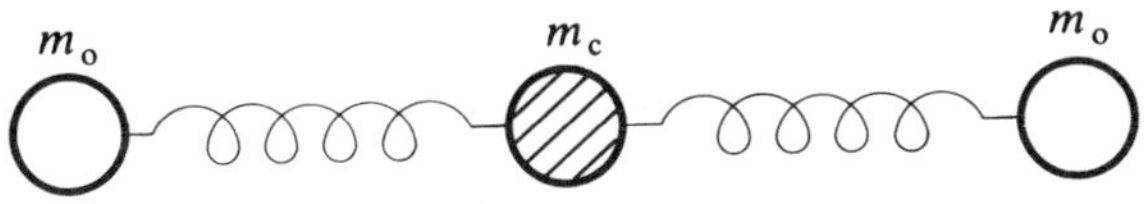

Figure 6.4

CO_2 is a linear molecule (Figure 6.4). The task is to find those normal modes involving only motion along the molecular axis.

Work through the following calculations for the longitudinal motion of this system:

(i) Obtain the Hooke's law equations of motion. (Take particular care over signs.)

(ii) Obtain equations for the α_{ji} of a trial solution:

$$\mathbf{disp}_i(t) = \sum_{j=1,2,3} \alpha_{ji} \cos(\omega_i t)\mathbf{d}_j.$$

(iii) Put (ii) in matrix form. Note that your matrix should be symmetric.

(iv) Solve for the three eigenfrequencies.

(v) For each eigenfrequency find the corresponding eigenvector. Sketch these vector displacements and comment on them.

(vi) Obtain the force and mass matrices for the system using (a) the atomic displacement vectors and (b) the eigenvectors. Comment appropriately.

6.4 *Bond bending forces*

In Section 6.3 only those contributions to the potential energy that are associated with changes in bond length were included. A small displacement $r_i \mathbf{d}_i$ of an atom in a direction normal to a bond does not lead to a bond length change proportional to r_i and therefore does not appear in the potential energy expression within the harmonic approximation. However, the change in any bond angles due to such a displacement is linear in r_i and quadratic changes in the potential energy due to bond bending will arise.

For the CO_2 molecule suppose that the displacements depicted give bond bending potential energies $\frac{1}{2}b(r_4+r_6)^2$ and

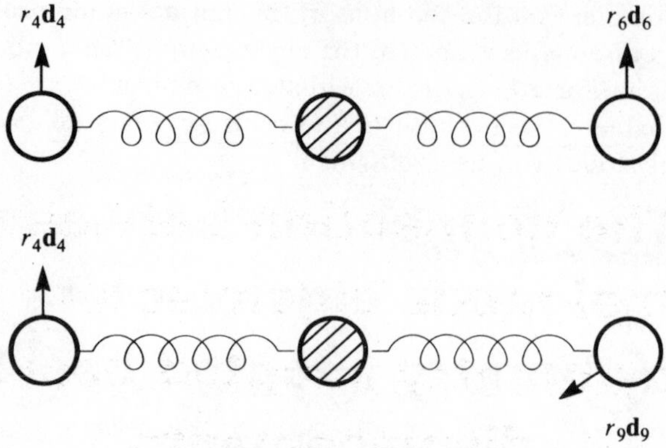

Figure 6.5

$\frac{1}{2}b(r_4^2+r_9^2)$ respectively. Write down the 9×9 effective force matrix of the molecule.

Discuss the eigenfrequencies and eigenvectors.

7

The connection between normal-mode eigenvectors and symmetry-adapted basis displacements

7.1 Wigner's theorem

The purpose of this section is to establish probably the most crucial result for applied group theory, a result which forms the link between abstract mathematical theorems and practical physical problems. This result sometimes appears under the name of Wigner's theorem, and we call it that here for convenience. In the context of normal modes:

> *Normal-mode eigenvectors are linear combinations of those symmetry-adapted displacements that belong to a specific row of a specific irreducible representation.*

We shall firstly establish this theorem by a verbal argument. Its consequences concerning the nature of the $[K-\omega^2 M]$ matrices will be discussed. Then we will state, without proof, a mathematical theorem called Schur's lemma, which leads to exactly the same results. An alternative statement of Wigner's theorem is given in section 7.2. Readers willing to accept the theorem for the present should move to that section, omitting the next subsection.

Proof of Wigner's theorem

The method of proof we shall employ here is to consider various situations, starting with the simplest, that could apply to the normal modes of systems. The important physical statement to be made is the following.

Consider some molecule oscillating in a specific normal mode of eigenvector $\mathfrak{D}_i$, with all the atoms vibrating at the same frequency ω_i. Then on application of a group symmetry operation (G_r) the atoms at different positions might vibrate or atoms might vibrate in different directions, but any vibrating atom must still oscillate at frequency ω_i. But under any operation G_r we know that $\mathfrak{D}_i$ transforms into some linear combination of the complete set of normal-mode eigenvectors:

$$G_r\mathfrak{D}_i = \sum_j \mathfrak{D}_j \Gamma^{\mathfrak{D}}(G_r)_{ji}.$$

Hence $\sum_j \mathfrak{D}_j \Gamma^{\mathfrak{D}}(G_r)_{ji}$ vibrates at frequency ω_i, and in turn every $\mathfrak{D}_j$ that appears in this sum with a non-zero coefficient must vibrate at ω_i. Thus if a set of l normal modes transform amongst themselves they must all have the same frequency eigenvalue—they form an l-fold degenerate set of modes. We now apply this statement to various situations.

(i) Given a mode ($\mathfrak{D}_1,\omega_1$) that is non-degenerate, then $\mathfrak{D}_1$ must transform only into plus or minus itself under all group operations. (If it transformed into a combination with any orthogonal mode, then this orthogonal mode would also have frequency ω_1). But this is just the property required of a symmetry-adapted basis displacement belonging to a one-dimensional irreducible representation. Hence $\mathfrak{D}_1$ must belong to a one-dimensional irrep. Furthermore, if in calculating the reduced representation of the group the irrep. correponding to $\mathfrak{D}_1$ was found to occur just once, then $\mathfrak{D}_1$ would have to be precisely the symmetry-adapted displacement $\mathfrak{D}_1$, belonging to this irrep. We can calculate such a displacement.

The symmetry-adapted displacement that forms the basis for a one-dimensional irrep. occurring just once in Γ^D must therefore itself be a normal-mode eigenvector.

(ii) For several normal modes, each transforming into themselves but each in accord with a *different* irrep., the above argument is easily extended.

For several irreps, each occurring once in Γ^D, each of their bases is a non-degenerate normal mode.

(iii) Now consider two non-degenerate modes ($\mathfrak{D}_1,\mathfrak{D}_2$) each transforming into itself in accord with the *same* one-dimensional irrep, Γ^j.

$$G_r\mathfrak{D}_1 = \mathfrak{D}_1\Gamma^j(G_r)_{11};\; G_r\mathfrak{D}_2 = \mathfrak{D}_2\Gamma^j(G_r)_{11}.$$

Hence

$$G_r[a\mathfrak{D}_1 + b\mathfrak{D}_2] = [a\mathfrak{D}_1 + b\mathfrak{D}_2]\Gamma^j(G_r)_{11};$$

any linear combination of $\mathfrak{D}_1$ and $\mathfrak{D}_2$ is also a basis for Γ^j, and the system has two such mutually orthogonal bases. Thus Γ^j must occur twice in Γ^D and the symmetry-adapted displacements obtained by calculation must be linear combinations of $\mathfrak{D}_1$ and $\mathfrak{D}_2$.

Conversely if a one-dimensional Γ^j occurs twice in Γ^D, then there are two non-degenerate normal modes, with eigenvectors that are linear combinations of any pair of symmetry-adapted basis displacements that we calculate. Note that because of the arbitrary choice of **Q** in the symmetry-adapted displacement formula there was always an associated arbitrariness in the $\mathbf{D}^j$ we calculated if the irrep. occurred more than once in Γ^D. We are now saying that there are particular choices of $\mathbf{D}^j$ that are also normal-mode eigenvectors.

(iv) Suppose two normal modes are degenerate; they transform into combinations of each other. Hence they form the basis for a two-dimensional irrep. If the corresponding irrep. occurs just once in Γ^D, then the calculated symmetry-adapted displacements must also be normal modes.

If a two-dimensional irrep. occurs once in Γ^D, then there is a two-fold degenerate pair of normal modes. The eigenvectors are the symmetry-adapted displacements. The fact that different **Q** choices or different choices of equivalent irreps led to different pairs of **D** vectors is easily handled. Any linear combination of degenerate normal modes is also a normal mode, at the same frequency. So we have the same arbitrariness in choosing $\mathfrak{D}$ vectors as we had for the **D** vectors.

(v) Finally consider the case where there are two pairs of two-fold degenerate normal modes, each pair transforming the same way (as Γ^k, say). Thus

$$G_r[\mathfrak{D}_1,\mathfrak{D}_2]=[\mathfrak{D}_1,\mathfrak{D}_2]\ \Gamma^k(G_r),$$

$$G_r[\mathfrak{D}_3,\mathfrak{D}_4]=[\mathfrak{D}_3,\mathfrak{D}_4]\ \Gamma^k(G_r).$$

The most general displacements to obey this form of equation are

$$G_r[a\mathfrak{D}_1+b\mathfrak{D}_3,\ a\mathfrak{D}_2+b\mathfrak{D}_4]=[a\mathfrak{D}_1+b\mathfrak{D}_3,\ a\mathfrak{D}_2+b\mathfrak{D}_4]\ \Gamma^k(G_r).$$

Then any symmetry-adapted displacements belonging to the two rows of Γ^k must be identical combinations of the mode eigenvectors belonging to the two rows. There must be two orthogonal combinations, corresponding to different a and b pairings.

Conversely, if a two-dimensional irrep. occurs twice in Γ^D, then the four normal-mode eigenvectors must be (a) one specific linear combination of the two symmetry-adapted displacements belonging to the first row of the irrep. and the same combination of the two displacements belonging to the second row; these two modes will be degenerate; (b) a second, mutually orthogonal pair of degenerate modes.

(vi) With the above argument as a basis it is relatively easy to extend to the case of a complicated, reduced representation.

7.2 Restatement of Wigner's theorem

> *If an l-dimensional irrep. occurs m times in Γ^D, then there are m sets of l-fold degenerate normal modes. Each mode eigenvector is a linear combination of the m displacements that belong to the same row of the irrep. (By analogy to case (v) there are further restrictions on the forms of the linear combinations.)*

This summarizes Wigner's theorem. The most immediate consequence concerns the degeneracies of the modes. Given the form of the reduced representation of a system (which is easy to obtain) we can immediately predict the number of distinct frequencies at which the system can oscillate and the degeneracies of the modes. Furthermore the modes have specific symmetry types; they can be labelled by the symmetry group irreducible representation labels.

In our ongoing example of the planar H vibrations of NH_3, we obtained

$$\Gamma^D = 1.A_1 \oplus 1.A_2 \oplus 2.E.$$

This tells us that there is one non-degenerate mode of symmetry type A_1, one of type A_2, and two pairs of doubly-degenerate modes of type E. Note that the presence of E tells us about the degeneracy (double); the multiplying factor tells us the number of such degenerate sets of modes. It is a coincidence that in this example the degeneracies and number of occurrences are equal. There are in summary just four possible distinct frequencies of oscillation of the system allowed by symmetry. (We shall see that in practice there are less, but this will be a consequence of the specific system under consideration rather than the symmetry.)

7.3 The form of $K^D - \omega^2 M^D$

It is now quite clear that the use of **D** vectors should help considerably in our search for normal modes eigenfrequencies and eigenvectors. In this section we want to extend the arguments of Section 7.1 in order to construct the form of the matrix β which will transform the calculable **D** vectors into normal-mode $\mathfrak{D}$ vectors:

$$[\mathfrak{D}_1 \ldots . \mathfrak{D}_{3N}] = [\mathbf{D}_1 \ldots . \mathbf{D}_{3N}]\beta.$$

This same β must diagonalize the force and mass matrices in accord with

$$\beta^{-1}[K^D - \omega^2 M^D]\beta = [K^{\mathfrak{D}} - \omega^2 M^{\mathfrak{D}}].$$

(Note that $\boldsymbol{K}^D$ is defined such that the potential energy under an arbitrary displacement $\Sigma_i R_i D_i$ is $\frac{1}{2}\tilde{R}K^D R$ etc.) Having obtained the form of β we shall then extract the form that $[K^D - \omega^2 M^D]$ must have. Thus, for example:

Case (i) of Section 7.1—$\Gamma^D = \Gamma^j$ (indicating a one-dimensional irrep.),

$$[\mathfrak{D}_1] = [\mathbf{D}_1]\,[1].$$

Case (ii)—$\Gamma^D = \Gamma^j \oplus \Gamma^{j'} \oplus \Gamma^{j''}$ (three one-dimensional irreps),

$$[\mathfrak{D}_1, \mathfrak{D}_2, \mathfrak{D}_3] = [\mathbf{D}_1, \mathbf{D}_2, \mathbf{D}_3] \begin{bmatrix} 1 & 0 & 0 \\ 0 & 1 & 0 \\ 0 & 0 & 1 \end{bmatrix}.$$

Case (iii)—$\Gamma^D = 2\Gamma^j$,

$$[\mathfrak{D}_1, \mathfrak{D}_2] = [\mathbf{D}_1, \mathbf{D}_2] \begin{bmatrix} \beta_{11} & \beta_{12} \\ \beta_{21} & \beta_{22} \end{bmatrix},$$

for a specific, orthonormal β matrix.

Case (iv)—$\Gamma^D = \Gamma^k$ (a two-dimensional irrep.),

$$[\mathfrak{D}_1, \mathfrak{D}_2] = [\mathbf{D}_1, \mathbf{D}_2] \begin{bmatrix} \beta_{11} & \beta_{12} \\ \beta_{21} & \beta_{22} \end{bmatrix},$$

but because the modes are degenerate any orthonormal β can be used. It is sensible to choose the unit diagonal matrix.

Case (v)—$\Gamma^D = 2\Gamma^k$,

$$[\mathfrak{D}_1, \mathfrak{D}_2, \mathfrak{D}_3, \mathfrak{D}_4] = [\mathbf{D}_1, \mathbf{D}_2, \mathbf{D}_3, \mathbf{D}_4] \begin{bmatrix} \beta_{11} & 0 & \beta_{13} & 0 \\ 0 & \beta_{11} & 0 & \beta_{13} \\ \beta_{31} & 0 & \beta_{33} & 0 \\ 0 & \beta_{31} & 0 & \beta_{33} \end{bmatrix}.$$

Here $\mathbf{D}_1$ and $\mathbf{D}_2$ are partner vectors, as are $\mathbf{D}_3$ and $\mathbf{D}_4$. The vectors $\mathbf{D}_1$ and $\mathbf{D}_3$ are assumed to belong to the first row of Γ^k and $\mathbf{D}_2$,$\mathbf{D}_4$ to the second row.

(vi) In our example, $\Gamma^D = A_1 \oplus A_2 \oplus 2E$,

$$[\mathfrak{D}_1, \mathfrak{D}_2, \mathfrak{D}_3, \mathfrak{D}_4, \mathfrak{D}_5, \mathfrak{D}_6] = [\mathbf{D}_1, \mathbf{D}_2, \mathbf{D}_3, \mathbf{D}_4, \mathbf{D}_5, \mathbf{D}_6] \begin{bmatrix} \beta_{11} & & & & & \\ & \beta_{22} & & & & \\ & & \beta_{33} & 0 & \beta_{35} & 0 \\ & & 0 & \beta_{33} & 0 & \beta_{35} \\ & & \beta_{53} & 0 & \beta_{55} & 0 \\ & & 0 & \beta_{53} & 0 & \beta_{55} \end{bmatrix}.$$

All non-marked elements are zero. In addition, the orthonormality conditions require

$$\beta_{11} = \beta_{22} = 1,\ \beta_{33}^2 + \beta_{53}^2 = \beta_{35}^2 + \beta_{55}^2 = 1 \text{ and } \beta_{33}\,\beta_{35} + \beta_{53}\,\beta_{55} = 0.$$

We are concerned therefore with only 6 independent transformation matrix elements, rather than the 36 that might relate the **d**-vectors to the eigenvectors.

(vii) Quite generally it is useful to notice that, if we reorder the symmetry-adapted displacements and corresponding normal-mode displacements such that those belonging to the same rows of each irrep. appear together in the **D** and $\mathfrak{D}$ vectors, then a block diagonalization of β is accomplished:

$$\Gamma^D = A_1 \oplus A_2 \oplus 2E,$$

$$[\mathfrak{D}_1,\mathfrak{D}_2,\mathfrak{D}_3,\mathfrak{D}_5,\mathfrak{D}_4,\mathfrak{D}_6] = [\mathbf{D}_1,\mathbf{D}_2,\mathbf{D}_3,\mathbf{D}_5,\mathbf{D}_4,\mathbf{D}_6]\begin{bmatrix} 1 & & & & & \\ & 1 & & & & \\ & & \beta_{33} & \beta_{35} & & \\ & & \beta_{53} & \beta_{55} & & \\ & & & & \beta_{33} & \beta_{35} \\ & & & & \beta_{53} & \beta_{55} \end{bmatrix}.$$

The 2×2 block

$$\begin{bmatrix} \beta_{33} & \beta_{35} \\ \beta_{53} & \beta_{55} \end{bmatrix}$$

originated from the factor of 2 multiplying E. The second appearance of this block, in exactly the same form, originated from the degeneracy exhibited by the two-dimensionality of E.

(viii) If $\Gamma^D = m\Gamma^l$ (m identical $l\times l$ blocks), then providing we reorder the displacements as indicated, $\beta = l\beta_m$ (l identical $m\times m$ blocks).

As well as β we are interested in the form of $[K^D - \omega^2 M^D]$. It is after all this matrix that we intend to manipulate. Its form is easy to see if we appreciate the matrix relationship (Problem 7.3),

$$[A\oplus B]\,[C\oplus D] = [AC\oplus BD],$$

where A and C are square matrices of the same dimension; and B,D are square and of some identical dimension. If we therefore consider $\beta^{-1}A\beta$, where the matrix A has the same blocked form as β, then the resultant product matrix is also of the same blocked form. (β^{-1} has the same blocked form as β). A more complicated A matrix results in elements in the product that fall away from the blocks. It is of course possible that the blocked form of the product may turn out to be entirely diagonal, and this is indeed what we require.

For the present the point to be made is that if the product is to be diagonal, then at least A must have blocked form identical to β. This is the required result. If the **D** vectors are ordered such that vectors belonging to the same row of the same irrep. appear together, then the matrix $[K^D - \omega^2 M^D]$ will be block diagonalized and we know that the β similarity transformation diagonalizes this matrix. The form of this block diagonalization is therefore the same as the form of the β matrices as discussed above and therefore complementary to the form of the reduced matrix representation of the group. Furthermore we can now see that for the case that Γ^D contains $m\Gamma^l$ there will be an $m \times m$ block in $[K^D - \omega^2 M^D]$ which appears l times. But the eigenvalues of ω are the roots of the determinant of this matrix (Section 6.3). Each of these blocks will lead to m roots (m different frequencies), but as the blocks are identical these m solutions must each appear l times; as expected then an l-fold irrep. in Γ^D corresponds to an l-fold degenerate set of normal modes. For our planar-H case, we must have the form:

$$[K^D - \omega^2 M^D]' = \begin{bmatrix} A_{11} & & & & & \\ & A_{22} & & & & \\ & & A_{33} & A_{35} & & \\ & & A_{35} & A_{55} & & \\ & & & & A_{33} & A_{35} \\ & & & & A_{35} & A_{55} \end{bmatrix}.$$

The prime indicates that the **D** vectors are reordered. Also we have chosen to define this physical matrix to be symmetric; that is $A_{53} = A_{35}$.

7.4 Schur's lemma

In the previous two sections, we have in effect proved Schur's lemma. Alternatively we could have stated Schur's result directly. Its consequence is precisely the form of $[K^D - \omega^2 M^D]$ that we have discussed. Two equivalent statements of the lemma are given.

Schur's lemma (statement 1) A matrix that commutes with all the matrices of an irreducible representation is a constant matrix (i.e., a scalar multiple of the unit matrix).

Schur's lemma (statement 2) If a group is represented by a set of completely reduced (block diagonal) matrices and some matrix A commutes with all the matrices in this set then:

(i) the element A_{ij} of A is zero unless the ith and jth elements of the total basis for the reduced rep. both belong to the same row of the same irrep.

(ii) if the ith and jth elements in the total basis both belong to the kth row of one of the irreps, Γ^j. And the i'th and j'th of the total basis both belong to the k'th row of the Γ^j, and $k' \neq k$.

Then $A_{i'j'} = A_{ij}$
(i' indicates a partner vector of i; etc.)
(one may have $i = j$, $i' = j'$ here)

Note: The form and order of the basis vectors of the irreducible reps must be such as to put the blocks which correspond to these reps into identical form (where the irrep. occurs more than once).

With a little persistence one can see that the second statement, applied to the case $\Gamma^D = A_1 \oplus A_2 \oplus 2E$, leads directly to the form of $[K^D - \omega^2 M^D]$ given in the previous section. It remains for us to verify that the force and mass matrices do commute with all the matrices Γ^D, so that we can be justified in applying Schur's lemma.

Proof that $K\Gamma = \Gamma K$, for any basis set. Consider an arbitrary displacement $\Sigma_i r_i \mathbf{d}_i$. The force matrix K^d is defined such that the potential energy for the displacements is $\frac{1}{2}\tilde{r}K^d r$. Rotate or otherwise operate on the displacement by some group operation G_r. The r_i amplitudes do not change but are now referred to the 'rotated' displacements $\mathbf{d}' = \mathbf{d}\Gamma^d(G_r)$. The original displacement has been changed to $\Sigma_i r_i \mathbf{d}'_i$. But we could equally write this as $\Sigma_i r'_i \mathbf{d}_i$, where $\mathbf{d}r' = \mathbf{d}'r = \mathbf{d}\Gamma^d(G_r)r$.

A displacement of amplitudes $\Gamma^d(G_r)r$ has an associated potential energy $\frac{1}{2}(\tilde{r}\tilde{\Gamma}^d(G_r))K^d(\Gamma^d(G_r)r)$. As G_r is a symmetry operation, the potential energy of the system cannot have been altered. Hence

$$\tilde{\Gamma}^d(G_r)K^d\Gamma^d(G_r) = K^d.$$

We know that Γ must be orthogonal. Pre-multiplying both sides by $\Gamma^d(G_r)$, we therefore obtain $K^d\Gamma^d = \Gamma^d K^d$.

Finally note that the choice of basis was not critical in this analysis and that the mass matrix, being associated with kinetic energy, is also invariant under G_r operations. Hence $K\Gamma = \Gamma K$ and $M\Gamma = \Gamma M$ for any representation and any G_r; and $[K^D - \omega M^D]$ will commute with all the Γ^D matrices. Schur's lemma can indeed be applied.

Problems

***7.1** Using matrix set (1) of the C_{3v} group representations (Table 3.1), demonstrate that a 2×2 matrix A that commutes with all of the 6 matrices in the set has to be of the form

$$\begin{bmatrix} A_{11} & 0 \\ 0 & A_{11} \end{bmatrix}.$$

(Do not appeal to Schur's lemma.)

7.2 The representation above was a two-dimensional irreducible one. Suppose instead you have a two-dimensional representation in a reduced form such as

$$\begin{bmatrix} \text{set (3)} & 0 \\ 0 & \text{set (3)} \end{bmatrix},$$

which equals

$$\begin{bmatrix} 1 & 0 \\ 0 & 1 \end{bmatrix}$$

for each G_r. What form will a matrix A take if it commutes with these $\Gamma(G_r)$?

7.3 Prove that

$$\begin{bmatrix} A & 0 \\ 0 & B \end{bmatrix}\begin{bmatrix} C & 0 \\ 0 & D \end{bmatrix} = \begin{bmatrix} AC & 0 \\ 0 & BD \end{bmatrix},$$

where A, C are square matrices of dimension n; B, D are matrices of dimension m, and all other elements outside the blocks are zeros.

***7.4** A molecule of T_d symmetry is found to have a representation of reduced form:

$$\Gamma^D = A_1 \oplus E \oplus T_1 \oplus 3T_2.$$

Write down the form of the $[K^D - \omega^2 M^D]$ matrix. Comment on the order of the **D** vectors to which your matrix refers. State the number of normal modes and the degeneracies of the eigenfrequencies.

8

The calculation of eigenfrequencies and eigenvectors

8.1 The NH_3 problem

In order to demonstrate how the conclusions of Chapter 7 are used in practice we will continue with the problem of the planar hydrogen motion of NH_3, for which the symmetry-adapted basis displacements were derived in Sections 5.3 and 5.4.

Given the form of the $[K^D - \omega^2 M^D]$ matrix, Section 7.3, there can be only five distinct matrix elements for the present problem; they are the elements 11, 22, 33, 35 and 55, referred to the symmetry-adapted displacements described in Figure 5.4. In contrast the matrix $[K^d - \omega^2 M^d]$ may have many non-zero elements that have no simple interrelations. It is useful now to make a few general comments on the evaluation of the force and mass matrix elements. Throughout it will be assumed that we can write:

$$\mathrm{KE} = \tfrac{1}{2} \sum_{\text{atoms}} m_{\text{atom}} (\text{velocity})^2,$$

$$\mathrm{PE} = \tfrac{1}{2} \sum_{\text{bonds}} k_{\text{bond}} |\text{extension}|^2.$$

In the basis displacement (**d**) representation, if we were to consider some arbitrary displacement $\Sigma r_i \mathbf{d}_i$, the atomic velocities and bond extensions can be formed in term of $\dot{r}_i^2$ and $r_i r_j$ respectively, and all elements of M^d and K^d are thereby obtained. If however, we only want specific elements of these matrices, it is simpler to consider either (i) $r_i\mathbf{d}_i$ alone to get diagonal elements or (ii) $(r_i\mathbf{d}_i + r_j\mathbf{d}_j)$ to get specific (ij) off-diagonal elements.

Example The planar-H problem.

(i) To find K^d_{16} in the planar-H problem. Consider $(r_1\mathbf{d}_1 + r_6\mathbf{d}_6)$, as shown in Figure 8.1.

The bond extensions are: a: $-r_6 \cos 60$,
b: $r_6 \cos 60 + r_1 \cos 30$,
c: $r_1 \cos 30$.

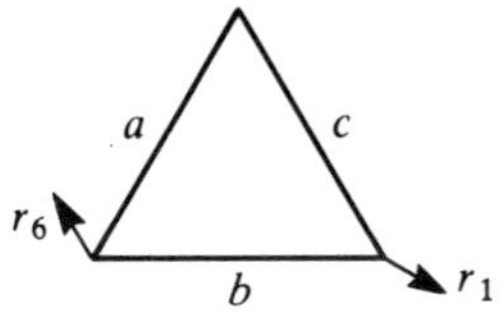

Figure 8.1 Schematic of $r_1\mathbf{d}_1 + r_6\mathbf{d}_6$, with the bonds labelled a, b, c.

Hence

$$\mathrm{PE} = \tfrac{1}{2}k[r_6^2/4 + (r_6/2 + r_1\sqrt{3}/2)^2 + r_1^2 3/4] = \tfrac{1}{2}[K^d_{11}r_1^2 + K^d_{66}r_6^2 + (K^d_{16} + K^d_{61})r_1r_6],$$

$$K^d_{11} = 3k/2, \qquad K^d_{66} = k/2, \qquad (K^d_{16} + K^d_{61}) = k\sqrt{3}/2.$$

We choose to define $K_{16} = K_{61}$, so finally $K^d_{16} = k\sqrt{3}/4$, which we note is non-zero.

Turning to the symmetry-adapted displacements:

(ii) To find K^D_{16} (Figure 8.2).

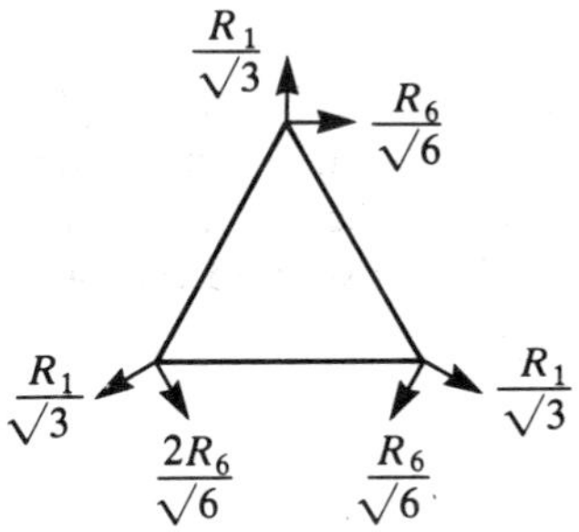

Figure 8.2 $R_1\mathbf{D}_1 + R_6\mathbf{D}_6$.

$$\mathbf{D}_1 = (\mathbf{d}_1 + \mathbf{d}_3 + \mathbf{d}_5)/\sqrt{3}, \ \mathbf{D}_6 = (\mathbf{d}_2 + \mathbf{d}_4 - 2\mathbf{d}_6)/\sqrt{6}.$$

Bond extensions: a: $\dfrac{R_1}{\sqrt{3}}\cos 30+\dfrac{R_6}{\sqrt{6}}\cos 60+\dfrac{R_1}{\sqrt{3}}\cos 30+\dfrac{2R_6}{\sqrt{6}}\cos 60,$

$$b: \frac{2R_1}{\sqrt{3}}\cos 30-\frac{3R_6}{\sqrt{6}}\cos 60,$$

$$c: \frac{2R_1}{\sqrt{3}}\cos 30.$$

Hence

$$\begin{aligned}\mathrm{PE}&=\tfrac{1}{2}k[(R_1+R_6\sqrt{(3/8)})^2+(R_1-R_6\sqrt{(3/8)})^2+R_1^2]\\&=\tfrac{1}{2}[K_{11}^DR_1^2+K_{66}^DR_6^2+(K_{16}^D+K_{61}^D)R_1R_6],\end{aligned}$$

$$K_{11}^D=3k,\qquad K_{66}^D=3k/4,\qquad (K_{16}^D+K_{61}^D)=0.$$

Examples (i) and (ii) are included to demonstrate that whilst K_{16}^d is non-zero (as are most other elements of K^d), we obtain, as required, that K_{16}^D is zero. [*Note*: Terms in the bond extension of higher orders in the R_i are ignored.] The non-zero elements of K^D and M^D are obtainable by considering just the three vectors $R_1\mathbf{D}_1$, $R_2\mathbf{D}_2$ and $(R_3\mathbf{D}_3+R_5\mathbf{D}_5)$.

(iii) K_{11}^D and M_{11}^D.

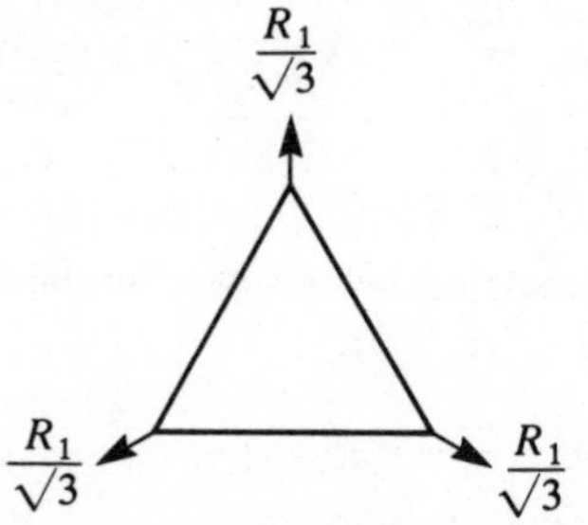

Figure 8.3 $R_1\mathbf{D}_1=\dfrac{R_1}{\sqrt{3}}(\mathbf{d}_1+\mathbf{d}_3+\mathbf{d}_5)$.

Each bond extension $=\dfrac{2R_1}{\sqrt{3}}\cos 30$:

$$\mathrm{PE}=3.\tfrac{1}{2}kR_1^2,\qquad K_{11}^D=3k.$$

Each mass moves at velocity $\dot{R}_1/\sqrt{3}$:

$$\mathrm{KE}=3.\tfrac{1}{2}m\dot{R}_1^2/3,\qquad M_{11}^D=m.$$

(iv) K_{22}^D and M_{22}^D.

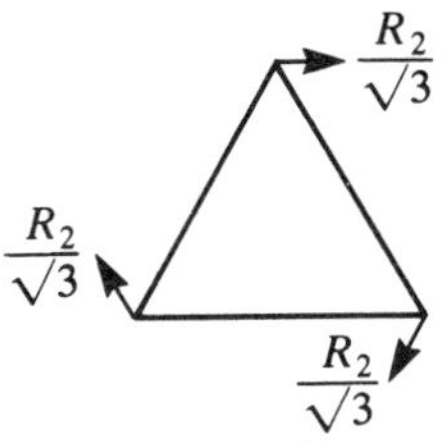

Figure 8.4 $R_2\mathbf{D}_2 = \frac{R_2}{\sqrt{3}}(\mathbf{d}_2 + \mathbf{d}_4 + \mathbf{d}_6)$.

There are no bond extensions: $K^D_{22} = 0$:

$$\text{KE} = 3.\tfrac{1}{2}m\dot{R}_2^2/3, \qquad M^D_{22} = m.$$

(v) K^D_{33}, K^D_{55}, K^D_{35}, etc...

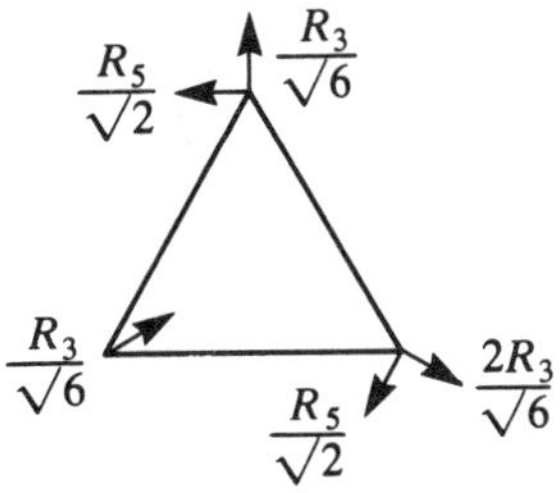

Figure 8.5 $(R_3\mathbf{D}_3 + R_5\mathbf{D}_5) = R_3(\mathbf{d}_1 + \mathbf{d}_3 - 2\mathbf{d}_5)/\sqrt{6} + R_5(\mathbf{d}_2 - \mathbf{d}_4)/\sqrt{2}$

Extension of bonds a, b: $-\frac{1}{2\sqrt{2}}(R_3 + R_5)$.

Extension of bond c: $\frac{1}{\sqrt{2}}(R_3 + R_5)$:

$$\text{PE} = \tfrac{1}{2}k[2/8(R_3 + R_5)^2 + 1/2(R_3 + R_5)^2];$$

$$K^D_{33} = 3k/4,\ K^D_{55} = 3k/4,\ K^D_{35} = K^D_{53} = 3k/4;$$

$$\text{KE} = \tfrac{1}{2}m[2(\dot{R}_3^2/6 + \dot{R}_5^2/2 + 4/6\dot{R}_5^2];$$

$$M^D_{33} = M^D_{55} = m, \qquad M^D_{35} = M^D_{53} = 0.$$

As anticipated, K^D_{55} is equal to the value of K^D_{66} obtained in case (ii). The equality of K_{33}, K_{55}, K_{35} is however a coincidence in that it is a consequence of the particular model chosen rather than its symmetry. It does make the resultant solutions simpler, however.

The full matrix

With the reordering, by rows of irreps, of the **D**-vectors, we now have, with respect to $\mathbf{D}_1$, $\mathbf{D}_2$, $\mathbf{D}_3$, $\mathbf{D}_5$, $\mathbf{D}_4$, $\mathbf{D}_6$:

$$[K^D-\omega^2 M^D]'=\begin{bmatrix} 3k-\omega^2 m &&&&& \\ & -\omega^2 m &&&& \\ && \frac{3k}{4}-\omega^2 m & \frac{3k}{4} && \\ && \frac{3k}{4} & \frac{3k}{4}-\omega^2 m && \\ &&&& \frac{3k}{4}-\omega^2 m & \frac{3k}{4} \\ &&&& \frac{3k}{4} & \frac{3k}{4}-\omega^2 m \end{bmatrix}$$

The prime on the left-hand-side is to indicate the new ordering.

The solutions are:

(i) $3k-\omega_1^2 M=0$; $\mathfrak{D}_1=\mathbf{D}_1$, $\omega_1=3k/m$.

(ii) $-\omega_2^2 m=0$; $\mathfrak{D}_2=\mathbf{D}_2$, $\omega_2=0$.

(iii) $\begin{vmatrix} \frac{3k}{4}-\omega^2 m & \frac{3k}{4} \\ \frac{3k}{4} & \frac{3k}{4}-\omega^2 m \end{vmatrix}=0$; $[\mathfrak{D}_3,\mathfrak{D}_5]=[\mathbf{D}_3,\mathbf{D}_5]\begin{bmatrix} \beta_{33} & \beta_{35} \\ \beta_{53} & \beta_{55} \end{bmatrix}$,

such that

$$\begin{bmatrix} \frac{3k}{4}-\omega_3^2 m & \frac{3k}{4} \\ \frac{3k}{4} & \frac{3k}{4}-\omega_3^2 m \end{bmatrix}\begin{bmatrix} \beta_{33} \\ \beta_{53} \end{bmatrix}=\begin{bmatrix} 0 \\ 0 \end{bmatrix}$$

and a similar solution for β_{35} and β_{55} using ω_5.

The roots of the determinant give $\omega_3=0$, $\omega_5=\sqrt{(3k/2m)}$. Substituting to find the β matrix and normalizing:

$$\beta_{53}=\beta_{33}=\frac{1}{\sqrt{2}};\ \beta_{35}=-\beta_{55}=\frac{1}{\sqrt{2}}.$$

Therefore

$$\mathfrak{D}_3=(\mathbf{D}_3+\mathbf{D}_5)/\sqrt{2}, \qquad \omega_3=0.$$
$$\mathfrak{D}_5=(\mathbf{D}_3-\mathbf{D}_5)/\sqrt{2}, \qquad \omega_5=\sqrt{(3k/2m)}.$$

(iv) Identical frequency solutions to case (iii), and identical β elements, but referred to $\mathbf{D}_4$ and $\mathbf{D}_6$:

$$\mathfrak{D}_4 = (\mathbf{D}_4 + \mathbf{D}_6)/\sqrt{2}, \qquad \omega_4 = 0.$$
$$\mathfrak{D}_6 = (\mathbf{D}_4 - \mathbf{D}_6)/\sqrt{2}, \qquad \omega_6 = \sqrt{(3k/2m)}.$$

This completes the solution of the problem; the results are summarized in Figure 8.6. Readers still uncertain as to how the above solution is obtained in detail may like to tackle Problem 8.1 at this stage.

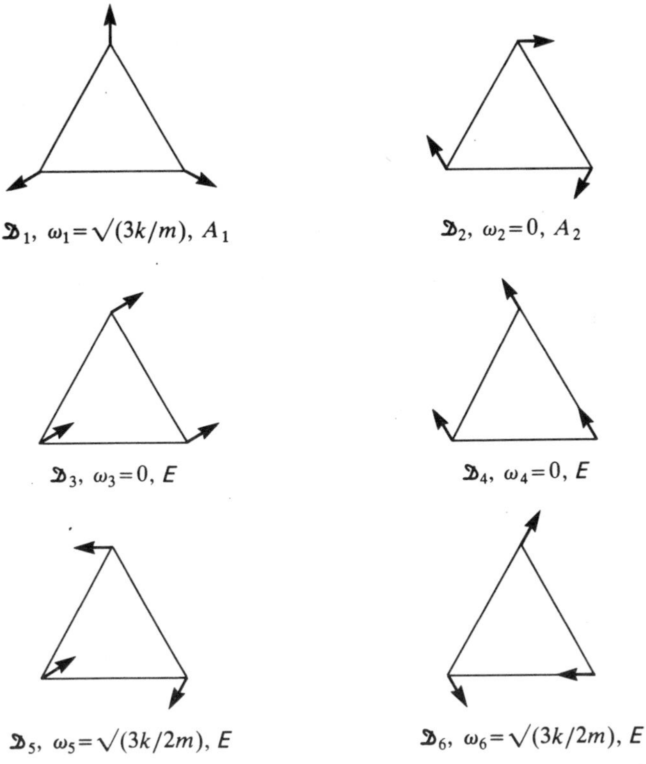

Figure 8.6 The normal modes of the planar-H model.

(i) As expected the A_1 mode is a 'breathing mode', of relatively high frequency.

(ii) The A_2 mode is a rotation mode; it must have zero frequency, because none of the bond lengths alter ($K_{22}^{D} = 0$).

(iii) The E modes $\mathfrak{D}_3$ and $\mathfrak{D}_4$ are the translational modes, at zero frequency; they are partner vectors.

(iv) $\mathfrak{D}_5$, $\mathfrak{D}_6$ are also partner vectors. They represent the two remaining vibrations of the system.

The full NH_3 problem

At this stage we should recall that we started with NH_3, with twelve degrees of freedom. The model evaluated completely above represents a far more restricted molecule. Moreover the modes obtained above are not in general going to be the normal modes of NH_3. To see this we need to obtain the characters of the twelve-dimensional displacement representation of NH_3, and reduce it.

	E	$2C_3$	3σ
Γ^d	12	0	2

These characters are obtained by considering three orthogonal displacements on each atom. Under C_3 only the nitrogen atom remains in position. Its displacements in the plane parallel to the H_3-plane transform like **i**, **j** with -1 total contribution to $\chi^d(C_3)$ (see matrix set (1)). The displacement perpendicular to this plane is unaltered and gives a $+1$ contribution. Thus $\chi^d(C_3)=0$. $\chi^d(\sigma)$ can be obtained likewise.

Using the C_{3v} character table we obtain

$$\Gamma^D = 3A_1 \oplus A_2 \oplus 4E.$$

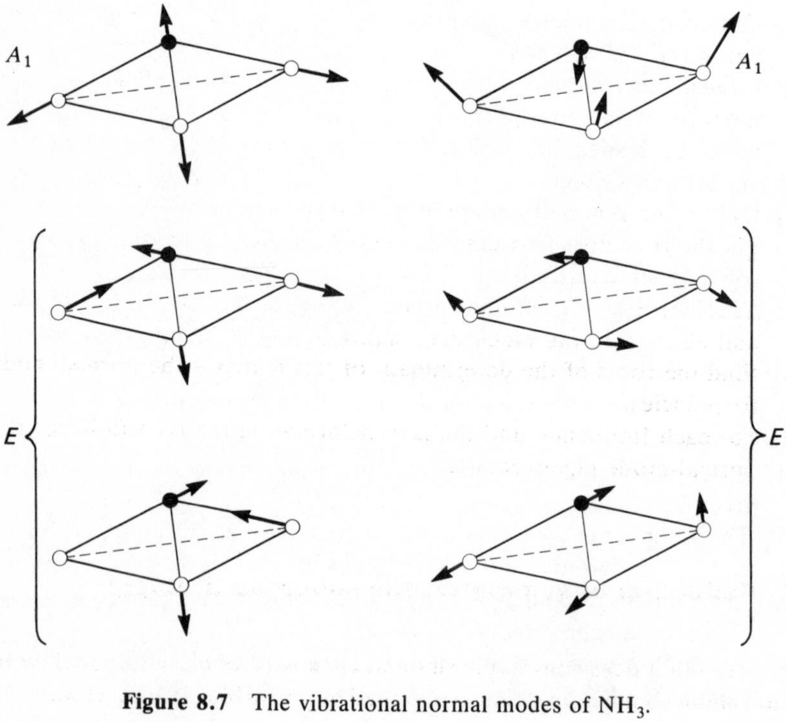

Figure 8.7 The vibrational normal modes of NH_3.

On application of Schur's lemma there must be four nondegenerate modes and four doubly degenerate. Also there are three symmetry-adapted displacements belonging to A_1, so that we have a 3×3 matrix problem to solve for the corresponding eigenvalues and vectors. The vectors will be linear combinations that contain $\mathbf{D}_1$ as found above, but there will not be a mode with $\mathfrak{D}$ identically equal to $\mathbf{D}_1$. In fact the breathing mode will involve all four atoms moving in phase, along axes through the centre-of-mass of the molecule, such that this centre is stationary. There must also now be six zero-frequency modes, the A_2 rotation already found, (with the N-atom unmoved) and a pair of rotations belonging to E. Of the translations there will be a pair similar to $\mathfrak{D}_3$, $\mathfrak{D}_4$ but involving N-atom translations also, and an A_1 mode—the translation perpendicular to the H_3-plane. This leaves two nondegenerate vibrations and two doubly-degenerate pairs, a total of $(3N-6)$ vibrations. (See Problem 5.6).

8.2 Summary of the classical solution for the normal modes of molecules

(i) Define a complete, orthonormal set of basis vectors, $\mathbf{d}$.
(ii) Obtain the symmetry operations of the group of the molecule and hence the characters χ^d.
(iii) Look up the character table for the group concerned and hence obtain the form of the reduced representation, Γ^D, using the a_j formula. This indicates the degeneracies and symmetry types of the normal modes, via Schur's lemma.
(iv) Obtain the symmetry-adapted basis displacements, $\mathbf{D}$.
(v) Put the $\mathbf{D}$ vectors in order, having obtained those belonging to each row of each irrep.
(vi) Consider $R_1\mathbf{D}_1$ etc. as appropriate, define force constants and masses, and hence find the block diagonalized $[K^D - \omega^2 M^D]$ matrix.
(vii) Find the roots of the determinant of this matrix—the normal-mode frequencies.
(viii) For each frequency find the β transformation matrix and hence the normal-mode eigenvectors.

8.3 Reduction of symmetry: the subgroup problem

There are often occasions in physics and chemistry where additional information about some sample can be obtained by perturbing it in some way. For example, by applying a uniform electric or magnetic field, or a constant stress. Other processes with similar effect from the point of view of group theory are the replacement of atoms or groups of atoms in a molecule or solid, by similar groups, different isotopes etc.

In this section we shall consider any such alteration which on the one hand is not expected to change the properties of the sample dramatically yet on the other hand clearly reduces the symmetry of the system. This is the *subgroup problem*; group theory has something very specific to say about this situation.

One specific problem is sufficient to exhibit the points we wish to make concerning the relationship between the normal modes of the original system and those of the perturbed system. The problem will be returned to in Section 10.3 in connection with quantum-mechanical eigenfunctions.

For the present we shall consider the normal modes of the molecule sulphur hexafluoride (SF_6), firstly with identical fluorine atoms and secondly with one atom replaced by a different fluorine isotope having, consequently, a slightly different mass. We should not expect any dramatic changes in the frequencies or eigenvectors of the normal modes in this reduced-symmetry problem.

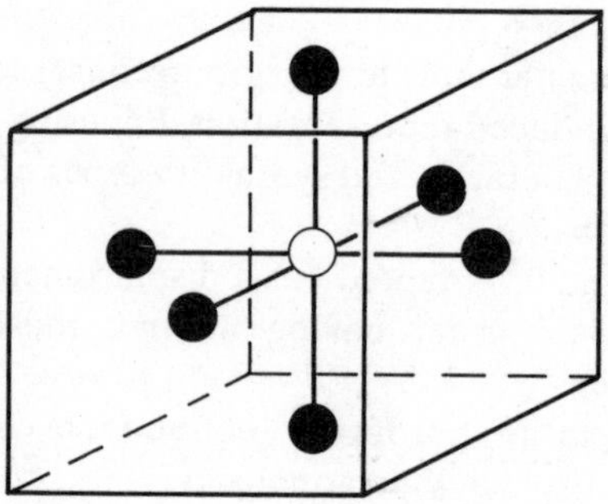

Figure 8.8 The structure of SF_6.

Figure 8.8 shows the structure of SF_6. It is useful to think of the sulphur atom at the body centre of a cube and the fluorine atoms at the six face centres. In this way it becomes clear that the molecule has cubic symmetry—the symmetry of the O_h group. The degeneracies and symmetry types of the normal modes are obtained by the techniques developed in this chapter, they are summarized briefly in Table 8.1.

Table 8.1 Reduction of the representation for the displacements of SF_6.

O_h	E	$8C_3$	$6C_4$	$3C_2$	$6C_2'$	i	$8S_6$	$6S_4$	$3\sigma_h$	$6\sigma_d$
Γ^d	21	0	3	−3	−1	−3	0	−1	5	3

$\Gamma^d \Rightarrow A_{1g} \oplus E_g \oplus T_{1g} \oplus T_{2g} \oplus 3T_{1u} \oplus T_{2u}$

There are seen to be eight distinct normal-mode frequencies for the molecule (barring accidental degeneracies). Only one normal mode, of A_{1g} type, is non-degenerate; there is one doubly-degenerate pair of modes and six triply-degenerate sets. If we were to pursue the problem beyond this stage we should need to construct only five distinct diagonal matrix elements of $[K^D - \omega^2 M^D]$ and one 3×3 block, the diagonalization of which gives the T_{1u} mode eigenvectors and frequencies.

Now replace one fluorine atom; the symmetry of the system SF_5F' reduces to C_{4v}. The operations that remain are just those that leave the F′ isotope in place, that is rotations about the SF′ axis or reflections with respect to planes that contain this axis.

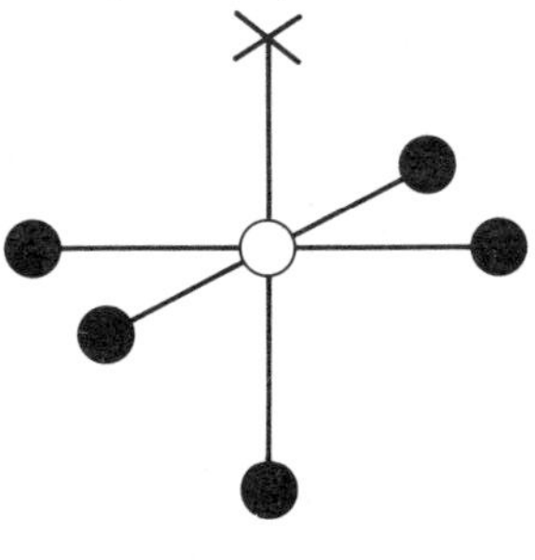

Figure 8.9 The structure of SF_5F'.

Table 8.2 Reduction of the SF_5F' displacement representation.

C_{4v}	E	$2C_4$	C_2	$2\sigma_v$	$2\sigma_d$
Γ^d	21	3	−3	5	3

$\Gamma^d \Rightarrow 5A_1 \oplus A_2 \oplus 2B_1 \oplus B_2 \oplus 6E$

One could use exactly the same methods as with the O_h group in order to find the degeneracies and normal modes of the new system (see Table 8.2). However, this takes no account of the fact that the original and new modes must be similar. Consider instead a particular degenerate set of modes of the

O_h case—say the T_{1g} set. The three eigenvectors must have the property that they transform into linear combinations of each other under all the O_h group operations; that is why they were degenerate. In particular, then, they will transform as a set under those O_h operations that remain in the reduced-symmetry C_{4v} subgroup. Hence they form a basis for a representation of C_{4v}. In general, then,

partner vector normal modes of the full group form a basis for a representation of the subgroup. This latter representation may not be an irreducible representation for the subgroup.

In our example there is no three-dimensional representation in C_{4v}, so the T_{1g} modes must be a basis for a reducible representation. Conversely the A_{1g} mode must be a basis for one of the one-dimensional (irreducible) representations of C_{4v}. Continuing with the T_{1g} modes, though, we may not have their eigenvectors but we do know the characters for their transformation under any of the O_h (or C_{4v}) operations and can therefore find the reduced form of the C_{4v} representation for which they do form a basis. We obtain $A_2 \oplus E$, meaning that from the original T_{1g} mode eigenvectors it is possible to construct three orthogonal linear combinations such that one combination is a symmetry-adapted function belonging to the A_2 irrep. of C_{4v} and such that the other two combinations belong to E. In terms of degeneracy the T_{1g} three-fold degenerate modes of the O_h system will split into one singly and one doubly-degenerate mode when the F atom is altered.

Table 8.3 Reduction of irreducible representations of the O_h group in terms of the irreps of the C_{4v} subgroup.

C_{4v}	E	$2C_4$	C_2	$2\sigma_v$	$2\sigma_d$	
$\Gamma(O_h)=A_{1g}$	1	1	1	1	1	$\Gamma(O_h) \Rightarrow A_1$
E_g	2	0	2	2	0	$A_1 \oplus B_1$
T_{1g}	3	1	−1	−1	−1	$A_2 \oplus E$
T_{2g}	3	−1	−1	−1	1	$B_2 \oplus E$
T_{1u}	3	1	−1	1	1	$A_1 \oplus E$
T_{2u}	3	−1	−1	1	−1	$B_1 \oplus E$

Figure 8.10 gives the resulting form of the frequency solutions (barring accidental degeneracies). The relative positions of the eigenfrequencies are of course not specified by group theory; the diagram is meant to indicate (i) the number of distinct frequencies and (ii) the presence of 'split-modes', modes of different but close frequency.

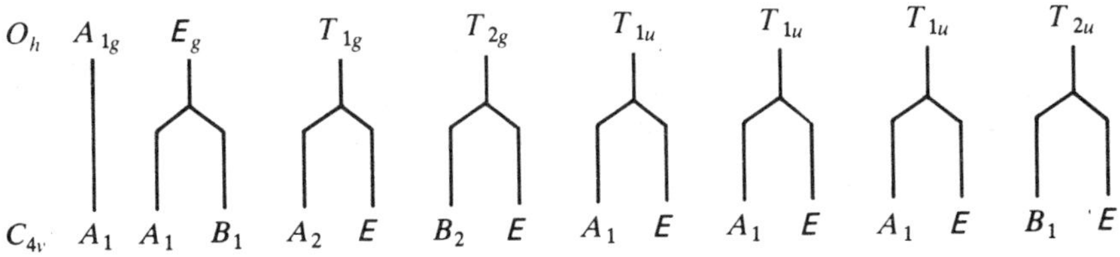

Figure 8.10 Splitting of the normal mode frequencies and degeneracies of SF_6.

As they must, the results of this analysis give the same number of distinct frequencies and the same symmetry types as obtained by the direct method of Table 8.2. In the C_{4v} group there are now fifteen distinct frequencies, eight corresponding to singly degenerate modes and seven to two-fold degenerate modes.

8.4 Hints on the use of character tables with respect to the normal-mode problem

Inversion groups

You will notice, and it will become apparent why in Chapter 13, that those groups that contain the inversion operation also contain one improper rotation (or reflection) for every proper rotation. The O_h group used in Section 8.3 is such an example. Now it can be relatively awkward to obtain characters for operations such as S_6 by comparison to those for the proper rotations, so consider what happens if we take the SF_6 normal-mode problem but ignore all the improper operations (Table 8.4). We find the same degeneracies and hence the same number of distinct mode frequencies (cf. Table 8.2). Quite generally, if we are only concerned to obtain the

Table 8.4 Reduction of the displacement representation of SF_6, ignoring improper operations.

$O_h \Rightarrow O$	E	$8C_3$	$6C_4$	$3C_2$	$6C'_2$
Γ^d	21	0	3	-3	-1

$\Gamma^d \Rightarrow A_1 \oplus E \oplus 4T_1 \oplus 2T_2$

degeneracies, then we can cut down by at least a factor of two on our effort if we ignore the improper rotations. However, if we wish to go on to solve the full normal-mode problem, we now have to diagonalize a matrix containing a 4×4 block and a 2×2 block compared to the single 3×3 block in the previous case. It will be fruitful to include the improper rotations, then, in order to simplify the diagonalization process. The point is that the symmetry-adapted displacements belonging to T_{1g} and T_{1u} respectively are not distinguished if the improper rotation operations are ignored.

The ordering of classes

It is often the case that in sorting out the specific symmetry operations of a group there is not a totally unambiguous correspondence between the classes written in the character table and the classes one writes down from looking at the symmetry. For example, How are C_2' and C_2'' distinguished in the D_4 group given in Table 4.11 or in the D_{6h} group given in Appendix A2? Although it may not be necessary to specify such operations exactly in achieving physical results it is essential to be consistent in the use of a given character table, and it is nice to write one's final answers in the same form as the author of the tables. When it does matter there are various hints in the tables that enable such consistency to be achieved.

(a) For those groups containing inversion the order of the improper rotations (described in terms of iC_n) is usually the order of the proper rotations C_n. In O_h, for example, $\sigma_d \equiv iC_2'$, the σ_d planes are normal to the C_2' axis. In this case there is no ambiguity, because the number of elements in each class is sufficient to define the form of the operations. An ambiguity does arise in the D_{6h} group, where one might define C_{2a} and C_{2b} and ask which is to correspond to C_2' and which to C_2'' in the character tables (see Figure 8.11).

Providing we recognize that the σ_d-plane is normal to C_2' etc. we will

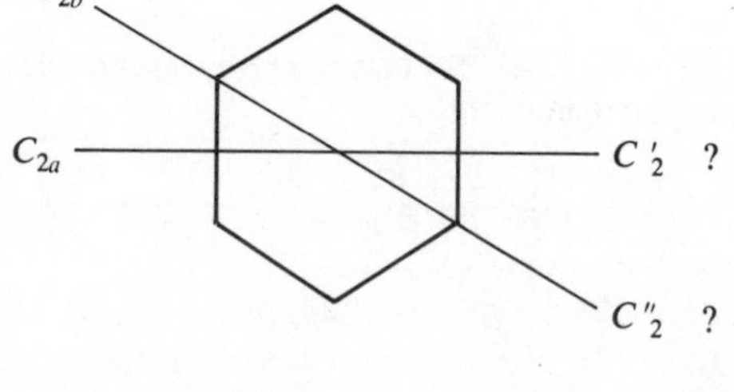

Figure 8.11 Schematic of C_2 operations of D_{6h} group.

not be led to incorrect physical results whichever choice we make. The only difference obtained by choosing $C_2' \equiv C_{2a}$, $C_2'' = C_{2b}$ rather than $C_2' = C_{2b}$, $C_2'' = C_{2a}$, is that different labels will end up being attached to the eigenvectors; B_{1g} and B_{2g} will interchange as will B_{1u} and B_{2u}. Similar statements apply to the D_4 group.

(b) Alongside most character tables one finds x, y, z, R_x, R_y, R_z. These are present in order to tell us how the vectors corresponding to x,y,z directions transform (or how functions of the symmetry of x,y,z transform) and how pseudo (axial) vectors transform (cf. Section 11.1). Thus in the cubic groups x, y, and z form the basis for a three-dimensional irrep. In D_2, however, the three directions are distinguishable and x, y, and z 'belong' to different irreps. These letters also give us a hint as to the author's intentions in labelling the classes. In D_2,x transforms as B_1 and we see that the character for B_1 is $(+1)$ for the C_2' operation and (-1) for the C_2'' and C'''_2 operations. Hence C_2' must be such that it transforms x into plus itself, etc. One is therefore meant to define the x-axis and C_2' axis to be the same; similarly $y \equiv C_2''$, $z \equiv C'''_2$.

Translations and rotations

Suppose that the displacements corresponding to the translation of a molecule in say the x-direction transform into a y-direction translation under a particular operation (for example a C_4 operation). The unit vector **i** at the point-group centre will therefore transform into **j** under the same operation. The symmetries of the translation modes of a molecule are therefore exactly the same as the symmetries of the polar vectors **i**, **j**, **k** centred at the point-group centre of the molecule. Equally the rotational symmetries are those of axial vectors $\mathbf{j} \wedge \mathbf{k}$ etc. One can then either determine the zero-frequency mode symmetries for any given molecule, or appeal directly to the character tables. The translational symmetries are the irreps. identified with x, y, z in the tables; the rotations are identified with R_x, R_y, R_z.

Uniform, breathing modes

In all of the examples used so far the symmetric irrep. (A_1, A_{1g}, etc.) appeared in the reduced, displacement representation Γ^D. This is a general result. For all molecules there is at least one symmetric mode, a uniform breathing mode in which all equivalently positioned atoms oscillate in and out together.

Problems

8.1 Consider the $[K^D - \omega M^D]$ matrix

$$A = \begin{bmatrix} k-\omega^2 m & k & 0 \\ k & k-\omega^2 m & 0 \\ 0 & 0 & k'-\omega^2 m \end{bmatrix}.$$

(i) Write down the form of the reduced Γ^D matrix.

(ii) Write out the three simultaneous equations corresponding to the matrix equation

$$A \begin{bmatrix} \beta_{1i} \\ \beta_{2i} \\ \beta_{3i} \end{bmatrix} = \begin{bmatrix} 0 \\ 0 \\ 0 \end{bmatrix}.$$

and find the frequency eigenvalues, ω_1, ω_2, ω_3.

(iii) Substitute each eigenfrequency back into the simultaneous equations and hence build up the form of the 3×3 β matrix. Write down the resulting normalized normal-mode eigenvectors in terms of $\mathbf{D}_1$, $\mathbf{D}_2$ and $\mathbf{D}_3$.

(iv) Comment on the form of β if $k'=2k$.

8.2** ***The water molecule

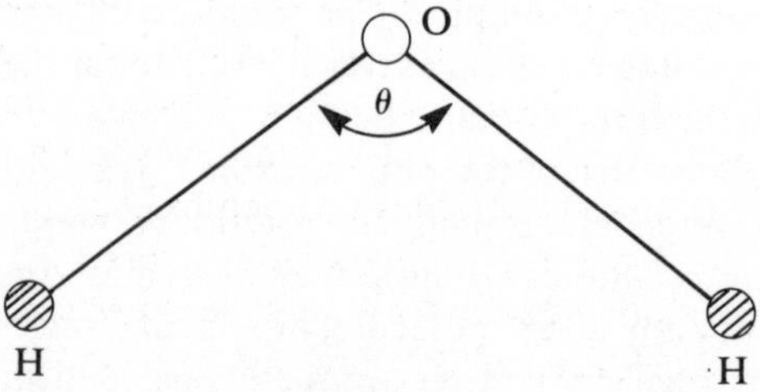

Figure 8.12

(i) What are the four rotation and reflection operations of the H_2O group?

(ii) How many distinct-frequency modes of vibration are there, barring accidental degeneracies?

(iii) Neglecting H–H bonding, find the A_1 symmetry-adapted displacements and the corresponding force and mass matrices. Verify that $\omega=0$ for one of the A_1 modes.

(iv) Obtain and sketch all the symmetry-adapted displacements. Can you pick out the six combinations that are the zero-frequency translations and rotations? By this means, or otherwise, state the symmetry types of the vibrational ($\omega \neq 0$) normal modes (you may even be able to sketch them roughly). Use the character tables as appropriate.

8.3 A molecule consists of four identical atoms in a plane, at the corners of a rhombus (Figure 8.13).

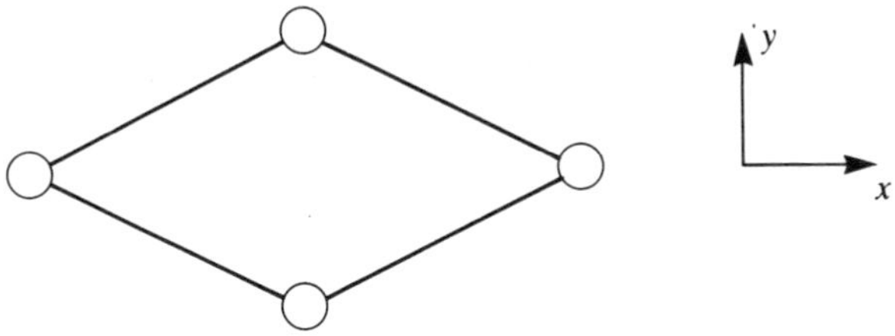

Figure 8.13

(i) Show that there are eight symmetry operations (including inversion and improper operations) in the group of the molecule, indicating with examples why each of them is in a separate class.
(ii) What dimensions and number of irreducible representations are therefore expected for the group?
(iii) Write down the characters for representations which have as their basis the twelve displacement vectors of the atoms.
(iv) The group with which you should be working should contain the irrep. labelled B_{1u}. Noting that an x-translation has B_{1u} symmetry, identify as far as possible the operations you have specified with those in the character tables (Appendix A2).
(v) Find all the B_{1u} normal-mode eigenvectors.
(vi) Introduce parameters appropriately and find expressions for the eigenfrequencies of the B_{1u} modes.

8.4 Determine the symmetry properties of the normal modes of a square planar molecule. Sketch as many modes as you can, without undertaking detailed calculations.

***8.5** The methane molecule consists of a carbon atom at the centre of a tetrahedron of hydrogen atoms (consider the H atoms to be at alternate corners of a cube).
(i) Given the character table for the methane symmetry group (Table 8.5) what are the degeneracies and symmetry types of the normal modes of CH_4?
(ii) What will the symmetry types of the pure translations be?
(iii) Suggest the form of a vibration of A_1 type.

Table 8.5

T_d	E	$8C_3$	$6S_4$	$3C_2$	$6\sigma_d$	
A_1	1	1	1	1	1	
A_2	1	1	-1	1	-1	
E	2	-1	0	2	0	
T_1	3	0	1	-1	-1	
T_2	3	0	-1	-1	1	$(\mathbf{i}, \mathbf{j}, \mathbf{k})$

8.6** ***The subgroup problem
For the molecular system considered in Problem 8.5, replace two of the H atoms by deuterium atoms.

(i) Which symmetry operations remain? Look up the character tables to determine the appropriate new symmetry group.
(ii) Determine the symmetry types and degeneracies of the normal modes of CH_2D_2 by considering the fifteen basis displacements of this molecule.
(iii) Use the information obtained in Problem 8.5 to verify the results of part (ii) and to relate the new degeneracies to those for CH_4.

8.7 The semiconductor gallium arsenide (GaAs) has a cell of tetrahedral symmetry (T_d). The $\mathbf{k}=0$ lattice vibrations are obtained by considering the displacements of one gallium and one arsenic atom: one obtains $\Gamma^D = 2T_2$.

Suppose a crystal of GaAs is subject to a uniaxial stress in the direction of one of the C_2 axes.

(i) Determine those T_d operations which do not alter the direction of this axis and so could be considered as the symmetry operations of the stressed system. Hence determine any splitting of the degeneracy of the lattice vibrational modes. Pick out and comment on the acoustic modes (cf. Section 6.7).
(ii) Would the same result be achieved if the stress was applied along a C_3 axis?

8.8 ***Site symmetry approximation in crystals***
In many crystals sets of atoms in any given cell behave much as isolated molecules. In particular the symmetry types of vibrations of these sets can be deduced from those of the equivalent molecule. Furthermore vibrational frequencies, deduced from the molecules in gases or liquids, may approximate closely to those in the crystal, depending on the strength of coupling of the molecule to the surrounding lattice.

For example, a free $(CO_3)^{2-}$ molecular ion has D_{3h} symmetry (the carbon is in the plane of the equilateral triangle of oxygens). In crystal calcite, $CaCO_3$, the CO_3 ions are positioned at sites of D_3 symmetry. Determine the splittings of the free ion normal modes in the crystal situation. $CaCO_3$ occurs in a second natural form, aragonite, in which the CO_3 ions are at C_{1h} sites. Discuss the modes in the aragonite example.

8.9 ***Bond lengths and angles***

Additional insight into the normal-mode vibrations of molecules may sometimes be obtained by considering bond-length changes and bond-to-bond angle changes, rather than atomic displacements, as basis entities. The water molecule H_2O provides a straightforward example.

(i) Treat the two O–H bond extensions as basis displacements. Obtain characters associated with their transformations with respect to the H_2O group of Problem 8.2. Reduce this representation and hence explain why only A_1 and vibrational modes involve bond extensions.

(ii) Use similar arguments to find those modes which involve a change in the bond angle.

(iii) Why can one only consider vibrations, rather than translations or rotations, using the above coordinates?

(iv) To summarize the symmetry properties of the H_2O molecule analyse the normal modes in terms of H displacements, O displacements, bond-length changes, angle changes, translations, rotations and vibrations, using only group-theoretical techniques.

8.10 For tetrahedral CH_4, consider the H–C bond lengths and the H–C–H bond angles as internal coordinates. Determine the symmetry types of the vibrational normal modes and comment on the number you obtain.

8.11 ***The benzene problem: normal-mode symmetries***

A continuation of the example stated in Problems 5.8, 5.9.

(i) From the results obtained for the reduced form of $\Gamma_c(12)$ and $\Gamma_c(6)$ in Problem 5.9, deduce the form of $\Gamma(36)$ for the entire C_6H_6 molecule. Hence state the degeneracies and the symmetry types of the benzene normal modes. State the symmetries of the zero-frequency, translation and rotation modes.

(ii) What deductions would have been made in part (i) if only the proper rotational operations had been considered?

(iii) Returning to the full group comment on the statement that no normal mode can involve motion both in and perpendicular to the molecular plane.

8.12 ***The benzene problem: normal-mode solution***

Consider carbon–carbon nearest-neighbour bonds to have a force constant, k_1, and carbon–hydrogen bonds to have a constant, k_2. Ignore interactions between non-adjacent atoms.

(i) Find expressions for the frequencies of the two A_{1g} modes of the entire C_6H_6 molecule.

(ii) Determine the corresponding eigenvectors.

(iii) What are the sizes of the matrix problems one must tackle to solve fully the benzene normal-mode problem (a) ignoring group-theoretical considerations, (b) as a consequence of accounting for symmetry?

8.13 ***The benzene problem: subgroups***

Replace a pair of oppositely positioned C-atoms by a different carbon isotope (C^{14}).

(i) Find the new symmetry group.

(ii) Use the irreps of the full benzene group to determine the degeneracies and symmetry types of the new normal modes and relate them to those of C_6H_6.

PART THREE

SYMMETRY APPLIED TO QUANTUM ENERGY LEVELS

9

Symmetry theory applied to functions

9.1 Basis functions

In the preceding sections a structure for the application of group theory to a physical problem has been constructed. The problem tackled was as simple as possible from an intuitive point in order to smooth as much as possible the introduction of the language of group theory. However, we always had in mind the development of symmetry theory for use with respect to solids and in particular to semiconductors. To this end we will be concerned with electronic (or phonon) eigenfunctions as opposed to nuclear displacements and with the Hamiltonian operator of a system rather than with the force and mass matrices.

Basis functions and the effect of a group operation

We will assume for the moment that we are concerned with the probability distribution for a single electron, and that some set of *basis functions* of the electronic coordinate, $\mathbf{r}$, are $f_i(\mathbf{r})$. These are supposed to be orthonormal functions; they will play the role in quantum mechanical problems that was played by the set of basis displacements $\mathbf{d}_i$ in the normal-mode problem. Their orthonormality is expressed by requiring

$$\int f_i(\mathbf{r})^* f_j(\mathbf{r}) \mathrm{d}\mathbf{r} = \delta_{ij}.$$

The integration is taken to be over all space. The $f_i(\mathbf{r})$ are required to have the same form of transformation properties as the displacement *vectors*; that is

$$G_r[f_1, f_2, \ldots] = [f'_1, f'_2, \ldots] = [f_1, f_2, \ldots]\ \Gamma(G_r).$$

If this is to be so it is necessary to define the effect of a group operation G_r on $f(\mathbf{r})$ to be

$$G_r f(\mathbf{r}) = f(G_r^{-1}\mathbf{r}).$$

This has the following interpretation. Consider a coordinate system with fixed axes. The electronic distribution described by $f_i(\mathbf{r})$ is thus dependent on the value of f_i at any given point $\mathbf{r}$. ($|f_1(\mathbf{r})|^2 d\mathbf{r}$ is interpreted as the probability of finding the electron in the region $\mathbf{r}$ to $(\mathbf{r}+d\mathbf{r})$ at some instant). The group operation G_r, applied to the physical system (the electron distribution), rotates/reflects it to some new distribution with respect to the fixed coordinates. $f_i(G_r^{-1}\mathbf{r})$ is the value of the distribution function which arrives at position $\mathbf{r}$ when the operation G_r is applied. With this definition the transformation properties of the basis functions are of the same form as those of basis vectors. The definition is also consistent with the idea introduced in Chapter 2 that operations are defined with respect to a fixed frame of reference and the object (in this case the electron probability distribution) is rotated etc. within the fixed frame.

Unitary transformation matrices

By analogy with the restriction that a matrix that transformed one set of orthonormal vectors/displacements to another had to be an *orthogonal* matrix, those matrices that transform sets of orthonormal functions must be *unitary matrices*, they obey the result:

$$\text{If } [f_1^N, f_2^N \ldots] = [f_1, f_2 \ldots]N, \qquad \text{then} \quad N^* = N^{-1}.$$

The proof follows that for orthogonal matrices, excepting that the presence of the complex conjugate of f_i in the function orthonormalization leads to the conjugate (*) in the above restriction.

Symmetry-adapted basis functions

The above results show us the analogy between basis functions and displacements. Given this analogy it follows directly that for any group we should be able to form matrix representations which have functions as their basis rather than conventional vectors or displacements. In particular the irreducible representations have symmetry-adapted basis functions, which can be obtained through the formula

$$F^j(Q) \propto \sum_{r=1} \chi^j(G_r)^* \; G_r Q.$$

Q is some chosen function of the general type of interest. For example, if we are looking for symmetry-adapted functions that are first-order powers of x,y,z we might choose $Q(\mathbf{r})$ to be the function x. We have introduced the notation F^j by analogy to the D^j.

9.2 The eigenfunction problem

Eigenfunctions and energy eigenvalues

The physical problem we are now attempting to solve is the time-independent Hamiltonian problem:

$$\mathscr{H}\psi = E\psi.$$

This is a differential equation, the solution of which gives the possible stationary states of the physical system,

$$\mathscr{H}\psi_n = E_n\psi_n.$$

Under the boundary condition that ψ_n tends to zero at infinity, the solutions will be denumerable (albeit they will in general be of infinite number). Unfortunately, though, we are seldom in a position to solve exactly the differential Hamiltonian equation. For the moment it is sufficient to think of the ψ as being one-particle (electron) distribution functions and for $\mathscr{H}$ to involve only the coordinates of this particle. The constants E_n are the allowed energies of the system. They are the *energy eigenvalues* of states of *eigenfunctions* ψ_n. Although we cannot solve for the E_n and ψ_n, let us suppose that there is another differential equation involving the same physical parameters ($\mathbf{r}$) and boundary conditions, which we can solve:

$$\mathscr{H}_0 f_j = \epsilon_j f_j.$$

There will be a one-to-one correspondence between the two sets of eigenfunction solutions; the entire sets of ψ_n or of f_j are said to be *complete sets of solutions* to the particular type of differential equations. They have the property that

$$\psi_n = \sum_j f_j a_{jn} \qquad \text{or} \qquad f_j = \sum_n \psi_n c_{nj},$$

where the a_{jn} and c_{nj} are constant coefficients.

As examples of complete sets of functions: the set of eigenfunctions of a three-dimensional harmonic oscillator can in principle be expanded in terms of the hydrogen-atom solutions or vice versa, providing only the three spatial coordinates are involved in each case. Clearly a function involving spin cannot be expanded in terms of functions that do not involve spin; nor can a three-dimensional function be expanded in terms of the solutions of a one-dimensional problem. A similar expansion technique with which some readers may be more familiar is the Fourier series method. Here *any* even periodic function (of period a) may be written as a linear combination of all cosine functions $\cos n\pi x/a$, with integer n:

$$\text{Periodic function} = \sum_{n=0}^{\infty} a_n \cos (n\pi x/a).$$

It may be that in practice a simple analytical expression can often be found for the a_n, or that the series can be curtailed at some n value. Nevertheless, provided all n are included, *any* periodic function may be expressed as such a sum. The cosine functions form a complete set for the periodic problem.

The point to be made is that given a complete set of functions f_j for the problem of interest then the required eigenfunctions ψ_n are related to the f_j by a matrix transformation.

$$[\psi_1, \psi_2 \ldots] = [f_1, f_2 \ldots]\, a.$$

The matrix a is equivalent to α in the normal mode case. However a is now quite likely, at least in principle, to be of infinite dimension; and a is now a unitary matrix.

A matrix diagonalization problem is arrived at from the following arguments. Given that we know the form of $\mathscr{H}$, and we know the form of the basis functions f_j; then we can evaluate Hamiltonian matrix elements, which are defined as

$$\mathscr{H}^f_{ij} = \int f_i^* \mathscr{H} f_j \mathrm{d}\mathbf{r}.$$

But

$$\psi_n = \sum_j f_j a_{jn} \qquad \text{and} \qquad \mathscr{H}\psi_n = E_n \psi_n.$$

Then

$$\sum_j (\mathscr{H} f_j) a_{jn} = E_n \sum_j f_j a_{jn},$$

and premultiplying by f_i^* and integrating gives

$$\sum_j \left(\int f_i^* \mathscr{H} f_j\right) a_{jn} = E_n \sum_j \left(\int f_i^* f_j\right) a_{jn}.$$

For orthonormal functions the right-hand-side simplifies and we obtain

$$\sum_j \mathscr{H}^f_{ij} a_{jn} = E_n a_{in}.$$

This can be expressed by either of the matrix relations

$$[\mathcal{H}^f - E_n I]\begin{bmatrix} a_{1n} \\ a_{2n} \\ \cdot \\ \cdot \\ \cdot \end{bmatrix} = \begin{bmatrix} 0 \\ 0 \\ \cdot \\ \cdot \\ \cdot \end{bmatrix} \quad \text{or} \quad a^{-1}[\mathcal{H}^f - EI]a = [\mathcal{E} - EI].$$

I is the unit diagonal matrix and $\mathcal{E}$ is a diagonal matrix of elements $\mathcal{E}_{nn} = E_n$.

These results are almost exactly of the form of the normal-mode problem. Note also that, had we started with the correct eigenfunctions, then

$$\mathcal{H}^{\psi}_{nm} = \int \psi_n^* \mathcal{H} \psi_m = \int \psi_n^* E_n \psi_m = E_n \delta_{mn}.$$

Hence $[\mathcal{H}^{\psi} - EI] = [\mathcal{E} - EI]$.

Once again, the transformation matrix we are seeking is the one that, through a similarity transformation, diagonalizes a matrix, in this case the Hamiltonian matrix.

Non-orthogonal bases

It is sometimes convenient to use non-orthogonal basis functions in the linear combination. For example, in molecular orbital theory a linear combination of all atomic orbitals centred at each atom is considered. But atomic orbitals based on different centres are not orthogonal to each other, though they may be nearly so. Define

$$S^f_{ij} = \int f_i^* f_j \, d\mathbf{r}.$$

S^f_{ij} is known as an overlap integral, for obvious reasons; we would hope to choose bases so that the S^f_{ij} are fairly small or zero. The matrix equations must be modified in such circumstances:

$$\sum_j \mathcal{H}^f_{ij} a_{jn} = E_n \sum_j S^f_{ij} a_{jn},$$

or

$$[\mathcal{H}^f - E_n S^f]\begin{bmatrix} a_{1n} \\ a_{2n} \\ \cdot \\ \cdot \\ \cdot \end{bmatrix} = \begin{bmatrix} 0 \\ 0 \\ \cdot \\ \cdot \\ \cdot \end{bmatrix} \quad \text{or} \quad a^{-1}[\mathcal{H}^f - ES^f]a = [\mathcal{E} - EI].$$

This final pair of equations is exactly of the mathematical form of

$$\alpha^{-1}[K^d - \omega^2 M^d]\alpha = [K^{\mathcal{D}} - \omega^2 M^{\mathcal{D}}],$$

and we can expect to apply group theory to it by direct analogy.

Wigner's theorem and Schur's lemma, applied to eigenfunctions

We know that symmetry-adapted functions can be constructed from the basis functions f_i. By the same arguments as used in Chapter 7 it follows that:

> *Eigenfunctions are linear combinations of symmetry-adapted basis functions belonging to the same row of the same irreducible matrix representations.*

In particular,

> *Eigenfunctions can be labelled by the irrep. to which they belong. The dimension of the irrep. tells us the degeneracy of the energy level.*

These results can be further substantiated by applying Schur's lemma. In order to be applicable the matrix $\mathscr{H}$ must now commute with Γ; this we now verify, for the case of a single particle. It is directly extendable to a general many-particle case.

Proof that $\mathscr{H}\Gamma = \Gamma\mathscr{H}$. Remember that the Hamiltonian is the sum of kinetic and potential energy operators:

$$\mathscr{H} = -\frac{\hbar^2}{2m}\nabla^2 + V(\mathbf{r}).$$

Now the group of the system is, by definition, the group of operations for which $V(\mathbf{r})$ is invariant. In addition, ∇^2 is invariant with respect to all point group operations, and certainly therefore for those in the group of interest. It is easily to demonstrate the invariance of ∇^2. In spherical polar coordinates,

$$\nabla^2 = \frac{1}{r^2}\left[\frac{\partial}{\partial r} r^2 \frac{\partial}{\partial r} + \frac{1}{\sin\theta}\frac{\partial}{\partial\theta}\sin\theta\frac{\partial}{\partial\theta} + \frac{1}{\sin^2\theta}\frac{\partial^2}{\partial\phi^2}\right].$$

Under a rotation by α about the arbitrarily selected z-axis, r and θ are unaltered, ϕ transforms to $(\phi+\alpha)$. Hence $\partial/\partial r' = \partial/\partial r$, $\partial/\partial\theta' = \partial/\partial\theta$, $\partial/\partial\phi' = \partial/\partial(\phi+\alpha) = \partial/\partial\phi$; ∇^2 is unaltered. We conclude then that $\mathscr{H}$ is invariant under group operations. Now consider the matrix elements of $\mathscr{H}$, $\mathscr{H}^f_{ij} = \int f_i^* \mathscr{H} f_j$ integrated over all space. Under some operation G_r both f_i^* and f_j change.

$$\mathscr{H}^{f'}_{ij} = \int f_i(G_r^{-1}\mathbf{r})^* \, \mathscr{H} f_j(G_r^{-1}\mathbf{r}).$$

But as the integration is over all space the coordinate system can be changed from $\mathbf{r}$ to $G_r^{-1}(\mathbf{r})$ with no effect. Thus

$$\mathscr{H}^{f'}_{ij} = \int f_i(\mathbf{r})^* \, \mathscr{H} f_j(\mathbf{r}),$$

where we have used the fact that $\mathscr{H}$ is unaltered by the change of coordinate systems. Thus

$$\mathscr{H}^{f'}_{ij} = \mathscr{H}^{f}_{ij}.$$

But

$$\mathcal{H}_{ij}^{f'} = \int f'_i \mathcal{H} f'_i = \int (\sum_k f_k \Gamma(G_r)_{ki})^* \mathcal{H} (\sum_l f_l \Gamma(G_r)_{lj})$$
$$= \sum_k \sum_l (\int f_k^* \mathcal{H} f_l) \Gamma_{ki}^* \Gamma_{lj}.$$

Using the unitary property of $\Gamma, \Gamma_{ki}^* = \Gamma_{ik}^{-1}$, we obtain

$$\therefore \mathcal{H}_{ij}^{f'} = \sum_k \sum_l \Gamma_{ik}^{-1} \mathcal{H}_{kl} \Gamma_{lj} = (\Gamma^{-1} \mathcal{H} \Gamma)_{ij}.$$

This must equal $\mathcal{H}_{ij}^{f}$ for all i,j, so that

$$\mathcal{H} = \Gamma^{-1} \mathcal{H} \Gamma$$

or

$$\Gamma \mathcal{H} = \mathcal{H} \Gamma.$$

We therefore expect that Schur's lemma can be applied to this problem both from the verbal argument and from a mathematical viewpoint. Therefore we may state

> *An l-dimensional irrep. occurring m times in Γ^F corresponds to m sets of l-fold degenerate energy levels.*

Wigner's theorem and Schur's lemma, applied to quantum mechanics, both support the following simple statement:

Given an eigenfunction ψ_n and a symmetry operation G_r, by definition G_r does not change the energy of the system, so that if $G_r(\psi_n)$ is distinct from ψ_n then the two must be degenerate eigenfunctions.

Time-reversal symmetry

If we assume that on reversing the direction of time all physical laws are maintained, then the energy of any state must be degenerate with that of its time-reversed version.

The time-dependent quantum mechanical problem is

$$\mathcal{H}\psi(r,t) = i\hbar\, \partial\psi(\mathbf{r},t)/\partial t.$$

The time-reversed function $\psi(\mathbf{r}, -t)$ therefore satisfies

$$\mathcal{H}\psi(\mathbf{r}, -t) = -i\hbar \partial\psi(\mathbf{r}, -t)/\partial t,$$

which is precisely the equation satisfied by $\psi(\mathbf{r},t)^*$. Hence time-reversal symmetry implies that states ψ_n and ψ_n^*, if they differ, are degenerate.

There are a number of symmetry groups (see Appendix A2) for which certain irreps have complex characters. However for any one such irrep. a second may be found with characters that are the complex conjugates of the first. As a consequence the symmetry-adapted functions for the two are complex conjugates and they will lead to conjugate (and hence degenerate) eigenfunctions. For this reason the conjugate irreps are placed together in the tables and may be treated as a single two-dimensional irrep., of real character $(\chi+\chi^*)$, for most purposes. [*Note*: In the presence of an external magnetic field time-reversal symmetry applies only if the field direction is also reversed.]

Problems

9.1** ***Symmetry-adapted functions

Draw up a table that shows how the functions x, y and z transform under the operations of the D_{2d} group (Table 4.12). From this table you will be able to deduce $G_r[x^2]$ etc. Hence apply the symmetry-adapted basis formula to obtain symmetry-adapted functions that are linear combinations of $x^2, y^2, z^2, xy, yz, zx$. Check your results with the character table in Appendix A2.

9.2 The orthonormal functions $f_1, f_2, \ldots, f_8$ transform as a set under the operation of a symmetry group. How many orthonormal symmetry adapted functions can be constructed as linear combinations of $f_1, \ldots, f_8$?

Symmetry adapted functions are obtained; F_1 and F_2 are found to transform into linear combinations of each other under all group operations—they are bases for a two-dimensional irrep. F_3 and F_4 are found to transform together in identical manner to F_1 and F_2. In group theory what do we call F_1 and F_2? How do we describe the similarity between F_1 and F_3? Write down the Hamiltonian matrix elements based on $F_1, \ldots, F_4$.

9.3** ***Invariant functions

A function that forms a basis for the identity representation of a group is known as an *invariant function*. Show that if the functions F^j_k belong to the kth row of the jth irrep. of a group, then $\Sigma|F^j_k|^2$ is an invariant function if the sum Σ is over all partner vectors. Look up the character tables in Appendix A2 and state why they support this result. Comment on the relevance of the result with regard to the symmetry of atoms.

Hint: Consider the effect of an operation G_r on $\Sigma_k(F^j_k)^*(F^j_k)$ in terms of the matrix elements of $\Gamma^j(G_r)$.

***9.4** For a given system of C_{4v} symmetry five functions form a basis for a reducible representation Γ^f. On reduction one obtains

$$\Gamma^f \Rightarrow 2A_1 \oplus B_2 \oplus E.$$

If five eigenstates of the system are constructed as linear combinations of the five functions, what are the symmetry properties of these states?

State the form of the Hamiltonian matrix, based on (i) the symmetry-adapted functions and (ii) the eigenfunctions. Repeat the problem for an initial basis of fifteen functions that have a reduced representation

$$\Gamma^f \Rightarrow 3A_1 \oplus A_2 \oplus B_1 \oplus 2B_2 \oplus 4E.$$

9.5 ***The invariance of ∇^2***

Verify that ∇^2 is invariant under rotations by considering

$$x' = x \cos\alpha + y \sin\alpha,$$

$$y' = -x \sin\alpha + y \cos\alpha, z' = z.$$

That is, demonstrate that for a rotation about the z-axis,

$$\frac{\partial^2}{\partial x'^2} + \frac{\partial^2}{\partial y'^2} + \frac{\partial^2}{\partial z'^2} \equiv \frac{\partial^2}{\partial x^2} + \frac{\partial^2}{\partial y^2} + \frac{\partial^2}{\partial z^2}.$$

10

The atomic Hamiltonian and the spherical-symmetry group

Fortunately the Schrödinger equation for the hydrogen atom, within certain simplifying approximations, is exactly soluble by analytic means. That is, the solutions can be written as power-series expansions of the simple functions r, $\cos\theta$, $\cos\phi$, $\sin\phi$. There are very few realistic Hamiltonian models that have this property. In turn, knowing that molecules and solids consist of arrays of atoms, it is hardly surprising that where possible we attempt to set up the problem of solving for the electronic motion of such systems in terms of hydrogenic states. The classic examples are the method for constructing molecular orbitals by linear combination of atomic orbitals, and the tight binding model for core electron states in solids.

In this section we wish to begin building a basis for the study of such electronic states by considering, from a group-theoretical point of view, the degeneracies and symmetries of the hydrogen states.

10.1 The hydrogen atom

For the present, spin effects and other relativistic effects will be ignored, so we are concerned with only the Coulombic contribution to the potential energy, and the nuclear and electronic kinetic terms, in this one-electron atom.†

†Expressions containing the Coulomb interaction are almost the only ones in this text where one must take care over units. Purely for simplicity, in these expressions the interaction is written as it would appear in cgs units. Readers wishing to obtain the SI version of the interaction itself or of any resulting expressions need only replace e^2 by $(e^2/4\pi\varepsilon_0)$ throughout.

$$\mathcal{H}_H = -\frac{\hbar^2}{2M_n}\nabla_n^2 - \frac{\hbar^2}{2m_e}\nabla_e^2 - \frac{e^2}{|\mathbf{r}_e - \mathbf{R}_n|},$$

where $\mathcal{H}_H\Psi(\mathbf{R}_e,\mathbf{r}_n) = E_H\Psi(\mathbf{R}_e,\mathbf{r}_n)$.

This differential equation in six variables (the coordinates of the two particles) is solved firstly by transforming to centre-of-mass and internal coordinates,

$$\mathbf{R} = (\mathbf{r}_e M_n + \mathbf{R}_n m_e)/(m_e + M_n), \qquad \mathbf{r} = \mathbf{r}_e - \mathbf{R}_n$$

and then switching to spherical polar coordinates for the internal motion, which is of primary concern. Thus

$$\Psi(\mathbf{R}_e,\mathbf{r}_n) = X(\mathbf{R})\psi(r,\theta,\phi), \qquad E_H = E_{cm} + E,$$

$$-\frac{\hbar^2}{2M}\nabla_{\mathbf{R}}^2 X(\mathbf{R}) = E_{cm}X(\mathbf{R}),$$

$$\left[-\frac{\hbar^2}{2\mu r^2}\left\{\frac{\partial}{\partial r}r^2\frac{\partial}{\partial r} + \frac{1}{\sin\theta}\frac{\partial}{\partial\theta}\sin\frac{\partial}{\partial\theta} + \frac{1}{\sin^2\theta}\frac{\partial}{\partial\phi^2}\right\} - \frac{e^2}{r}\right]\psi(r,\theta,\phi) = E\psi(r,\theta,\phi).$$

M is the total atomic mass and μ is in principal the reduced mass $M_n m_e/(M_n + m_e)$ but is sufficiently close to the electronic mass for the distinction to be of minor interest.

From a group theoretical point of view the important statement about the internal motion is the obvious one—the system and its Hamiltonian are spherically symmetric. We have already, in Chapter 9, established that the ∇^2 operator is invariant under all symmetry group operations; clearly the Coulombic interaction, a central-force interaction, has full spherical symmetry ($-e^2/r$ is invariant under rotations and reflections).

The rotation–inversion group

Spelling out the operations for the group of the internal H-atom problem, they are:

(i) The identity, E.

(ii) Rotation, by any angle α up to 2π, about any axis ξ through the origin, $R(\alpha,\xi)$. Rotations by different angles must fall into distinct classes. However, as each axis is physically equivalent, rotations by the same angle but about different axes will be in the same class.

Rotations by α and $(2\pi - \alpha)$ about some specific axis will fall in the same class either from the point of view that physically the sense of rotation is of no consequence or, and it amounts to the same statement, because $R((2\pi-\alpha),\xi)$ *is* a rotation by α about the axis ξ' pointing along ξ but in the opposite direction.

(iii) The inversion operation, i.
(iv) Improper rotations $iR(\alpha,\xi)$.

Groups such as this, which contain operations involving continuously varying parameters (α or the orientation of ξ) are known as *continuous* or *Lie groups*. This spherical symmetry group we label $R_3 \otimes S_2$ for reasons that will become apparent in Chapter 13.

As far as the character table is concerned, because there are infinitely many classes of operation (denoted by subscripts in Table 10.1) we anticipate infinitely many irreducible representations, and because there are an infinite number of operations (denoted by premultipliers) in each class then we might expect any degree of degeneracy for the irreps. Table 10.1 confirms these comments.

Table 10.1 The character table for the group $R_3 \otimes S_2$.

	E	$\infty R(\alpha,\xi)_\infty$	i	$\infty iR(\alpha,\xi)_\infty$
D^{0g}	1	1	1	1
D^{0u}	1	1	-1	-1
D^{1g}	3	$\frac{\sin(3\alpha/2)}{\sin(\alpha/2)}$	3	$\frac{\sin(3\alpha/2)}{\sin(\alpha/2)}$
$\vdots$	$\vdots$		$\vdots$	
D^{lg}	$2l+1$	$\frac{\sin(l+1/2)\alpha}{\sin \alpha/2}$	$2l+1$	$\frac{\sin(l+1/2)\alpha}{\sin \alpha/2}$
D^{lu}	$2l+1$	$\frac{\sin(l+1/2)\alpha}{\sin \alpha/2}$	$-(2l+1)$	$-\frac{\sin(l+1/2)\alpha}{\sin \alpha/2}$

Now suppose that we wished to find the eigenfunctions of the system, by a similar technique to the generation of eigenvectors, using group theory. The first question to ask would be, 'what possible form can the symmetry-adapted basis functions for the various irreps take?' In principle one can pick a function $f(r,\theta,\phi)$ and run it through the projection operator formula,

$$F^j(f) \propto \sum_r \chi^j(G_r)^* G_r f.$$

The summation needs now to be replaced by integrations, one for the angle of rotation and one for the orientation of the axis. Nevertheless from any initial complete set of basis functions a set of symmetry-adapted functions can be constructed and the form of the eigenfunctions (linear combinations of F^j_k) can be found. Obviously to find the precise eigenfunctions the Hamiltonian matrix elements would need to be determined and the diagonalization of $\mathcal{H}$ carried out. At this stage we would have established an infinite (but denumerable) set of symmetry-adapted basis functions for each irrep. Hence the Hamiltonian, although block-diagonalized, would have infinite blocks and we could not pursue the full solution of the problem to an exact conclusion. What we would have established though are the possible symmetries of states that are based on spatial wave functions and the degeneracies of such states. We will however leave the construction of symmetry-adapted functions of continuous groups until we deal with the axial symmetry problem (Chapter 11) and instead consider the H-atom problem from the reverse point of view. We will demonstrate that the form of the functions obtained by integration of the angular part of the Schrödinger equation are indeed symmetry-adapted functions of the $R_3 \otimes S_2$ irreps and that consequently the eigenfunctions of the full problem, regardless of their precise form, must also belong to specific irreps. Firstly separate the radial and angular parts of the problem:

$$\psi(r,\theta,\phi)=R(r)\ Y(\theta,\phi)$$

We assume a solution of form

$$\left[-\frac{1}{\sin\theta}\frac{\partial}{\partial\theta}(\sin\theta\frac{\partial}{\partial\theta})+\frac{1}{\sin^2\theta}\frac{\partial^2}{\partial\phi^2}\right]\ Y(\theta,\phi)=cY(\theta,\phi),$$

so that

$$\left[-\frac{\hbar^2}{2\mu}\frac{1}{r^2}\{\frac{\partial}{\partial r}r^2\frac{\partial}{\partial r}-c\}-\frac{e^2}{r}\right]\ R(r)=ER(r).$$

The point of this separation, facilitated by the introduction of a separation constant c, is that the exact form of the potential energy term does not affect the solution for Y. Providing we are dealing with a spherically symmetric potential the conclusions we draw here from the Y-equation must be valid. But this is a well-known mathematical differential equation; its solutions are the spherical harmonic functions:

$$\{\frac{1}{\sin\theta}\frac{\partial}{\partial\theta}\sin\theta\frac{\partial}{\partial\theta}+\frac{1}{\sin^2\theta}\frac{\partial^2}{\partial\phi^2}\}\,Y^l_m(\theta,\phi)=l(l+1)Y^l_m(\theta,\phi).$$

$$Y^l_m(\theta,\phi)\propto P^l_{|m|}(\cos\theta)\ e^{im\phi}=\Theta^l_{|m|}(\theta)\ e^{im\phi}$$

say, and the $P^l_{|m|}(\cos\theta)$ are associated Legendre functions (polynomials in $\cos\theta$). The physical boundary conditions that the eigenfunctions (and hence the Y factors) must be everywhere continuous, finite and single-valued place restrictions on the values of l and m; namely.

$$l=0,\ 1,\ 2,\ldots \qquad \text{with} \qquad m=-l,\ -l+1,\ldots+l$$

for a given l value. For each l there are thus $(2l+1)$ values that m may take. The fact that m must be an integer is seen easily from the single valuedness of Y^l_m; it requires that $e^{im\phi}=e^{im(\phi+2\pi)}$.

Table 10.2 The lowest-order spherical harmonic functions $Y_m^l \equiv \Theta^l_{|m|}(\theta)e^{im\phi}$, normalized so that $\int |Y|^2 \sin\theta \, d\theta \, d\phi = 1$.

Y_0^0	$\sqrt{(1/4\pi)}$
Y_0^1	$\sqrt{(3/4\pi)} \cos\theta$
$Y_{\pm 1}^1$	$\mp\sqrt{(3/8\pi)} \sin\theta \exp(\pm i\phi)$
Y_0^2	$\sqrt{(5/16\pi)} (3\cos^2\theta - 1)$
$Y_{\pm 1}^2$	$\mp\sqrt{(15/8\pi)} \sin\theta \cos\theta \exp(\pm i\phi)$
$Y_{\pm 2}^2$	$\sqrt{(15/32\pi)} \sin^2\theta \exp(\pm i2\phi)$

Given the solution for Y, the radial equation takes the form

$$\left[-\frac{\hbar^2}{2\mu}\frac{1}{r^2}\frac{\partial}{\partial r}r^2\frac{\partial}{\partial r}+\frac{\hbar^2 l(l+1)}{2\mu r^2}-\frac{e^2}{r}\right]R(r)=ER(r).$$

Without solving for E and R we can certainly make the statement that they are independent of m. In practice for this particular form of potential energy the solutions that obey the boundary condition that the eigenfunction must be zero at infinity (in order to avoid the electron spending all its time at infinite separation from the proton) happen to have energy eigenvalues that are also independent of l; this is not a consequence of symmetry, it is an accidental degeneracy. In general one must obtain bound solutions of the form

$$\psi_{nlm}(r,\theta,\phi)=R_{nl}(r)\,\Theta^l_{|m|}(\theta)\,e^{im\phi};\; E_{nlm}=E_{nl}.$$

For any given n,l combination there must be $(2l+1)$ different eigenfunctions having identical energies. In turn these $(2l+1)$ eigenfunctions must transform into linear combinations of each other under the group symmetry operations—they must form a basis for a $(2l+1)$-dimensional irrep. of the $R_3 \otimes S_2$ group. This being so we should be able to generate the characters for the irrep.:

(i) There are $(2l+1)$ basis functions, hence $\chi^l(E)=2l+1$.

(ii) Rotations by an angle α about different axes fall in the same class and therefore have the same characters for their matrix representations. In attempting to find $\chi^l(R(\alpha,\xi))$ we are therefore at liberty to choose the axis ξ at will. The choice to make is obvious; we consider the same axis that we have already chosen to be special (arbitrarily) in defining the coordinate system—the z-axis. The functions $Y_m^l(\theta,\phi)$ have simple properties with respect to $R(\alpha,z)$.

Remembering that $G_r(f(\mathbf{r}))$ is defined as the function that arrives at $\mathbf{r}$ under the group operation G_r, (see Fig. 10.1):

$$R(\alpha,z)\,f(r,\theta,\phi)=f(r,\theta,\phi-\alpha).$$

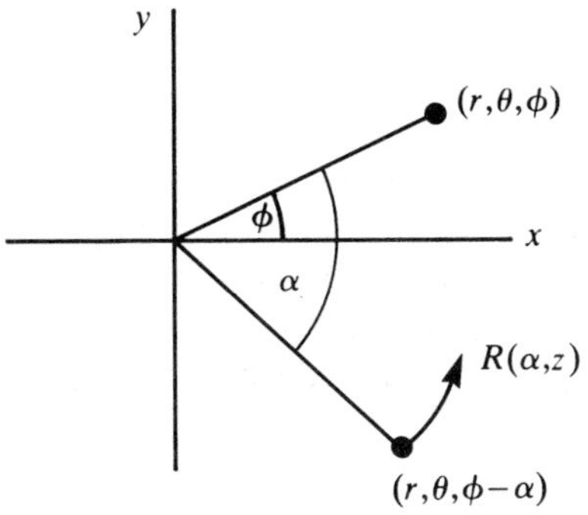

Figure 10.1 To show the effect of the operation $R(\alpha,z)$ on a function $f(r,\theta,\phi)$.

Hence the function $R_{nl}\theta^{l}_{|m|}\ e^{im\phi}$ is transformed to $R_{nl}\theta^{l}_{|m|}\ e^{im(\phi-\alpha)}$. The choice of rotational axis has been such that ψ_{nlm} transforms to $e^{-im\alpha}\ \psi_{nlm}$; for a different choice it will transform to some linear combination of all $(2l+1)$ functions $\psi_{nl(-l)}$ to $\psi_{nl(l)}$. We have, for this set of $(2l+1)$ functions,

$$R(\alpha,z)\ [\psi_{nl(-l)},\ \psi_{nl(-l+1)},\ldots,\ \psi_{nl(l)}]$$

$$=[\psi_{nl(-l)},\ \psi_{nl(-l+1)},\ldots\ \psi_{nl(l)}]\begin{bmatrix} e^{il\alpha} & & & & \\ & e^{i(l-1)\alpha} & & & \\ & & \cdot & & \\ & & & \cdot & \\ & & & & e^{-il\alpha} \end{bmatrix},$$

so

$$\chi^{l}(R(\alpha,\xi))=\chi^{l}(R(\alpha,z))=(e^{il\alpha}+e^{i(l-1)}\alpha+\cdots+e^{-il\alpha}).$$

We will find it useful to remember this form for the character for future reference. As far as tables are concerned though a neater form is always presented. From the geometric progression format of χ^{l},

$$\chi^{l}=\sum_{m=-l}^{l}e^{-im\alpha}=\frac{e^{-il\alpha}-e^{il\alpha}\ e^{\alpha}}{1-e^{i\alpha}}$$

$$=e^{i\alpha/2}(e^{-i(l+1/2)\alpha}-e^{i(l+1/2)\alpha})/(e^{i\alpha/2}[e^{i\alpha/2}-e^{-i\alpha/2}]);$$

$$\chi^{l}(R(\alpha,\xi))=\frac{\sin\ (l+1/2)\alpha}{\sin\ \alpha/2}.$$

Note that this is consistent with $\chi^{l}(E)=(2l+1)$, as in the limit $\alpha\to 0$ the sine ratio tends to $(2l+1)$.

(iii) Under inversion

$$if(r,\theta,\phi)=f(r,\ \theta-\pi,\ \phi+\pi),$$

so that $iY_m^l(\theta,\phi)\propto P^l_{|m|}(\cos\ (\theta-\pi))\ e^{im(\phi+\pi)}$.

The Legendre functions invert to $(-1)^{l-|m|}P$, so

$$iY_m^l(\theta,\phi)=(-1)^l Y_m^l(\theta,\phi).$$

The character for the irrep. based on the $(2l+1)$ degenerate functions must therefore be

$$\chi^l(i)=(2l+1)\ (-1)^l.$$

(iv) Under improper rotations in general:

$$iRf(r,\theta,\phi)=f(r,\theta-\pi,\ \phi+\pi-\alpha),$$

giving $\chi^l(iR(\alpha,\xi))=(-1)^l\dfrac{\sin\ (l+1/2)\alpha}{\sin\ \alpha/2}$.

We conclude that the $(2l+1)$-fold degenerate eigenfunctions $R_{nl}Y_m^l$ form the basis for those irreducible representations of $R_3 \otimes S_2$ which, in Table 10.1, are labelled as D^{lg} (if l is even) or D^{lu} (if l is odd.) Note that only half of the possible irreps are covered by such functions. This implies that there might be eigenfunctions of different symmetry types, but that these could not be functions of the spatial coordinates r,θ,ϕ alone. We will come across such functions in connection with many-particle states.

The superscripts g,u refer to the *parity* of the states. An eigenfunction that is unaffected by inversion is said to have even parity (g stands for *gerade*, meaning 'even' in German); a function which changes sign under inversion has odd parity (i.e. u for *ungerade*, uneven).

10.2 Many-electron atoms

The treatment of many-electron atoms can be dealt with very briefly. Starting from the atomic Hamiltonian,

$$\mathcal{H}_{\text{atom}}=\frac{-\hbar^2}{2M_n}\nabla_n^2+\sum_i\frac{-\hbar^2}{2m_e}\nabla_i^2+\sum_i\frac{Ze^2}{|\mathbf{r}_i-\mathbf{R}_n|}+\tfrac{1}{2}\sum_i\sum_{j\neq i}\frac{e^2}{|\mathbf{r}_i-\mathbf{r}_j|},$$

which contains in order the nuclear kinetic term, the total electron kinetic energy (for each electron labelled i), the electron–proton Coulombic interactions, and the electron–electron Coulombic repulsions. (Note that the factor $\frac{1}{2}$ in the interaction term takes care of the fact that we must only count the interaction between any given pair of electrons once in $\Sigma_i\ \Sigma_{j\neq i}$.)

Z is the atomic number of the element. Again the centre-of-mass motion is separated out. This, however, leaves now an insoluble problem in the coordinates of each electron (with respect to the nucleus) that includes an awkward coupling term—the electron–electron repulsion:

$$\mathscr{H}_{\text{elect}} = \sum_i \left(-\frac{\hbar^2}{2\mu}\nabla_i^2 - \frac{Ze^2}{r_i} + \tfrac{1}{2}\sum_{j\neq i}\frac{e^2}{r_{ij}} \right).$$

As one cannot specify positions for the electrons j, then $\mathscr{H}_{\text{elect}}$ cannot be split into a sum over single-electron Hamiltonians. The final term must therefore be handled is some approximate way. We could treat it as an average—the average Coulomb energy of electron i in the field of all the other electrons. Suppose to start with though we ignore the term completely. Now the problem decouples into a set of H-atom equations:

$$\Psi(\mathbf{R}_n, \mathbf{r}_{e1}, \mathbf{r}_{e2}, \ldots) = X(\mathbf{R})\psi(\mathbf{r}_1)\psi(\mathbf{r}_2)\cdots,$$

with

$$\left(-\frac{\hbar^2}{2\mu}\nabla_i^2 - \frac{Ze^2}{r_i}\right)\psi(r_i,\theta_i,\phi_i) = E_i\psi(r_i,\theta_i,\phi_i),$$

and

$$E_{\text{atom}} = E_{\text{cm}} + \sum_i E_i.$$

If we know the solutions E_i and $\psi(r_i)$ we can appeal to the Pauli exclusion principle and allot the electrons in the atom to the lowest energy states. This gives us sufficient information to have an intelligent guess at $\frac{1}{2}\sum_{j\neq i} e^2/r_{ij}$. That is, we can calculate

$$\bar{V}_e(\mathbf{r}_i) = \tfrac{1}{2}\sum_{j\neq i}\int \psi_j^* \psi_j \frac{e^2}{r_{ij}}\, \mathrm{d}\mathbf{r}_j,$$

which contains integrals of the potential at $\mathbf{r}_i$ due to the electron distribution implied by ψ_j. Reinserting this, we obtain

$$\left[-\frac{\hbar^2}{2\mu}\nabla_i^2 - \frac{Ze^2}{r_i} + \bar{V}_e(\mathbf{r}_i)\right]\psi(\mathbf{r}_i) = E_i\,\psi(\mathbf{r}_i).$$

In principle numerical solutions are again possible, giving a new set of E_i values and a new set of eigenfunctions. This cycle can be carried out as many times as required until successive solutions agree to within the required accuracy. This technique is known as the Hartree self-consistent field method. It is a numerical solution and can clearly become tedious. It should also be noted that the solutions considered do not automatically obey Pauli exclusion; neither are they orthogonal, as different $\bar{V}_e(\mathbf{r}_i)$ are used for each electron. A modification of the theory allows this to be catered for. The electron product wavefunction used in calculating $\bar{V}_e(\mathbf{r}_i)$ is of the form

$$\prod_i \psi_n(\mathbf{r}_i),$$

where for each electron (i) we must choose a different state (n), thereby imposing Pauli exclusion. But there is no reason that we should have to put a particular electron into a given state. That is, having determined all the states that are to be occupied, we could distribute the electrons amongst them in any permutation. For two electrons the function $\psi_{n=1}(\mathbf{r}_i)\ \psi_{n=2}(\mathbf{r}_2)$ and $\psi_{n=2}(\mathbf{r}_1)\ \psi_{n=1}(\mathbf{r}_2)$ are equally valid, as is any combination of them. In particular consider

$$\psi(\mathbf{r}_1,\mathbf{r}_2)=\frac{1}{\sqrt{2}}\,[\psi_1(\mathbf{r}_1)\ \psi_2(\mathbf{r}_2)-\psi_2(\mathbf{r}_1)\ \psi_1(\mathbf{r}_2)].$$

This normalized function has the property that if we pick identical one-particle states ($n=1$ and $n=2$ the same state), then $\psi(\mathbf{r}_1,\mathbf{r}_2)$ vanishes; this function automatically obeys Pauli exclusion. In general there is such a 'Slater determinant' function for any N-electron system,

$$\psi(\mathbf{r}_1\ldots\mathbf{r}_n)=\frac{1}{(N!)^{\frac{1}{2}}}\begin{vmatrix} \psi_1(\mathbf{r}_1) & \psi_1(\mathbf{r}_2) & \ldots & \psi_1(\mathbf{r}_N) \\ \psi_2(\mathbf{r}_1) & \psi_2(\mathbf{r}_2) & \ldots & \\ \vdots & \vdots & & \\ \psi_N(\mathbf{r}_1) & & & \psi_N(\mathbf{r}_N) \end{vmatrix},$$

which we should use. $\bar{V}_e(\mathbf{r}_i)$ can still be set up in terms of

$$\int \psi'(\mathbf{r}_1\ldots\mathbf{r}_N)^*\frac{e^2}{2r_{ij}}\psi'(\mathbf{r}_1\ldots\mathbf{r}_N)\ \mathrm{d}\mathbf{r}_1\ldots\mathrm{d}\mathbf{r}_N,$$

where all electrons bar the ith appear in ψ', but you can see that what was a minor irritation is now becoming a serious problem. The many-electron atom problem is not elementary.

But now look at the problem from a group theory point of view. The one-electron Hamiltonian

$$\mathscr{H}_i=-\frac{\hbar^2}{2\mu}\nabla_i^2-\frac{Ze^2}{r_i}+\tfrac{1}{2}\sum_{j\neq i}\frac{e^2}{r_{ij}}$$

is, to a first approximation, still a spherically symmetric operator. Any departure from sphericity is removed from the initial problem and reintroduced later as a perturbation. Within this approximation the average Coulombic interaction takes the form $\bar{V}_e(r_i)$, depending only on the radial coordinate as opposed to the vector $\mathbf{r}_i$.

We can therefore then make the same group-theoretical statements as for the one-electron H atom. The one-particle eigenfunctions belong to the $(2l+1)$-fold degenerate irreps of $R_3 \otimes S_2$.

$$\psi_{nlm}\ (\mathbf{r}_i) = R_n(r_i)\Theta^l_{|m|}\ (\theta_i)\ \mathrm{e}^{im\phi_i}.$$

The only difference lies in the radial functions $R_{nl}\ (r_i)$ and in the fact that the accidental l-degeneracy of hydrogen is no longer present.

10.3 Atoms in crystal fields

Having established the symmetry types and degeneracies of the levels of free atoms and, in Chapter 8, having shown that the effect of a reduction of symmetry on normal-mode degeneracies was to cause a splitting of various degeneracies, it is now a suitable time to consider the splitting of atomic levels on reduction of symmetry.

As a first example consider the effect of a crystal field on the degeneracies of states of an impurity atom. We have in mind, for example, the rare earth and transition metal atoms, which have incomplete electronic shells and are strongly affected by the crystal field. It is sufficient, however, to think of a hydrogen atom in order to demonstrate the principles of the problem.

Crystal sites have at most three equivalent axes running through them. This is manifested by the fact that the crystallographic point groups contain only one, two and three-dimensional irreps (the latter are present only in the cubic groups). All states must therefore have degeneracy of three or less in a crystallographic environment. As a consequence atomic states of l value greater than 1 must split regardless of the exact symmetry at the atomic site, because they have degeneracy greater than three in the free atom.

As a specific example we will consider the s, p, and d states of a hydrogen atom ($l = 0, 1, 2$) and ask, 'what degeneracies and symmetry types would these states take up when the atom is placed in a cubic lattice?' Two cases will be studied. Taking the crystal lattice to be body-centred, the impurity will be assumed firstly to be substitutional at the body-centre and secondly interstitial at a position half way between the body-centre and a lattice position (Figures 10.2 a,b respectively).

For the substitutional case we are dealing with a site of O_h symmetry. To find the degeneracies we need to obtain the characters for the representations based on the s,p and d-state functions and reduce the representations. In turn the characters might be obtained by picking an operation in each class of the O_h group and seeing how d-states transform, etc. However, we

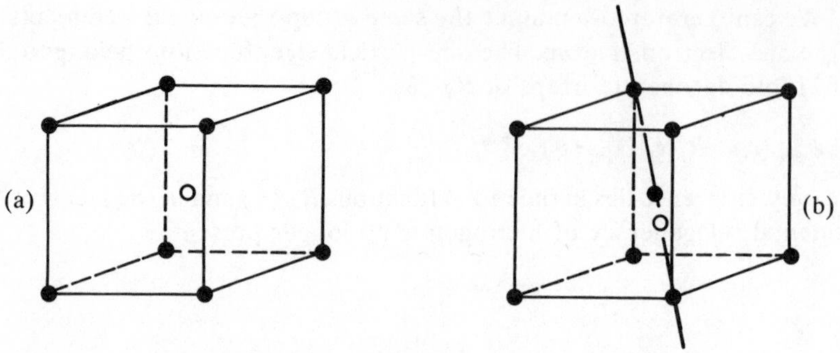

Figure 10.2 The site symmetry of an impurity in a body-centred cubic lattice: (a) substitional (b) interstitial.

already know the answer to this question because we know the transformation properties of free-atom degenerate states, under all the pure or improper rotations;

$$[\sin (l+1/2)\alpha \;/\sin \alpha/2] \qquad \text{and} \qquad [(-1)^l \sin (l+1/2)\alpha \;/\sin \alpha/2]$$

respectively. The free-atom states are just those for which the axis of rotation is inconsequential, only the angle being significant, so that for example $\chi^l(C_2)$ must equal $\chi^l(C_2')$ for the different classes of the O_h group C_2 operations. In order to analyse this type of problem it is clearly useful to tabulate some of the values of $\sin(l+\frac{1}{2})\alpha/\sin\alpha/2$. For the crystallographic groups we are interested only in rotations C_2, C_3, C_4, C_6 (Chapter 17) and in improper rotations that can be written in the form $iC_2 \equiv S_2$ etc. The required sine ratios are given in Table 10.3 for $l = 0, 1, 2, 3$. This information, together with the fact that $\chi(iR) = (-1)^l \chi(R)$, is all that is required to solve the crystal field problem.

Table 10.3 Values of $\sin(l+\frac{1}{2})\alpha/\sin\alpha/2$ for the crystal rotations.

α \ l	0	1	2	3
π	1	−1	1	−1
$2\pi/3$	1	0	−1	1
$\pi/2$	1	1	−1	−1
$\pi/3$	1	2	1	−1

Tables 10.4 and 10.5 contain the required information for the substitutional case. The O_h characters are reproduced in Table 10.4. Table 10.5 shows the characters for representations based on the $l = 0$, 1 and 2 atomic

eigenfunctions. These characters are taken from Table 10.3; the labels D^{0g}, etc. refer to the irreps of the full rotational group. Note that the improper rotations have been written in both the conventional form (S_n or σ) and in iC_n form in Table 10.5. This enables us to use expressions for $\chi(iR)$ in order to determine the appropriate characters.

Table 10.4 The O_h group character table.

O_h	E	$8C_3$	$6C_4$	$3C_2$	$6C_2$	i	$8S_6$	$6S_4$	$3\sigma_h$	$6\sigma_d$
A_{1g}	1	1	1	1	1	1	1	1	1	1
A_{2g}	1	1	−1	1	−1	1	1	−1	1	−1
E_g	2	−1	0	2	0	2	−1	0	2	0
T_{1g}	3	0	1	−1	−1	3	0	1	−1	−1
T_{2g}	3	0	−1	−1	1	3	0	−1	−1	1
A_{1u}	1	1	1	1	1	−1	−1	−1	−1	−1
A_{2u}	1	1	−1	1	−1	−1	−1	1	−1	1
E_u	2	−1	0	2	0	−2	1	0	−2	0
T_{1u}	3	0	1	−1	−1	−3	0	−1	1	1
T_{2u}	3	0	−1	−1	1	−3	0	1	1	−1

Table 10.5 Characters for $l=0$, 1, 2 states, under the O_h group symmetry operations.

O_h	E	$8C_3$	$6C_4$	$3C_2$	$6C'_2$	i	$8S_6 \equiv iC_3$	$6S_4 \equiv iC_4$	$3\sigma_h \equiv iC_2$	$6\sigma_d \equiv iC'_2$
D^{0g}	1	1	1	1	1	1	1	1	1	1
D^{1u}	3	0	1	−1	−1	−3	0	−1	1	1
D^{2g}	5	−1	−1	1	1	5	−1	−1	1	1

The reductions of the above irreps of the full rotation group $R_3 \otimes S_2$, in the O_h group are:

$D^{0g} \Rightarrow A_{1g}$,
$D^{1u} \Rightarrow T_{2u}$,
$D^{2g} \Rightarrow E_g \oplus T_{2g}$.

From these results we therefore conclude that atomic $l=0$ states have symmetry of type A_{1g} at a site of O_h symmetry. They must remain non-degenerate of course, and have to be bases for the symmetric irrep. $l=1$ states remain three-fold degenerate, now of symmetry type T_{2u}. However, $l=2$ states, which have to split as they are five-fold degenerate, split into a pair of levels—a doubly degenerate E_g state and a triply-degenerate T_{2g} state. If the crystal field is relatively small, compared to the Coulombic field of the atom itself, the energies in the crystal are not too dissimilar to those of the atom, and the splittings are small.

Turning to the interstitial case, the site is now of C_{3v} symmetry. The body diagonal, shown in Figure 10.2(b), is the C_3 axis, the three σ_d planes of O_h that contain this axis are now the σ_v planes of C_{3v}. Table 10.6 summarizes the group-theoretical analysis.

Table 10.6 C_{3v} character table; and characters for the bases of D^{0g} etc., for the C_{3v} operations.

C_{3v}	E	$2C_3$	$3\sigma_v \equiv iC_2$
A_1	1	1	1
A_2	1	1	−1
E	2	−1	0
D^{0g}	1	1	1
D^{1u}	3	0	1
D^{2g}	5	−1	1

On reduction of the full rotation group representations, we obtain

$$D^{0g} \Rightarrow A_1,$$
$$D^{1u} \Rightarrow A_1 \oplus E,$$
$$D^{2g} \Rightarrow A_1 \oplus 2E.$$

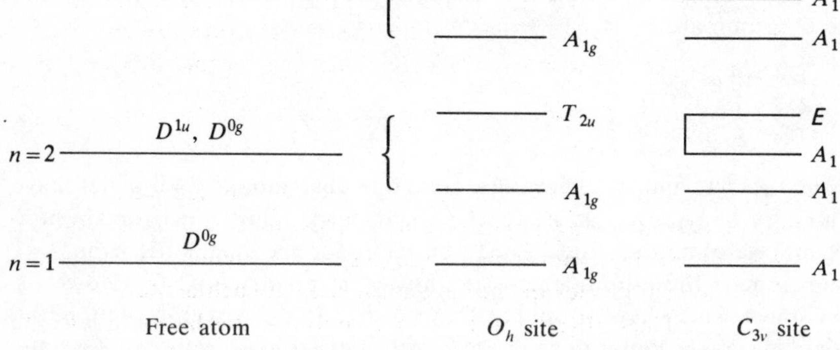

Figure 10.3 Schematic for the degeneracies and symmetry types of the $n=1$, 2, 3 H-atom levels of the free atom, and at sites of O_h and C_{3v} symmetry.

In this case the group is non-cubic, the greatest irrep. dimensionality is two, and so the $l=1$ states must also split. The fact that different numbers of energy levels and different symmetry types pertain in the two cases is of great use physically. As we shall discover in Chapter 15, the symmetry types will tell us which radiation selection rules are obtained. We will be able to predict the number of absorption frequencies that should be observable (and polarization selection rules) for the transitions between the H-atom levels. This will differ in the two cases, so that despite the absence of any quantitative information we should, by measuring the form of the absorption spectrum near a free-atom absorption frequency, be able to say whether the atom took up a substitutional or interstitial position in the lattice.

In Figure 10.3 the degeneracies and groupings of the levels are shown schematically. We can say nothing about the relative positions of levels in any groupings. Different symmetry types for the free-atom levels correspond to accidental degeneracies; there may of course be such degeneracies even in the crystal field cases.

Problems

***10.1** The p-state eigenfunctions of the hydrogen atom may be written as $xf_p(r)$, $yf_p(r)$, $zf_p(r)$. Verify that these functions form a basis for the irrep. D^{1u} of the $R_3 \otimes S_2$ group.

***10.2** Extend Table 10.3 to include the $l=4$ states.

10.3 We know that energy levels of a free atom split in crystal fields. In particular we obtained the degeneracies and symmetry types of various l-states at site of C_{3v} symmetry. Given that the set of atomic orbitals ψ_{nlm} form a complete set for the crystal field situation, obtain the most general forms for the eigenfunctions in the C_{3v} case.

10.4 Verify that the Slater determinant eigenfunction automatically satisfies Pauli exclusion.

***10.5** (i) The outer electron of the Ce^{3+} ion is a $4f$-electron, all other electrons being in closed shells. This rare earth ion forms an interstitial impurity in CaF_2 at a site of O_h symmetry. What are the symmetries of the seven $4f$-states at such a site?

The CaF_2 structure is a face-centred-cubic lattice of calcium atoms, with fluorine atoms displaced by $(\frac{1}{4}, \frac{1}{4}, \frac{1}{4})$ and $(\frac{3}{4}, \frac{3}{4}, \frac{3}{4})$ with respect to each calcium. Satisfy yourself that the site of O_h symmetry referred to above is the body-centre of the Ca lattice, by considering some example operations.

(ii) Substitute an F^- ion for one of the F atoms closest to the Ce impurity in order to compensate for one of the Ce charges. What is the new symmetry of the Ce site and how are the 4*f*-states changed?

(From knowledge of the energy-level symmetries and their absorption selection rules it is possible to determine the fraction of Ce^{3+} sites which achieve partial compensation, by making absorption spectroscopy measurements.)

(iii) The sample is now subjected to a static electric field in the [0,0,1] direction. Sketch the symmetries and degeneracies of the 4*f*-states, for both site symmetries, for a weak field such that the electric-field splittings are far less than the crystal-field splittings.

11

The axial groups

11.1 Group character tables

There are a number of interesting points to be made concerning those groups that allow for any rotations but about a *specific* axis. In order to construct the elements of these *axial groups*, five characteristic systems will be considered:

(i) the H_2 molecule;
(ii) the HCl molecule or
(iii) an atom in an electric field;
(iv) an atom in a magnetic field;
(v) an HCl molecule in a magnetic field.

The group of the H_2 molecule

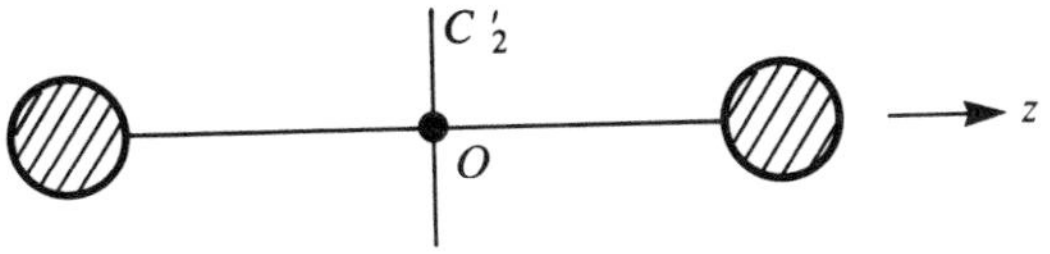

Figure 11.1 Schematic H_2 molecule.

The operations of the H_2 symmetry group consists of:
(i) the identity, E;

(ii) rotations by any angle about the molecule axis. Each different angle of rotation is in a different class excepting that $R(\alpha, z)$ and $R(2\pi - \alpha, z)$ are physically indistinct. Thus each rotational class will contain two operations $2R(\alpha, z)$;

(iii) rotations by π about any axis such as that depicted, which passes through 0, normal to the molecular axis, C'_2;

(iv) inversion;

(v) improper rotations, $2iR(\alpha, z)$;

(vi) reflections with respect to any plane perpendicular to a C'_2 axis, for example the plane of the paper, σ_v.

The group is labelled $D_{\infty h}$.

Table 11.1 Character table for the $D_{\infty h}$ group. Molecular symmetry labels are included in brackets. In the final line $p = g$ or u. The positive signs apply if $p = g$, the negative if $p = u$.

	E	$2R(\alpha,z)_\infty$	$\infty C'_2$	i	$2iR(\alpha,z)_\infty$	$\infty\sigma_v$
$A_{1g}(\Sigma_g^+)$	1	1	1	1	1	1
$A_{1u}(\Sigma_u^-)$	1	1	1	-1	-1	-1
$A_{2g}(\Sigma_g^-)$	1	1	-1	1	1	-1
$A_{2u}(\Sigma_u^+)$	1	1	-1	-1	-1	1
$E_{1g}(\pi_g)$	2	$2\cos\alpha$	0	2	$2\cos\alpha$	0
$E_{1u}(\pi_u)$	2	$2\cos\alpha$	0	-2	$-2\cos\alpha$	0
.						
.						
.						
E_{Mp}	2	$2\cos M\alpha$	0	± 2	$\pm 2\cos M\alpha$	0

There are an infinite number of classes and hence an infinite number of irreps. Almost all of the irreps are two-dimensional, a consequence of the two operations per $R(\alpha, z)$ class and the infinite number of C'_2 and σ_v operations in each class. (Note that the σ operation with respect to the plane that contains the C'_2 axes is contained as $iR(\pi, z)$.)

The HCl molecule

For diatomic molecules with distinguishable atoms certain of the $D_{\infty h}$ operations no longer apply. There is clearly no centre of inversion or improper rotation. The C'_2 axes are no longer symmetry axes, but the σ_v operations remain. This is the $C_{\infty v}$ group.

Figure 11.2 Schematic HCl molecule.

Table 11.2 Character table for the $C_{\infty v}$ group.

	E	$2R(\alpha,z)_\infty$	$\infty\sigma_v$
$A_1(\Sigma^+)$	1	1	1
$A_2(\Sigma^-)$	1	1	−1
$E_1(\pi)$	2	2 cos α	0
E_M	2	2 cos $M\alpha$	0

Atom in a uniform electric field

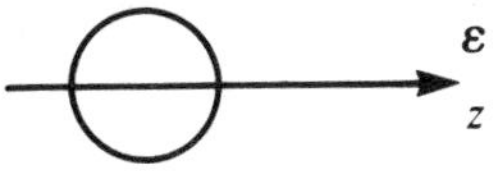

Figure 11.3 Schematic of an atom in an electric field.

An electric field has a direction associated with it, and it is described by a *polar vector* $\boldsymbol{\varepsilon}$ which changes sign on inversion. Physically, a charged particle placed in an $\boldsymbol{\varepsilon}$ field will accelerate in a particular direction. Reverse the field and this direction will reverse. An atom in an electric field does not, as a consequence, have inversion symmetry. It has the same $C_{\infty v}$ symmetry as the HCl molecule.

Atom in a magnetic field

In contrast to the above example a magnetic field is not associated with a polar vector. We can see this mathematically by noting the definition $\mathbf{B} = \nabla \wedge \mathbf{A}$, where $\mathbf{B}$ indicates the magnetic field vector. $\mathbf{A}$ is the vector potential, and can be chosen to be proportional to $\boldsymbol{\varepsilon}$; it must be a polar vector field. Equally ∇ is a polar vector ($\partial/\partial x$ becomes $-\partial/\partial x$ if x is inverted to $-x$, etc.). Hence $\mathbf{B}$ does not change under inversion; it is a *pseudo* or *axial vector*, and its symmetry can be thought of as that of a rotational vector around the field axis.

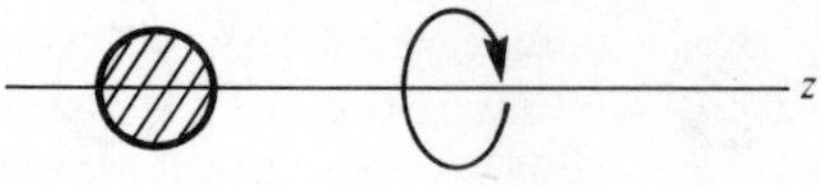

Figure 11.4 Schematic of an atom in a magnetic field.

One can think alternatively of $\nabla \wedge \mathbf{A} = \text{curl}\, \mathbf{A}$; the use of the word curl is not accidental. Finally from a physical point of view a charged particle placed in a **B** field does not accelerate in either direction. However, a charged particle of finite velocity perpendicular to the **B** axis performs circular motion. Regardless of the direction of the velocity this motion is of the same *sense*. This indicates that rotations of opposite sense must be physically distinguishable in this system. We no longer expect $R(\alpha,z)$ and $R(2\pi-\alpha,z)$ to be in the same class. Inversion is a group operation (as **B** is invariant on inversion) and so are the improper rotations. However, C'_2 and σ_v operations are not allowed because they reverse the sense of rotation. This is the group $C_\infty \otimes S_2$. Note that as all classes now contain only one operation, all of the irreps must be one-dimensional and hence no non-accidental degeneracies exist.

Table 11.3 Character table for the group $C_\infty \otimes S_2$.

	E	$R(\alpha,z)$	i	$iR(\alpha,z)$
A_g	1	1	1	1
A_u	1	1	-1	-1
E_{+1g}	1	$e^{i\alpha}$	1	$e^{i\alpha}$
E_{+1u}	1	$e^{i\alpha}$	-1	$-e^{i\alpha}$
E_{-1g}	1	$e^{-i\alpha}$	1	$e^{-i\alpha}$
E_{-1u}	1	$e^{-i\alpha}$	-1	$-e^{-i\alpha}$
$\vdots$				
$E_{\pm Mg}$	1	$e^{\pm iM\alpha}$	1	$e^{\pm iM\alpha}$
$E_{\pm Mu}$	1	$e^{\pm iM\alpha}$	-1	$-e^{\pm iM\alpha}$

HCl in a magnetic field

Only those elements that are common to the HCl group and the group of an atom in a magnetic field remain in this group. It is called C_∞ and again contains only one-dimensional irreps.

Various notions appear in the literature for the classes of these groups, and for $R(\alpha,\xi)$ of the spherical symmetry group. The operational angle α is

Table 11.4 Character table for the C_∞ group.

	E	$R(\alpha,z)$
A	1	1
E_{+1}	1	$e^{i\alpha}$
E_{-1}	1	$e^{-i\alpha}$
$\vdots$		
$E_{\pm M}$	1	$e^{\pm iM\alpha}$

often replaced by ϕ. However, as ϕ is reserved for the variable spherical polar coordinate we prefer to use a different notation here. As will be seen in the next section, both α and ϕ can appear in the same equations. We will now use the $C_{\infty v}$ and $C_\infty \otimes S_2$ groups in order to discuss the Stark and Zeeman effects. The H_2 molecular orbitals will be discussed, using the $D_{\infty h}$ group, in Chapter 12.

11.2 The Stark and Zeeman effects

Example (a): The Stark effect

We wish to use the example of an atom in a uniform applied electric field in order firstly to demonstrate how the angular part of eigenfunctions can be derived from the character tables, and secondly to demonstrate the splitting of atomic energy levels in a gradually increasing (perturbing) field.

In Section 10.1 we verified that the states ψ_{nlm} formed bases for the irreps of $R_3 \otimes S_2$. Now we wish to consider the converse problem. Given a set of basis functions, can we find those that belong to each irrep. of a continuous group? That is, can we adapt the formula for symmetry-adapted basis functions:

$$F^j(Q) \propto \sum_r \chi^j(G_r)^* G_r Q.$$

There are two G_r types of concern, $R(\alpha,z)$ and σ_v. For the latter we need to define the planes of reflection. Looking down the z-axis at the x–y plane a single angle β is sufficient to define a σ_v plane. The angle β ranges from 0 to π.

The sum over all rotations (including the identity $R(0,z)$) and all reflections can now be replaced by a pair of integrations:

$$F^j(Q) = \int_0^{2\pi} \chi^j(R(\alpha,z))^* R(\alpha,z)\, Q \, \mathrm{d}\alpha + \int_0^{\pi} \chi^j(\sigma_v(\beta))^* \sigma_v(\beta)\, Q \, \mathrm{d}\beta.$$

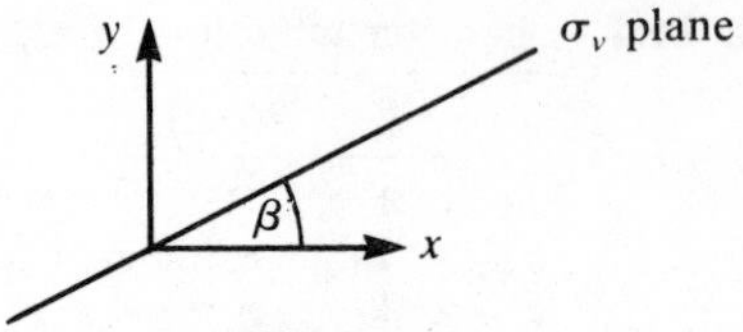

Figure 11.5

We need to use

$$R(\alpha,z)\ Q(r,\theta,\phi)=Q(R^{-1}(r,\theta,\phi))=Q(r,\theta,\phi-\alpha),$$
$$\sigma_v(\beta)\ Q(r,\theta,\phi)\ =Q(\sigma_v^{-1}(r,\theta,\phi))=Q(r,\theta,2\beta-\phi),$$

and select various Q possibilities. We are concerned with the irreps (Γ^j) of the $C_{\infty v}$ group (Table 11.2).

$$Q=f_1(r):\ F^{A_1}(f_1)\propto\left[\int_0^{2\pi}1.f_1\ \mathrm{d}\alpha+\int_0^{\pi}1.f_1\ \mathrm{d}\beta\right]\propto f_1;$$

$$Q=xf_2(r):\ F^{A_1}(xf_2)\propto\left[\int_0^{2\pi}\sin\theta\ \cos(\phi-\alpha)\ f_2\ \mathrm{d}\alpha+\int_0^{\pi}\sin\theta\ \cos(2\beta-\phi)f_2\ \mathrm{d}\beta\right]=0;$$

$$Q=xf_2(r):\ F^{E_1}(xf_2)\propto\left[\int_0^{2\pi}2\ \cos\alpha\ r\ \sin\theta\ \cos(\phi-\alpha)\ f_2\ \mathrm{d}\alpha+0\right]\propto xf_2.$$

Table 11.5 Symmetry adapted functions for the $C_{\infty v}$ group.

function	f_1	xf_2, yf_2	zf_2	$(x^2+y^2)f_3$	z^2f_3	xzf_2, yzf_3	$(x^2-y^2)f_3, xyf_3$
irrep.	A_1	E_1	A_1	A_1	A_1	E_1	E_2

By the same token the assignments shown in Table 11.5 can be made. Hence from any set of polynomials $x^ay^bz^cf(r)$ we can construct symmetry-adapted functions belonging to each irrep. Further, for the one-dimensional irrep. A_1, the non-degenerate eigenfunctions must have the general form

$$\psi^{A_1}=a_1f_1+a_2zf_2+a_3(x^2+y^2)f_3+a_3z^2f_3+\cdots$$
$$=R(r)\Theta(\theta)\ \mathrm{e}^{\mathrm{i}0\phi}.$$

It is not surprising that the A_1 eigenfunctions are independent of ϕ, because only ϕ is altered under the group operations. The E symmetry-adapted functions can easily be formed into pairs belonging to different rows of E_M.

$$\psi_{+}^{E_1}=[a_1(x+\mathrm{i}y)f_2+a_2(x+\mathrm{i}y)zf_3+\cdots]\propto \mathrm{e}^{\mathrm{i}\phi},$$

$$\psi_{-}^{E_1}=[a_1(x-\mathrm{i}y)f_2+a_2(x-\mathrm{i}y)zf_3+\cdots]\propto \mathrm{e}^{-\mathrm{i}\phi}.$$

All the terms in $\psi_{\pm}^{E_1}$ transform into $\mathrm{e}^{\mp \mathrm{i}\alpha}$ times themselves under $R(\alpha,z)$ etc., confirming that they belong to the same row of the same form of the E_1 irrep. Thus starting from a position of no knowledge except the character table, we can induce and confirm the general results:

$$F^{E_M}(R\Theta \mathrm{e}^{im\phi})\propto\int_0^{2\pi} 2\cos M\alpha\, \mathrm{e}^{-im\alpha}\, R\,\theta\, \mathrm{e}^{\mathrm{i}m\phi}\, \mathrm{d}\alpha$$

$$\propto R\Theta \mathrm{e}^{\mathrm{i}m\phi} \qquad \text{providing } m=\pm M.$$

Also, because eigenfunctions are linear combinations of symmetry-adapted functions,

$$\psi_{\pm}^{E_{|m|}}=R\theta \mathrm{e}^{\pm im\phi},$$

with $\pm m$ functions being degenerate. The energy levels can be labelled by the $|m|$ values; this is a 'good quantum number' for a $C_{\infty v}$ system. The $R\theta$ functions cannot be specified without resort to the Hamiltonian of the problem. However, if the electric field is sufficiently weak, then the eigenfunctions and energies must closely resemble the zero-field atomic functions. Figure 11.6 summarizes, for $n=1,2$ atomic states, the conclusions that can be made *from symmetry alone*.

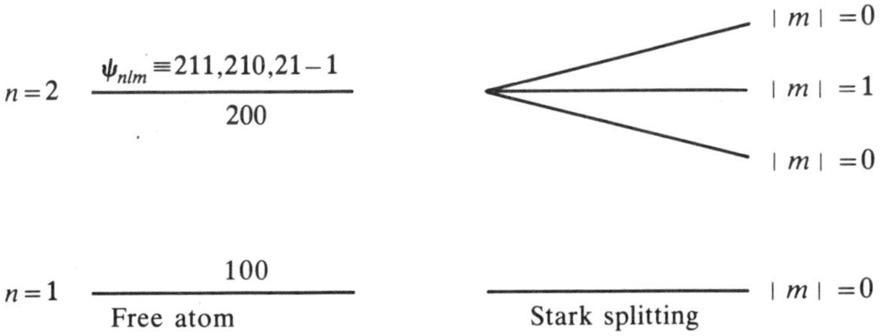

Figure 11.6 Schematic of group-theoretic conclusions with regard to Stark splitting of the H-atom states.

The $|m|=1$ state is doubly degenerate; the $n=2$, $|m|=0$ states are basically mixtures of the states ψ_{200} and ψ_{210} (see Section 15.3). Their splitting gives the familiar linear Stark effect; it is linear in the field strength.

Expanding the states in the electric-field case as linear combinations of the complete set of atomic orbitals, most generally:

$$\psi_{\pm}^{E|m|} = \sum_{n,l} a_{nl} \psi_{nl\pm m}.$$

Example (b): the Zeeman effect

In direct contrast to the above, for an atom placed in a magnetic field, use of the $C_\infty \otimes S_2$ group character table gives such results as:

$$F^{E_{+1u}}(xf_2) \propto (x+iy)f_2 \text{ etc.,}$$

indicating that a function $R\Theta\, e^{i\phi}$ is an eigenfunction of E_{+1u} type if $\Theta e^{i\phi}$ has odd parity, etc. Using the $C_\infty \otimes S_2$ character table,

$$\psi^{A_g} = R\Theta,$$

$$\psi^{E_{mg}} = R\Theta e^{-im\phi} \qquad \text{for even parity functions,}$$

$$\psi^{E_{mu}} = R\Theta e^{+im\phi} \qquad \text{for odd parity functions.}$$

None of the states are degenerate, the 'good quantum number' is now m, and it can take positive or negative values. In terms of atomic orbitals,

$$\psi^{E_{mg}} = \sum_{n,l} a_{nl}\, \psi_{nlm} \qquad (l \text{ even}),$$

$$\psi^{E_{mu}} = \sum_{n,l} a_{nl}\, \psi_{nlm} \qquad (l \text{ odd}).$$

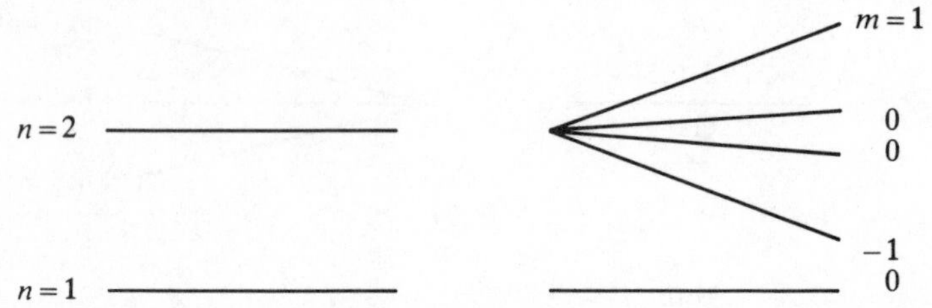

Figure 11.7 Schematic of group-theoretic conclusions for orbital Zeeman splitting of the H-atom.

These results contrast nicely the different consequences of the polar and axial natures of electric and magnetic field vectors. In the schematic figures the right-hand lines are drawn sloped to indicate the increasing effect of the fields. The order of states is of course uncertain and there may be additional accidental degeneracies (cf. the $m = 0$ states at the $n = 2$ level in the magnetic field case).

Problems

***11.1** Find the combinations of (i) x, y and z, (ii) $x^2, y^2, z^2, xy, yz, zx$ which form symmetry-adapted bases for irreps of (a) the C_∞ group and (b) the $D_{\infty h}$ group.

11.2 For a linear molecule containing N atoms, and having no centre of inversion, show that there are $(N-1)$ vibrational modes of A_1 symmetry and $(N-2)$ vibrational modes of E_1 symmetry.

***11.3** Discuss the symmetry properties of the normal modes of the linear molecules (i) CO_2 and (ii) OCS.

11.4 The core states of atoms may be only slightly perturbed in the configuration of a molecule. Discuss the likely changes in the $n=1,2$ states of iodine in potassium iodide, KI. (The outer occupied electronic states contribute to the molecular bonds; they are discussed in the following chapter.)

12

Molecular orbital theory

12.1 The molecular Hamiltonian

The half way stage between the hydrogen atomic problem (which is soluble) and the solid-state problem (which most definitely is not exactly soluble) is the molecular system. It is useful to introduce some of the quantum-mechanical and group-theoretical ideas that will be used in the solid-state case in the more tractable, and finite, molecular situation. The beautiful symmetries of many molecules lends this branch of physics to group-theory methods. First though, one needs to set up the problem such that group theory may best be applied; starting with the molecular Hamiltonian:

$$\mathcal{H}_{\mathrm{mol}} = \sum_n \frac{-\hbar^2}{2M_n}\nabla_n^2 + \sum_i \frac{-\hbar^2}{2m}\nabla_i^2 + \sum_n \sum_i \frac{-Z_n e^2}{|\mathbf{r}_i - \mathbf{R}_n|}$$
$$+ \tfrac{1}{2}\sum_{n,}\sum_{p\neq n} \frac{Z_n Z_p e^2}{|\mathbf{R}_n - \mathbf{R}_p|} + \tfrac{1}{2}\sum_{i,}\sum_{j\neq i} \frac{e^2}{|\mathbf{r}_i - \mathbf{r}_j|}.$$

Again we should like to reduce this to a sum of terms, each for a separate coordinate.

The Born–Oppenheimer approximation

The use of centre-of-mass coordinates no longer enables the nuclear kinetic term to be removed. Instead we appeal to the *Born–Oppenheimer adiabatic approximation*.

If we could hold the nuclei rigid, then the nuclear kinetic term would be eliminated from $\mathscr{H}_{\text{mol}}$ and the nuclear Coulombic repulsion term would be a constant and therefore trivial to handle. The remaining Hamiltonian, in the electronic coordinates, would contain the coordinates of the rigid nuclear configuration only in the electron–nuclear attraction energy, and the many-electron eigenfunctions would be of form

$$\psi_{\text{elect}} = \psi_{e\mathbf{R}}\,(\mathbf{r}_1, \mathbf{r}_2, \mathbf{r}_3, \ldots).$$

Here e is meant to refer to one of the set of eigenfunctions, $\mathbf{R}$ manifests the nuclear configuration and $\mathbf{r}_i$ runs over all the electrons. For the molecule as a whole,

$$\psi = X(\mathbf{R})\psi_{e\mathbf{R}}(\mathbf{r}),$$

where the X problem for the nuclear distribution is trivial (nucleus-1 is fixed at $\mathbf{R}_1$ etc). Now consider what happens if this form of eigenfunction is used in the presence of nuclear motion:

$$\begin{aligned}\mathscr{H}_{\text{mol}}X(\mathbf{R})\psi_{e\mathbf{R}}(\mathbf{r}) = {} & X(\mathbf{R})\left[\sum_i \frac{-\hbar^2}{2m}\nabla_i^2 + \tfrac{1}{2}\sum_{i,j}\frac{e^2}{r_{ij}} + \sum_{i,n}\frac{-Z_n e^2}{r_{in}} + \tfrac{1}{2}\sum_{n,p}\frac{Z_n Z_p e^2}{R_{np}}\right]\psi_{e\mathbf{R}}(\mathbf{r}) \\ & + \psi_{e\mathbf{R}}(r)\left[\sum_n \frac{-\hbar^2}{2M_n}\nabla_n^2\right] X(\mathbf{R}) \\ & - \sum_n \frac{\hbar^2}{2M_n}\left[2\nabla_n X(\mathbf{R})\nabla_n\psi_{e\mathbf{R}}(\mathbf{r}) + X(\mathbf{R})\nabla_n^2\psi_{e\mathbf{R}}(\mathbf{r})\right] \\ = {} & E_{\text{mol}}X(\mathbf{R})\psi_{e\mathbf{R}}(\mathbf{r}).\end{aligned}$$

Thus the nuclear and electronic parts can be separated provided that the final term can be ignored. That is,

$$\left[\sum_i \frac{-\hbar^2}{2m}\nabla_i^2 + \tfrac{1}{2}\sum_{i,j}\frac{e^2}{r_{ij}} + \sum_{i,n}\frac{-Z_n e^2}{r_{in}} + \tfrac{1}{2}\sum_{n,p}\frac{Z_n Z_p e^2}{R_{np}}\right]\psi_{e\mathbf{R}}(\mathbf{r}) = E_e(\mathbf{R})\psi_{e\mathbf{R}}(\mathbf{r})$$

and

$$\left[\sum_n \frac{-\hbar^2}{2M_n}\nabla_n^2 + E_e(\mathbf{R})\right] X(\mathbf{R}) = E_{\text{mol}}X(\mathbf{R}),$$

providing

$$[2\nabla_n X(R)\,\nabla_n\psi_{e\mathbf{R}}(\mathbf{r}) + X(\mathbf{R})\,\nabla_n^2\psi_{e\mathbf{R}}(\mathbf{r})] \ll \psi_{e\mathbf{R}}(\mathbf{r})\,\nabla_n^2 X(\mathbf{R}).$$

The criterion for this separation is that the electronic eigenfunctions depend less rapidly on the nuclear coordinates than do the nuclear functions themselves. The result is that in the electronic problem the nuclei are taken to be rigid whilst in the nuclear problem the electronic motion is averaged out in some way; it is manifested by the presence of $E_e(\mathbf{R})$ $(=\int\psi_{e\mathbf{R}}\mathscr{H}_{\text{elect}}\psi_{e\mathbf{R}})$. There is a non-rigorous physical argument that supports this statement. The nuclei, being 2000 times more heavy than the electrons, have a far smaller acceleration in response to any given force. Thus, whilst the electrons can respond rapidly to a change of the nuclear configuration, the converse is not true; the nuclei are unable to 'follow' electronic redistribution. When we talk above of the electronic eigenfunctions for a rigid set of nuclei, we can interpret this to mean that as the nuclei move, the electrons are able to adapt 'instantly' to any new configuration; they see each configuration the nuclei pass through and will attain equilibrium distributions $\psi_{e\mathbf{R}}(\mathbf{r})$ of energies $E_{e\mathbf{R}}(\mathbf{r})$ at any given instant of nuclear configuration $\mathbf{R}$. The interaction of the electrons with *motion* of the nuclei can be treated as a perturbation in most situations.

We shall take it then that the final term in the molecular Hamiltonian can be dropped initially. The nuclear motion itself will be dealt with in Chapter 16; it is of course anticipated that there is a connection between the nuclear motion considered quantum mechanically and the classical normal mode approach of Chapters 6–8. This leaves us with the electronic problem

$$\mathcal{H}_{\text{elect}} = \sum_i \left[\frac{-\hbar^2}{2m}\nabla_i^2 + \sum_n \frac{-Z_n e^2}{r_{in}} + \tfrac{1}{2}\sum_{j\neq i}\frac{e^2}{r_{ij}}\right] + \tfrac{1}{2}\sum_{n,p}\frac{Z_n Z_p e^2}{R_{np}} = \sum_i \mathcal{H}_i + \text{constant}.$$

This is now analogous to the many-electron atomic problem, but the spherical symmetry of the atom is replaced by whatever symmetry pertains for the nuclear configuration of the molecule. The approach is also the same. The electron–electron interaction, or at least that part of it that does not bear the symmetry of the molecule, must be held back until the one-electron problem is solved. The interaction is then reintroduced as a perturbation:

$$\mathcal{H}_i \approx \frac{-\hbar^2}{2m}\nabla_i^2 + \sum_n \frac{-Z_n e^2}{r_{in}} + \bar{V}_e \text{ (symmetric)}.$$

The resulting one-electron eigenfunctions are known as *molecular orbitals*. The one-electron Hamiltonian problem is still insoluble. Here we wish to discuss an approximate solution which accounts for the symmetry of the system—the approximation by linear combination of atomic orbitals (LCAO).

12.2 The LCAO–MO method

For electronic positions close to any one nucleus the electronic distribution must be very close to the corresponding atomic distribution. Thus a reasonable molecular orbital function is of the form

$$\psi_i = \sum_{nlm,j} a_{(nlm,j),i}\, \psi_{nlm,j},$$

where $\psi_{nlm,j}$ is an atomic eigenfunction centred at the jth nucleus. One might also anticipate that fairly few such functions will appear with large coefficients in the summation. In particular, one would expect that only those functions with rather similar energies would mix. Otherwise one would be saying that a given electron has a strong probability of scattering from a low energy to a high energy as it moves about the molecule. We will now demonstrate that further restrictions on the presence of different states in ψ_i are imposed by symmetry. After all, the ψ_i must be bases for the irreps of the group of the molecule.

Example (a): benzene

An excellent example of the LCAO–MO method is provided by the π-orbitals of the benzene molecule. Amongst these is the highest-energy populated state of benzene, a state which is therefore readily accessible to absorption spectroscopy at relatively low frequencies.

Carbon atoms have six electrons, and hydrogen one. It is therefore reasonable to expect that the low-energy molecular orbitals of benzene (C_6H_6) are closely approximated by linear combinations of the carbon $1s$, $2s$ and all the $2p$ atomic states centred at each C-atom and the hydrogen $1s$ states at each H-atom.

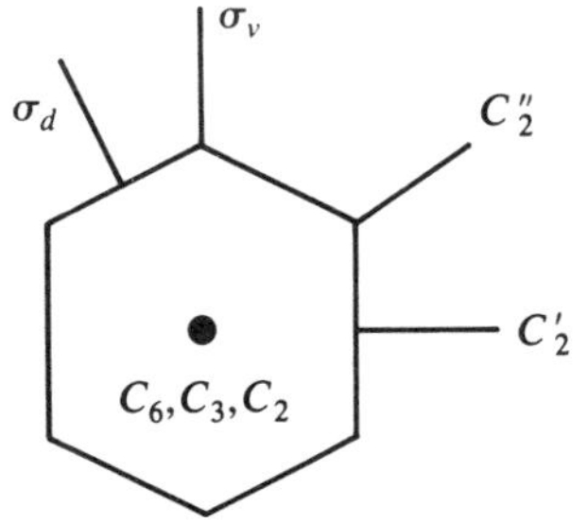

Figure 12.1 The benzene symmetry types—for use with Table 12.1.

The first task of group theory is to determine to which irreps of the benzene symmetry group linear combinations of such atomic orbitals can belong. The full group for benzene is D_{6h} (Table 12.1).

Table 12.1 The benzene symmetry group character table; characters for representations based on atomic s and p orbitals; and the effect of the operations on an s-state orbital.

D_{6h}	E	$2C_6$	$2C_3$	C_2	$3C_2'$	$3C_2''$	i	$2S_3$	$2S_6$	σ_h	$3\sigma_v$	$3\sigma_d$
A_{1g}	1	1	1	1	1	1	1	1	1	1	1	1
A_{2g}	1	1	1	1	−1	−1	1	1	1	1	−1	−1
B_{1g}	1	−1	1	−1	1	−1	1	−1	1	−1	1	−1
B_{2g}	1	−1	1	−1	−1	1	1	−1	1	−1	−1	1
E_{1g}	2	1	−1	−2	0	0	2	1	−1	−2	0	0
E_{2g}	2	−1	−1	−2	0	0	2	−1	−1	−2	0	0
A_{1u}	1	1	1	1	1	1	−1	−1	−1	−1	−1	−1
A_{2u}	1	1	1	1	−1	−1	−1	−1	−1	−1	1	1
B_{1u}	1	−1	1	−1	1	−1	−1	1	−1	1	−1	1
B_{2u}	1	−1	1	−1	−1	1	−1	1	−1	1	1	−1
E_{1u}	2	1	−1	−2	0	0	−2	−1	1	2	0	0
E_{2u}	2	−1	−1	−2	0	0	−2	1	1	2	0	0
Γ^s	6	0	0	0	0	2	0	0	0	6	2	0
Γ^{p_r}	6	0	0	0	0	2	0	0	0	6	2	0
Γ^{p_t}	6	0	0	0	0	−2	0	0	0	6	−2	0
Γ^{p_z}	6	0	0	0	0	−2	0	0	0	−6	2	0
$G_r f_1$	1	2,6	3,5	4	2,4,6	1,3,5	4	2,6	3,5	1	1,3,5	2,4,6

Consider firstly the carbon 1s states. Each of these six states transforms into either itself or one of the other of the six, under each D_{6h} group operation. The characters of the six-dimensional matrices that represent the transformations are given in Table 12.1 alongside the label Γ^s. For example, by reference to Figure 12.1, each atom moves to a new position under C_6, C_3, or C_2 rotations about the central axis normal to the molecule; hence for these operations $\chi^s=0$. Under C_2'' two atoms are unmoved although they rotate. But spherically symmetric functions are clearly invariant on rotation, so each of the two atomic functions contributes $(+1)$ to χ^s for the C_2'' operations, etc. On reduction of the Γ^s representation one obtains

$$\Gamma^s \Rightarrow A_{1g} \oplus B_{2u} \oplus E_{2g} \oplus E_{1u}.$$

Thus any eigenfunctions that contain contributions from the carbon atom 1s states must be of A_{1g}, B_{2u}, E_{2g} or E_{1u} symmetry types. The same is obviously true for the carbon 2s and hydrogen 1s states.

In order to discuss the 2p-states it is useful to pick specific p-orbitals. For a given C-atom take those orthogonal combinations of the spherical harmonics Y_1^1, Y_0^1, and Y_{-1}^1 which are directed along the benzene radius, tangential to it, and in the direction normal to the benzene plane. If you like, these can be thought of as the x,y, and z directions for the particular atom. In this way the three types of orbital transform independently. Using the notations p_r, p_t and p_z respectively, a p_r orbital transforms into itself or into one of the other p_r

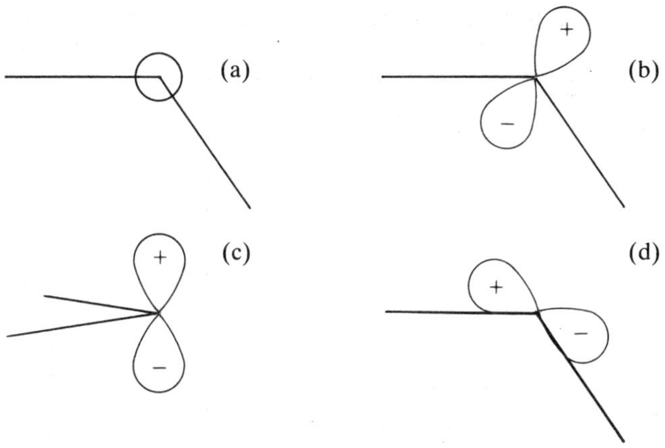

Figure 12.2 Schematic of s-states: (a) and orthogonal p-states (*b–d*) for benzene atom positions; (b) is a typical radial p-state p_r; (c), a p_z state, is directed normal to the molecular plane and (d) is a tangential state p_t.

orbitals centred at a different atom, but never into a p_t or p_z orbital. The characters for the three Γ^p representations can now be found. It is necessary only to note that a p-state has the symmetry of a polar vector; the wave-function changes sign under inversion. Reducing these representations:

$$\Gamma^{p_r} \Rightarrow A_{1g} \oplus B_{2u} \oplus E_{2g} \oplus E_{1u},$$

$$\Gamma^{p_t} \Rightarrow A_{2g} \oplus B_{1u} \oplus E_{2g} \oplus E_{1u},$$

$$\Gamma^{p_z} \Rightarrow B_{1g} \oplus A_{2u} \oplus E_{1g} \oplus E_{2u}.$$

Of particular note is that all the irreps appearing in Γ^{p_z} are different from those in Γ^{p_r}, Γ^{p_t}, or Γ^s. From all the states considered one can therefore construct six eigenfunctions that are combinations *only* of the p_z orbitals. Of these six, two are non-degenerate and there are two degenerate pairs. The six orbitals are known as the π-molecular orbitals,

$$\psi_\pi = \sum_{j=1}^{6} a_j f_j(\mathbf{r}_i),$$

where $f_j(\mathbf{r}_i)$ is a $2p_z$ orbital centred at the jth carbon atom, numbered cyclically around the benzene ring.

Because each irrep. in the reduction of Γ^{p_z} appears only once, their symmetry-adapted basis functions are also eigenfunctions (cf. Wigner's theorem). In this particular case, then, the π-orbital eigenfunctions are of the form

$$\psi^j \propto \sum_r \chi^j(G_r)^* G_r Q,$$

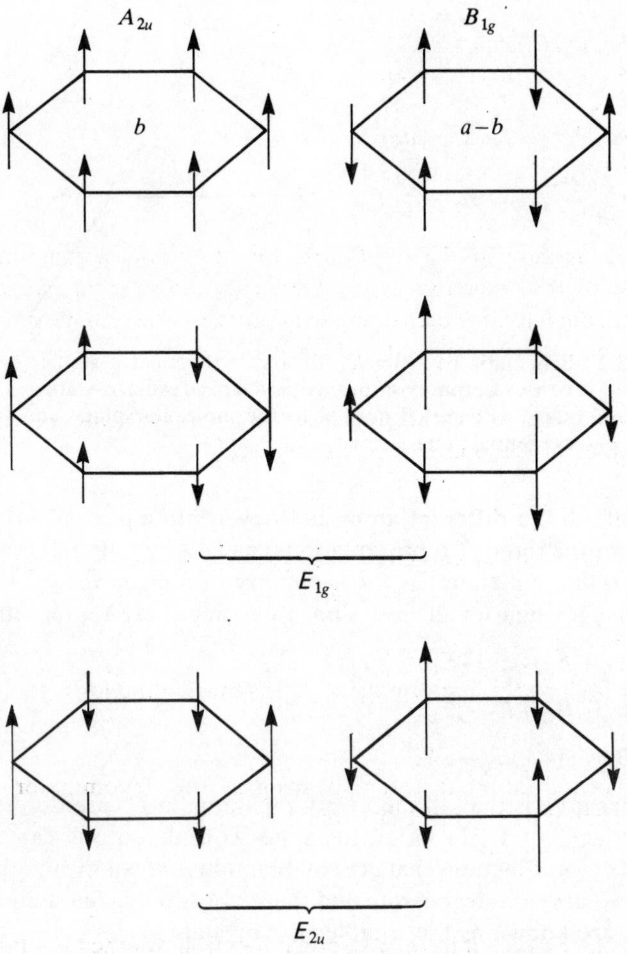

Figure 12.3 The π-orbitals of benzene.

where the obvious candidate for the function Q is one of the p_z-orbitals, f_1 say. The functions $G_r f_1$ are included in Table 12.1. Application of the formula gives:

$$\psi^{B_{1g}} \propto (f_1-f_2+f_3-f_4+f_5-f_6).$$
$$\psi^{A_{2u}} \propto (f_1+f_2+f_3+f_4+f_5+f_6).$$
$$\psi^{E_{1g}} \propto (2f_1+f_2-f_3-f_4-f_5+f_6).$$
$$\propto (2f_2+f_3-f_4-2f_5-f_6+f_1).$$
$$\psi^{E_{2u}} \propto (2f_1-f_3-f_3+2f_4-f_5-f_6).$$
$$\propto (2f_2-f_3-f_4+2f_5-f_6-f_1).$$

The second functions of E_{1g} and E_{2u} are found by cyclic permutation; they correspond to the selection $Q=f_2$. Obviously orthogonal pairs could be constructed, but it is not worth doing so at present. The symmetry inherent in the eigenfunctions is displayed in Figure 12.3.

It is now possible, still within the bounds of group theory, to determine relative values of the one-electron energies.

$$E_n=\mathcal{H}^{\psi}_{nn}=\int\psi_n^*\mathcal{H}_i\psi_n.$$

(i) As we have not yet normalized the eigenfunctions it is useful to express E_n as

$$E_n=\int\psi_n'^*\mathcal{H}_i\psi_n'/\int\psi_n'^*\psi_n',$$

where ψ_n' are the unnormalized, but simple, functions

$$(f_1-f_2+f_3-f_4+f_5-f_6) \text{ etc.}$$

(ii) The normalization factor, appearing as the denominator $\int\psi_n'^*\psi_n'$ is complicated by the fact that atomic orbitals centred at different atoms are not orthogonal:

$$\int f_i^* f_j \neq 0.$$

(iii) Although we do not have orthogonal functions for the two-dimensional irreps, E_{1g} say, it is sufficient to choose just one of the partner functions in order to determine the energy. This is because any linear combination of degenerate eigenfunctions is also an eigenfunction of the same energy. The only apparent anomaly arising from the use of non-orthogonal degenerate eigenfunctions would be the appearance of off-diagonal elements in $\mathcal{H}$.

(iv) It is necessary to introduce diagonal and off-diagonal elements of $\mathcal{H}$, with respect to the basis functions f_i. For simplicity it is conventional to restrict such elements to lie between identical or adjacent-atom functions only. This is known as the Hückel approximation:

$$\int f_j^* \mathcal{H}_i f_j = \alpha; \qquad \int f_j^* f_j = 1.$$

$$\int f_j^* \mathcal{H}_i f_{j\pm 1} = -\beta; \qquad \int f_j^* f_{j\pm 1} = S.$$

All other elements are set equal to zero. The precise nature of $\mathcal{H}_i$ has not been used here; these matrix elements are introduced purely from a phenomenological point of view. The reason for using $-\beta$ is given below. The energies can now be calculated in terms of α,β,S. For example, for $n=1$, the state of symmetry type B_{1g}, we have

$$E_1 = \frac{\int (f_1-f_2+f_3-f_4+f_5-f_6)^* \mathcal{H}_i (f_1-f_2+f_3-f_4+f_5-f_6)}{\int (f_1-f_2+f_3-f_4+f_5-f_6)^* (f_1-f_2+f_3-f_4+f_5-f_6)}$$

$$= \frac{6\alpha+12\beta}{6-12S}.$$

B_{1g}	$(\alpha+2\beta)/(1-2S)$
E_{2u}	$(\alpha+\beta)/(1-S)$
E_{1g}	$(\alpha-\beta)/(1+S)$
A_{2u}	$(\alpha-2\beta)/(1+2S)$

Figure 12.4 The π-orbitals of benzene.

For the six states one obtains an energy structure as given in Figure 12.4. Even though we do not know the values of α,β or S there is a large amount of information in this diagram:

(i) The state symmetries and degeneracies are shown.
(ii) S is likely to be small so the *relative* splittings of the levels are known; they are dependent primarily on the coupling matrix-element β.
(iii) In Chapter 15 we shall see that absorption selection rules between states can be found from group theory. By inspection of an experimental spectrum, containing perhaps many lines in addition to those corresponding to π-electron transitions, from the relative splitting and the selection rules it should be possible to identify these levels and thus obtain β and S from the spectral splittings. It is emphasized again that α, β, S have been introduced as parameters; their sizes and hence the separations and orderings of the four levels in Figure 12.4 are not defined by group theory.

The populations of the states have not yet been discussed, nor the energies of the π-orbitals by comparison to other orbitals constructed from s,p_r, and p_t functions. A clue to the ordering is obtained by considering the

A_{2u} and B_{1g} π-states (Figure 12.3). The point is that for a molecule to be stable the repulsion between the nuclei must be shielded by building up a concentration of charge between them. This is what we understand by the concept of a bond. A state with such an electron distribution is on the whole more stable and has a lower energy than one in which charge is removed from the bond region. Thus, for example, the A_{2u} state has a build up of charge just above and below the centre of each bond. Conversely the B_{1g} state has zero charge everywhere along such a line (See Figure 12.5). The A_{2u} orbital is a weak *bonding* state and has a lower energy than the weak *anti-bonding* B_{1g} orbital. Hence the use of $-\beta$ in the matrix element definitions; α, β, and S are all expected to be positive.

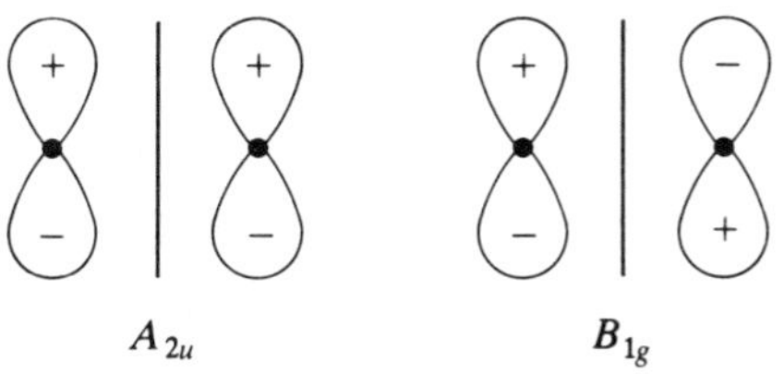

Figure 12.5 Schematic to show the build up of charge midway between the bounds for the A_{2u} π-orbital and the destructive interference of the wavefunctions for the B_{1g} orbital.

Let us turn to the other low-lying benzene orbitals. It is anticipated that the carbon 1*s* states form tightly bound, low-energy core orbitals with little admixture from the other states. From the C−2*s*, $2p_r$, $2p_t$ and H−1*s* states are created the C–H bonds and the strong σ-bonds between carbon atoms. In the latter the charge is concentrated directly along the bond (rather than off-axis as in the π-bond case) by a suitable combination of s, p_r, and p_t functions. In addition there are a number of anti-bonding high-energy molecular orbitals based on these functions. When the seven electrons per C–H pair are distributed amongst these one-electron orbitals, two take up core 1*s* positions, two are associated with the C–H bonds and two contribute to the σ-bonds. The remaining one-electron per C–H pair goes into the π-orbitals, which therefore hold six electrons. Two (of opposite spins) will be in the A_{2u} state and the remaining four in the E_{1g} (Figure 12.6). These are the uppermost populated states of benzene. The π-E_{2u} and B_{1g} states are the lowest empty states, followed by the σ-antibonding orbitals. The allowed transitions are discussed in Chapter 15.

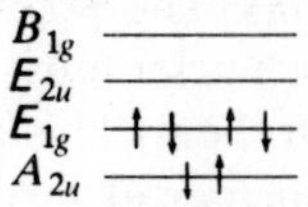

Figure 12.6 The electronic distribution in the benzene π-orbitals.

12.3 Example (b): the hydrogen molecule

The idea of bonding and anti-bonding orbitals is most usefully described using the H_2 molecule as an example. There have been several methods applied to this problem, and the LCAO–MO method is not the most accurate. Nevertheless from the group-theory point of view it does provide us with an example that can be carried through fully and the method is extremely useful for the more complex molecules. The lowest energy orbitals of H_2 will be assumed to be linear combinations of $1s$ atomic functions centred at the two atoms:

$$\mathscr{H}_{\text{mol}} = -\frac{\hbar^2}{2M}(\nabla_a^2 + \nabla_b^2) - \frac{\hbar^2}{2m}(\nabla_1^2 + \nabla_2^2) + \frac{e^2}{R_{ab}} - \frac{e^2}{r_{1a}} - \frac{e^2}{r_{1b}} - \frac{e^2}{r_{2a}} - \frac{e^2}{r_{2b}} + \frac{e^2}{r_{12}}.$$

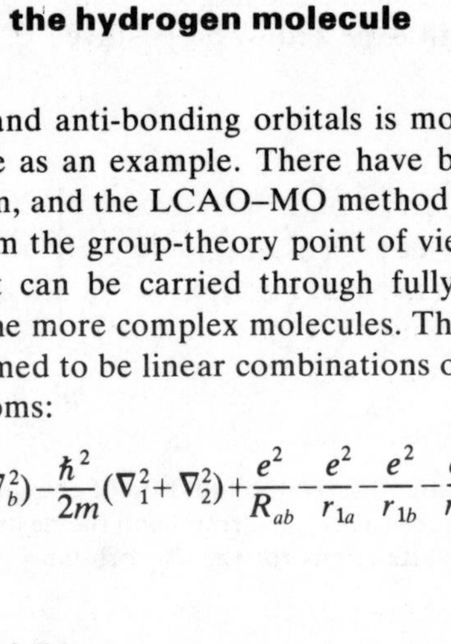

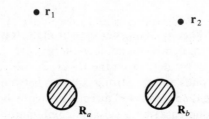

Figure 12.7 Schematic of the hydrogen molecule.

Applying the Born–Oppenheimer approximation, the electronic motion satisfies

$$\mathscr{H}_{\text{elect}} = \sum_{i=1,2} \left(-\frac{\hbar^2}{2m}\nabla_i^2 - \frac{e^2}{r_{ia}} - \frac{e^2}{r_{ib}}\right) + \frac{e^2}{r_{12}} + \frac{e^2}{R_{ab}}.$$

The Coulomb electron–electron interaction will be removed in determining the molecular orbitals, which then reduce to the form of an H_2^+ ion problem:

$$\mathcal{H}_i = \frac{-\hbar^2}{2m}\nabla_i^2 - \frac{e^2}{r_{ia}} - \frac{e^2}{r_{ib}}.$$

The trial functions have form $f_a(\mathbf{r}_i)$ and $f_b(\mathbf{r}_i)$, where f_a is a $1s$ function centred at nucleus a, etc. The group theory can be handled easily in this case. H_2 has $D_{\infty h}$ symmetry (see Chapter 11 or Appendix A2).

	E	$2R(\alpha,z)$	C'_2	i	$2iR(\alpha,z)$	σ_v
Γ^s	2	2	0	0	0	2

The reduction of the representation based on the two $1s$ states is

$$\Gamma^s \Rightarrow A_{1g} \oplus A_{2u},$$
$$\equiv \sigma_g^+ \oplus \sigma_u^+.$$

Note that σ has been used to label one-electron orbitals. By analogy to the atomic case, lower-case letters are used for one-electron and upper-case for many-electron orbitals (s,σ and S,Σ, etc.).

The formula may be used for the symmetry-adapted functions although in this case it is obvious by inspection that

$$F^{\sigma_g^+} \propto (f_a + f_b),$$
$$F^{\sigma_u^+} \propto (f_a - f_b).$$

Define the overlap integral

$$\int f_a(\mathbf{r}_i)^* f_b(\mathbf{r}_i) = \int f_b(\mathbf{r}_i)^* f_a(\mathbf{r}_i) = S,$$

and again note that as each irrep. appears just once in Γ^s, the symmetry adapted-functions are eigenfunctions (within the approximations of the model). Thus

$$\psi^{\sigma_g^+} = (f_a + f_b)/\sqrt{(2+2S)},$$
$$\psi^{\sigma_u^+} = (f_a + f_b)/\sqrt{(2+2S)},$$

It is clear that the σ_g^+ orbital is such that the charge is concentrated between the two repelling protons. This is the bonding, low-energy orbital. Still ignoring the electron–electron interaction, the energies of the one-electron orbitals are

$$E_{\sigma_p^+} = \int (f_a(\mathbf{r}_i) \pm f_b(\mathbf{r}_i))^* \left(\frac{-\hbar}{2m}\nabla_i^2 - \frac{e^2}{r_{ia}} - \frac{e^2}{r_{ib}}\right)(f_a(\mathbf{r}_i) \pm f_b(\mathbf{r}_i))/(2 \pm 2S)$$

(the + signs apply if $p = g$, the − signs if $p = u$). But

$$\left(\frac{-\hbar^2}{2m}\nabla_i^2 - \frac{e^2}{r_{ia}}\right) f_a(\mathbf{r}_i) = E_s f_a(\mathbf{r}_i),$$

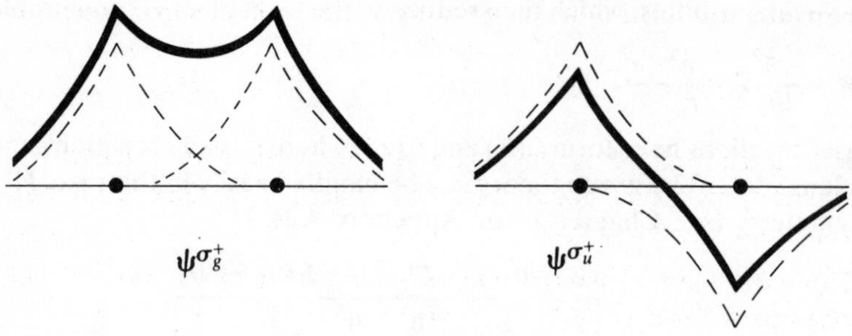

Figure 12.8 The molecular orbitals of H_2.

where E_s is the energy of a 1s electron in an H-atom. There are only two further parameters in the energy equation:

$$\int f_a(\mathbf{r}_i)^* \frac{-e^2}{r_{ib}} f_a(\mathbf{r}_i) = -A,$$

$$\int f_a(\mathbf{r}_i)^* \frac{-e^2}{r_{ia}} f_b(\mathbf{r}_i) = -B.$$

The former is the Coulomb energy of an electron sited at atom a, with respect to the nucleus at b. The latter does not have a direct classical interpretation:

$$E_{\sigma_p^+} = E_s - \frac{(A \pm B)}{(1 \pm S)}.$$

The bonding orbital has the lower energy, by approximately $2B$. The procedure for completing the problem is to allot the two H_2 electrons to the lowest one-electron states—in this case they will both enter the σ_g^+ state, with opposing spins. The two-electron ground state eigenfunction is thus

$$\psi(\mathbf{r}_1,\mathbf{r}_2) = \frac{1}{\sqrt{2}} \begin{vmatrix} \psi^{\sigma_g^+}(\mathbf{r}_1)\alpha_1 & \psi^{\sigma_g^+}(\mathbf{r}_2)\alpha_2 \\ \psi^{\sigma_g^+}(\mathbf{r}_1)\beta_1 & \psi^{\sigma_g^+}(\mathbf{r}_2)\beta_2 \end{vmatrix}.$$

It is appreciated that we have not discussed spin here; α and β can be considered purely as additional labels for any state:

$$\psi(\mathbf{r}_1,\mathbf{r}_2) \equiv (\psi_\alpha(\mathbf{r}_1)\psi_\beta(\mathbf{r}_2) - \psi_\beta(\mathbf{r}_1)\psi_\alpha(\mathbf{r}_2))/\sqrt{2},$$

using an obvious compressed notation for the one-electron functions. This two-electron function has the symmetry Σ_g^+.

Re-introducing the electron–electron interaction now, the two-electron, combined energy is

$$E_e = 2E_{\sigma_g^+} + \int \psi(\mathbf{r}_1,\mathbf{r}_2)^* \frac{e^2}{r_{12}} \psi(\mathbf{r}_1,\mathbf{r}_2).$$

The perturbation has the form, in terms of the one-electron functions,

$$E'_e = \int \psi(\mathbf{r}_1)^* \psi(\mathbf{r}_2)^* \frac{e^2}{r_{12}} \psi(\mathbf{r}_1) \psi(\mathbf{r}_2),$$

where the α,β have been dropped once account is taken of the fact that states of opposite spins are orthogonal (in spin space) regardless of their spatial wavefunctions.

This is not an elementary integration; it involves direct Coulomb integrals of the form

$$\int |f_c(\mathbf{r}_1)|^2 |f_{c'}(\mathbf{r}_2)|^2 e^2/r_{12},$$

where c,c' can take values a or b. These integrals represent the Coulomb interaction of the charge distributions associated with the atomic problem. In addition, E'_e involves integrals of type

$$\int f_a(\mathbf{r}_1)^* f_b(\mathbf{r}_2)^* f_b(\mathbf{r}_1) f_a(\mathbf{r}_2) e^2/r_{12},$$

which are called exchange interactions. The exchange-charge interactions in such double integrals and the one-electron integrals that rely on charge overlap (S and B) are responsible for the stability of the molecule. The two energy levels split by an amount determined by the magnitude of these interactions, and the decrease in energy experienced by the ground state determines the equilibrium separation of the nuclei. The molecular energy at equilibrium,

$$E_{\text{mol}} \approx E_e(R_{\text{eq}}) + e^2/R_{\text{eq}},$$

must be less than the energy of two separated atoms, $2E_s$. Also

$$E_{\text{mol}} \approx 2E_s + C \pm D,$$

where C contains direct integrals and D exchange integrals. Clearly D must exceed C in magnitude if the molecule is to be stable. It is a continual source of vexation that the stability of the H_2 molecule, a well-established fact, has no classical interpretation for its cause.

Given that the exchange interaction, associated with charge overlap, determines the molecular binding energy, it is not surprising that the presence in the molecular orbitals of p-like atomic functions directed along a bond can, by increasing the overlap, reduce the bonding-state energy. In many cases the bonds are dominated by the atomic p orbitals. Thus the oxygen atom has a $1s^2 2s^2 2p^4$ electronic configuration and the p electrons can form orthogonal bonds (each containing two electrons of opposite spins). It is partly for this reason that the bond angle of H_2O is fairly close to 90°. A similar statement applies to NH_3, where the $1s^2 2s^2 2p^3$ configuration of nitrogen leads to three nearly orthogonal bonds. In carbon bonding, however, the additional overlap obtained by use of the p orbitals produces a greater reduction in bond energy than the energy required to promote an s electron to a carbon atom p state. As a consequence, in many molecules carbon essentially takes on a $1s^2 2s 2p^3$ configuration and behaves like a tetravalent atom. Similar promotion is responsible for the tetrahedral bonding common amongst many group IV, III–V and II–VI semiconductors (Chapter 19).

The two molecules considered in this section, C_6H_6 and H_2, are exceptional in that one can attempt to obtain expressions for the energies of the states described in terms of a few parameters (α, β, S; A, B, S, E_s). In general, however, the advantages of a group-theoretical analysis lie in the irrep. labels that describe the *symmetries* of eigenstates. From these labels such properties as the radiation absorption and Raman scattering selection rules can be obtained. The allocation of the order of energy levels is in general possible only after experimental information has been used to compare these selection rules and the splitting of degeneracies under reduction of symmetry.

Problems

***12.1** Consider a hypothetical H_3^{++} molecular ion. The electronic ground state is, to a first approximation, a linear combination of atomic $1s$ orbitals centred on the three (equidistant) hydrogen nuclei (call these orbitals f_1, f_2, f_3).

(i) Obtain the full group of the molecule.

(ii) Obtain the symmetry-adapted functions in terms of f_1, f_2, f_3.

(iii) Write down the molecular Hamiltonian. Apply the Born–Oppenheimer approximation and hence find the electron Hamiltonian for the one electron that we have in this problem.

(iv) Define

$$\int f_i^* \left(\frac{-\hbar^2}{2m} \nabla^2 - \frac{e^2}{|\mathbf{r}-\mathbf{R}_i|} \right) f_i = E_s \text{ (the atomic } s \text{ energy).}$$

$$\int f_i^* \left(\frac{-e^2}{|\mathbf{r}-\mathbf{R}_j|} \right) f_i = -A.$$

$$\int f_i^* \left(\frac{-e^2}{|\mathbf{r}-\mathbf{R}_j|} + \frac{-e^2}{|\mathbf{r}-\mathbf{R}_k|} \right) f_j = -B.$$

$$\int f_i^* f_j = S, \qquad i \neq j.$$

Hence find the approximate energies of the three lowest levels of the ion.

(v) Given that E_s, A, B, S are positive, consider the ground state. What changes in eigenfunction would you expect if you allowed for all the $n=2$ atomic orbitals in your calculations?

12.2 Consider the water molecule, as shown in Figure 12.9.

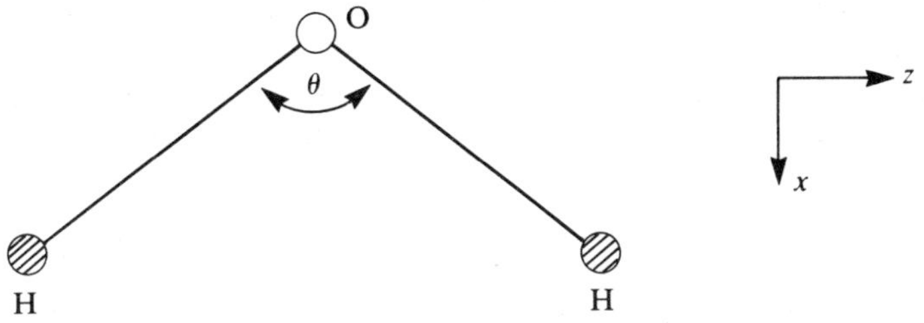

(i) Suppose that the electrons which form bonding orbitals are $1s^1$ on each H-atom and $2s^2$, $2p^4$ from the O-atom. In the Hückel LCAO approximation, that only nearest-neighbour atoms have non-zero matrix elements of $\mathscr{H}$ and S, write down a suitable form of a one-electron molecular orbital wavefunction, and the matrix equation which will determine the molecular orbitals and their eigenvalues. (Assume spin effects merely make each level doubly degenerate.)

(ii) Suggest how you might reduce the total number of parameters in your matrix, either by symmetry arguments, introducing the angle θ, or on the basis of dropping terms of small magnitude.

(iii) Use group theory to find a simpler form for the molecular orbitals than you had in (i).

12.3 BCl_3 is a planar molecule with the boron atom at the centre of an equilateral triangle of chlorine atoms. By considering the full group of the molecule, obtain the symmetry types of those molecular orbitals that are formed as linear combinations of p_z atomic orbitals on each of the four atoms. (The z-direction is defined to be normal to the molecular plane.)

Obtain the symmetry-adapted functions and state briefly how you would proceed to find the eigenfunctions for the four molecular orbitals. Sketch the expected form of the lowest energy (π-bonding) orbital.

***12.4** Within the LCAO approximation show that no molecular orbital of CH_4 (see Problem 8.5) contains both s-type and p-type carbon atomic orbitals.

12.5 ***Hybrid orbitals***

In order to form strong bonds it is necessary that the electrons in molecules and solids are slightly more concentrated along the lines between adjacent, positive nuclei. The core electrons are relatively unaffected by other than their central nucleus. The bonding charge comes from the outermost atomic occupied energy levels; the bond energy comes from the overlap of charge distributions. One technique for *visualizing* the bonds is to consider combinations of atomic orbitals that produce maximum electron distribution along the bond. For example, function $(s+p_z)$ concentrates the charge in the positive z direction and could be considered as a hybrid bonding orbital for a second atom situated in this direction.

In general, hybrid orbitals are not eigenfunctions.

(i) Consider four equivalent orbitals each of which concentrates the electron charge along one of the C–H bonds of methane (Problem 8.5). By reducing the representation based on these orbitals (or otherwise) show that they are not eigenfunctions.

(ii) Using your answer to Problem 12.4, show that possible tetrahedral–CH_4 hybrid orbitals could be linear combinations of the carbon s and p states. These are known as sp^3 orbitals; would you expect the outer carbon electrons to be in such states?

12.6 Suggest which Xe atomic orbitals would be used to construct the four hybrid orbitals of XeF_4 (a square, planar molecule), and show that the s and p states are insufficient. Make full use of the character table in Appendix A2.

12.7 Discuss the symmetry properties of those molecular orbitals of CO_2 that involve s and p-type atomic orbitals.

12.8 ***The correlation diagram of H_2***

(i) We have shown how the lowest energy *one-electron* states of H_2 can be constructed from linear combinations of 1s states of separated atoms. Find the symmetry type of orbitals that are combinations of 2s and 2p atomic states of separated atoms respectively.

(ii) Consider the H_2 levels to be those of a *single* atom perturbed by a potential of $D_{\infty h}$ symmetry. Determine any degeneracy splitting of the 1s, 2s and 2p atomic states.

(iii) As the proton separation is increased (hypothetically) from zero to infinity, the molecular energy levels must gradually change from those of model (ii) to those of model (i). The symmetry types however cannot change.

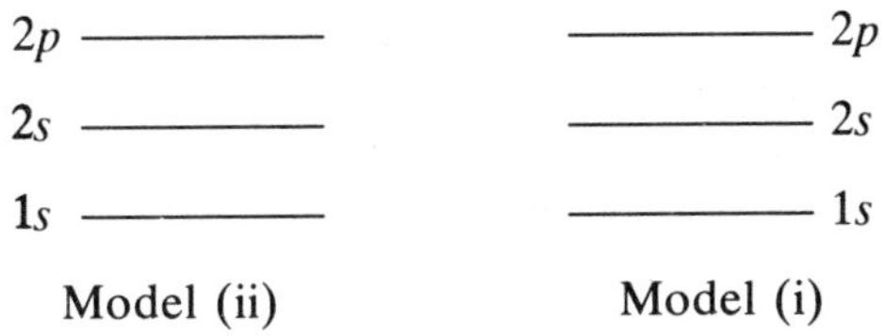

Starting with the above figure, join up the lowest energy states of each symmetry type. This is the beginning of the correlation diagram for H_2, and for all homonuclear diatomic molecules. For many-electron molecules these one-electron states are filled up with electrons, such that Pauli exclusion is obeyed, in similar manner to the filling of H-atom like states on many-electron atoms. The existence of energy crossing points on the correlation diagram indicates that the actual atomic separation does influence the state occupancies.

(iv) Verify that the H_2 ground state has Σ_g^+ spatial symmetry.

12.9 Obtain the B_{1u} symmetry-adapted orbitals, based on s and p atomic orbitals, of benzene.

PART FOUR

FURTHER SYMMETRY IDEAS AND APPLICATION TO THE RADIATION–MATTER INTERACTION

13

Direct product groups

The theme of Part 4 of this text is selection rules, particularly those for radiation absorption and scattering. Selection rules for electronic states are discussed in Chapter 15. These require an understanding of the symmetry properties of products of functions and these properties are described in Chapter 14 and applied to the vector model of the atom. In turn the direct product of matrices is required in order to represent the transformation properties of product functions. As this type of matrix multiplication is also needed in the treatment of direct product groups these groups are discussed in the present chapter, although they have no immediate bearing on selection rules. In Chapter 16 the ideas of normal modes of vibration, energy-level quantization and selection rules are brought together in a discussion of molecular vibration spectroscopy.

13.1 Outer direct products

It will have become apparent from inspection of the set of character tables that those groups which contain the inversion operation have equal numbers of improper and proper symmetry operations in them. Thus for example the O_h group consists of the O group rotation operations plus the same number of improper rotations.

$$O_h \begin{cases} E & 8C_3 & 6C'_2 & 6C_4 & 3C_2 \\ i & 8S_6 & 6\sigma_d & 6S_4 & 3\sigma_h \end{cases}$$

Furthermore it may be noticed that the structure of the O_h character table has a form predictable from the O group table. These facts are no coincidence, they occur because O_h can be described as the *outer direct product* of the two groups O and S_2. It is the purpose of this section to investigate the properties of this type of group, a type we shall also meet later in the crystal problem. First, though, we need a general definition.

Definition of the outer product group

Consider two groups, $\mathcal{A}$ and $\mathcal{B}$, such that they have only one common element (the identity) and such that all the operations in $\mathcal{B}$ commute with all the operations in $\mathcal{A}$. The set of operations formed by the product of each operation in $\mathcal{B}$ with each in $\mathcal{A}$ will be shown to have the properties of a group. It is called the direct product group of $\mathcal{A}$ and $\mathcal{B}$, for which we shall employ the notation $\mathcal{A}\otimes\mathcal{B}$.

Proof that the outer direct product group obeys the group postulates

Taking each of the group postulates of Chapter 2:

(i) The set of elements of $\mathcal{A}\otimes\mathcal{B}$ is the set $\mathcal{G}_{ij}\equiv\mathcal{A}_i\,\mathcal{B}_j$, where $\mathcal{A}_i$ is a particular element of $\mathcal{A}$, etc. The total number of elements is the product of the orders of the two groups $g_{\mathcal{A}\otimes\mathcal{B}}=g_{\mathcal{A}}g_{\mathcal{B}}$.

(ii) The product of two elements of $\mathcal{A}\otimes\mathcal{B}$ has a straightforward meaning

$$\mathcal{G}_{ij}\mathcal{G}_{kl}\equiv(\mathcal{A}_i\mathcal{B}_j)(\mathcal{A}_k\mathcal{B}_l)\equiv\mathcal{A}_i\mathcal{B}_j\mathcal{A}_k\mathcal{B}_l.$$

(iii) Because the operations in the two groups are required to commute, $\mathcal{B}_j\mathcal{A}_k\equiv\mathcal{A}_k\mathcal{B}_j$, so that

$$\mathcal{G}_{ij}\mathcal{G}_{kl}\equiv(\mathcal{A}_i\mathcal{A}_k)(\mathcal{B}_k\mathcal{B}_l).$$

But $\mathcal{A}_i\mathcal{A}_k$ is just one of the operations in $\mathcal{A}$, etc. Hence $\mathcal{G}_{ij}\mathcal{G}_{kl}=\mathcal{A}_m\mathcal{B}_n=\mathcal{G}_{mn}$, for some m and n. The closure relation is obeyed.

(iv) The associative law follows just as for ordinary groups, because we are dealing with symmetry operations.

(v) The identity element is clearly $\mathcal{A}_1\,\mathcal{B}_1$ where $\mathcal{A}_1\equiv\mathcal{B}_1\equiv E$.

(vi) The operation $\mathcal{G}_{ij}$ has an inverse $(\mathcal{G}_{ij})^{-1}\equiv\mathcal{A}_i^{-1}\mathcal{B}_j^{-1}$, as, using the commutation property,

$$(\mathcal{A}_i\mathcal{B}_j)(\mathcal{A}_i^{-1}\mathcal{B}_j^{-1})\equiv(\mathcal{A}_i\mathcal{A}_i^{-1})(\mathcal{B}_j\mathcal{B}_j^{-1})\equiv E\,E\equiv E.$$

It is necessary only to prove that this inverse is unique. It will be so if in general each operation in the group is unique.

Proof that elements of the direct product group are unique

Suppose that $\mathcal{G}_{ij} \equiv \mathcal{G}_{kl}$. If this is true, then pre-operating by $\mathcal{A}_i^{-1}$,

$$\mathcal{A}_i^{-1}\mathcal{A}_i\mathcal{B}_j \equiv \mathcal{A}_i^{-1}\mathcal{A}_k\mathcal{B}_l.$$

But $(\mathcal{A}_i^{-1}\mathcal{A}_k) \equiv \mathcal{A}_m$ for some m. Therefore $\mathcal{B}_j \equiv \mathcal{A}_m\mathcal{B}_l$.

However there must be a member of the group $\mathcal{B}$ such that $\mathcal{B}_j \equiv \mathcal{B}_p\ \mathcal{B}_l$,

so that $\mathcal{A}_m \equiv \mathcal{B}_p$.

Finally, because the only common element of the two groups is the identity, then $\mathcal{A}_m \equiv \mathcal{B}_p \equiv E$, which in turn demands that $\mathcal{A}_k = \mathcal{A}_i$ and $\mathcal{B}_l \equiv \mathcal{B}_j$. We have thus shown that each element of the double group is unique:

$$\mathcal{G}_{ij} \equiv \mathcal{G}_{kl} \qquad \text{only if} \quad k=i,\ l=j.$$

Thus $\mathcal{A} \otimes \mathcal{B}$ obeys all the groups postulates; it is a genuine group.

Classes of the direct product groups

We will be interested in the character table for a direct product group, and need to know the number of classes and irreps in it. To find an element in the same class as $\mathcal{A}_i\mathcal{B}_j$, consider

$$(\mathcal{A}_k\mathcal{B}_l)^{-1}\mathcal{A}_i\mathcal{B}_j(\mathcal{A}_k\mathcal{B}_l) \equiv (\mathcal{A}_k^{-1}\mathcal{A}_i\mathcal{A}_k)(\mathcal{B}_l^{-1}\mathcal{B}_j\mathcal{B}_l).$$

This demonstrates that for a given class of operation of $\mathcal{A}$ and a class of $\mathcal{B}$ one generates a single class of $\mathcal{A} \otimes \mathcal{B}$. The number of classes of the direct product group is therefore equal to the product of the numbers of classes in the individual groups. The same statement must apply to the numbers of irreps.

The irreps of direct product groups

We shall state, without giving a full proof, that the irreps of $\mathcal{A} \otimes \mathcal{B}$ are the set of matrices obtained by making the *direct matrix products* (defined below), of the irreps of the individual groups. Thus if $\Gamma^a(\mathcal{A}_i)$ is one of the irreducible matrices of group $\mathcal{A}$, representing the particular operation $\mathcal{A}_i$, and $\Gamma^b(\mathcal{B}_j)$ is the bth irrep. for $\mathcal{B}_j$, then $\Gamma^a(\mathcal{A}_i) \otimes \Gamma^b(\mathcal{B}_j)$ is an irreducible matrix representing the operation $\mathcal{A}_i\mathcal{B}_j$ in the direct product group. We employ the notation $\Gamma^{a \otimes b}$ to label the irreps of $\mathcal{A} \otimes \mathcal{B}$.

Direct matrix products

The definition of a direct product between two matrices Γ^a and Γ^b is the following:

$$\Gamma^a \otimes \Gamma^b = \begin{bmatrix} \Gamma^a_{11}[\Gamma^b] & \Gamma^a_{12}[\Gamma^b] & \cdots \\ \Gamma^a_{21}[\Gamma^b] & \Gamma^a_{22}[\Gamma^b] & \cdots \\ \vdots & \vdots & \end{bmatrix}.$$

That is, each Γ^a element multiplies the full Γ^b matrix.

$$\Gamma^a \otimes \Gamma^b = \begin{bmatrix} \Gamma^a_{11}\Gamma^b_{11} & \Gamma^a_{11}\Gamma^b_{12} & \cdots\ \Gamma^a_{12}\Gamma^b_{11} & \cdots \\ \Gamma^a_{11}\Gamma^b_{11} & \Gamma^a_{11}\Gamma^b_{22} & \cdots\ \Gamma^a_{12}\Gamma^b_{21} & \cdots \\ \vdots & & & \\ \Gamma^a_{21}\Gamma^b_{11} & \Gamma^a_{21}\Gamma^b_{12} & \cdots\ \Gamma^a_{22}\Gamma^b_{11} & \cdots \\ \vdots & & & \end{bmatrix} \begin{matrix} \text{row } 1,1 \\ \text{row } 1,2 \\ \\ \text{row } 2,1 \\ \\ \end{matrix}$$

$$\begin{matrix} \text{column } 1,1 & \text{column } 1,2 & \text{column } 2,1 \end{matrix}$$

The individual elements require four suffices to label them

$$(\Gamma^a \otimes \Gamma^b)_{ii'jj'} = \Gamma^a_{ij}\Gamma^b_{i'j'},$$

where (ii') labels the particular row and (jj') the column. Unlike conventional matrix products, there is no restriction on the form of the individual matrices Γ^a and Γ^b. In the cases we have in mind, the Γ-matrices are square, but in general Γ^a and Γ^b do not have identical dimensionality. For such matrices conventional products could not be formed.

Proof that $\Gamma^a \otimes \Gamma^b$ represents the direct product group

Whilst it is a complicated matter to prove that $\Gamma^a \otimes \Gamma^b$ are irreps of $\mathcal{A} \otimes \mathcal{B}$ we can offer a partial proof by demonstrating that they do at least form a representation of the direct product group.

A lemma is needed in order to facilitate the proof, namely:

$$[AB] \otimes [CD] = [A \otimes C][B \otimes D]. \tag{13.1}$$

This is proved by considering the $ii'jj'$ elements of the left-hand side of equation (13.1):

$$\begin{aligned} ([AB] \otimes [CD])_{ii'jj'} = (AB)_{ij}(CD)_{i'j'} &= \sum_{kk'} A_{ik}B_{kj}C_{i'k'}D_{k'j'} \\ &= \sum_{kk'} A_{ik}C_{i'k'}B_{kj}D_{k'j'} \\ &= \sum_{kk'} (A \otimes C)_{ii'kk'}(B \otimes D)_{kk'jj'}. \end{aligned}$$

This is precisely the $ii'jj'$ element of the right-hand side of equation (13.1).

Using the above lemma,

$$[\Gamma^a(\mathcal{A}_i)\otimes\Gamma^b(\mathcal{B}_j)][\Gamma^a(\mathcal{A}_k)\otimes\Gamma^b(\mathcal{B}_l)]=[\Gamma^a(\mathcal{A}_i)\,\Gamma^a(\mathcal{A}_k)]\otimes[\Gamma^b(\mathcal{B}_j)\Gamma^b(\mathcal{B}_l)]. \tag{13.2}$$

As Γ^a and Γ^b are matrix representations, the right-hand side of equation (13.2) is equal to

$$\Gamma^a(\mathcal{A}_i\mathcal{A}_k)\otimes\Gamma^b(\mathcal{B}_j\mathcal{B}_l)=\Gamma^{a\otimes b}(\mathcal{A}_i\mathcal{A}_k\mathcal{B}_j\mathcal{B}_l)$$

Hence a matrix $\Gamma^{a\otimes b}(\mathcal{A}_i\mathcal{B}_j)$ defined by $\Gamma^a(\mathcal{A}_i)\otimes\Gamma^b(\mathcal{B}_j)$ does obey the relation,

$$[\Gamma^{a\otimes b}(\mathcal{A}_i\mathcal{B}_j)][\Gamma^{a\otimes b}(\mathcal{A}_k\mathcal{B}_l)]=\Gamma^{a\otimes b}(\mathcal{A}_i\mathcal{B}_j\mathcal{A}_k\mathcal{B}_l),$$

which is the required relation for a representation of the group $\mathcal{A}\otimes\mathcal{B}$. That is,

$$\Gamma^{a\otimes b}(\mathcal{G}_{ij})\,\Gamma^{a\otimes b}(\mathcal{G}_{kl})=\Gamma^{a\otimes b}(\mathcal{G}_{ij}\mathcal{G}_{kl}).$$

The characters of direct product irreps

From the form of the direct product matrices, with trace equal to

$$\Gamma^a_{11}\text{ trace }\Gamma^b+\Gamma^a_{22}\text{ trace }\Gamma^b+\ldots,$$

the character of the matrix representation $\Gamma^{a\otimes b}(\mathcal{G}_{ij})$ is

$$\chi^{a\otimes b}(\mathcal{G}_{ij})=\sum_m\Gamma^a_{mm}(\mathcal{A}_i)\sum_n\Gamma^b_{nn}(\mathcal{B}_j)$$
$$=\chi^a(\mathcal{A}_i)\chi^b(\mathcal{B}_j).$$

That is, the characters of a direct product group are the products of those of the individual groups.

13.2 Rotation–inversion groups

The inversion operation commutes with all rotations. Hence any group containing only proper rotations forms a direct product group with the S_2 group. This latter contains only the identity and the inverse as its elements.

Table 13.1 The character table of the S_2 group.

	E	i
A_g	1	1
A_u	1	−1

The irreducible representations of S_2 are labelled A_g, A_u, where the g stands for '*gerade*', meaning even in German. The bases for A_g are even on inversion, as can be seen from the characters under the operation i. In contrast the bases of A_u are odd (*ungerade*). The bases are said to have even and odd *parity* respectively.

If the direct product of S_2 with a rotation group is considered, the number of operations must double, as must the number of classes and irreps. The character table must contain two types of irrep.: one with the same character for both the proper rotations R and the corresponding improper rotations iR, and one that has a sign reversal. For the full rotation–inversion group $R_3 \otimes S_2$ it is useful to write the character table as in Table 13.2.

Table 13.2 Character table for $R_3 \otimes S_2$, demonstrating the structure of even; and odd-parity irreps. The notation D^{0g} indicates that the particular irrep. has the form $D^0 \otimes A_g$, etc.

	E	$R(\alpha,\xi)$	i	$iR(\alpha,\xi)$
D^{0g}	1	1	1	1
D^{1g}	$(2l+1)$	$\dfrac{\sin(l+1/2)\alpha}{\sin \alpha/2}$	$(2l+1)$	$\dfrac{\sin(l+1/2)\alpha}{\sin \alpha/2}$
$\vdots$				
D^{0u}	1	1	-1	-1
D^{1u}	$(2l+1)$	$\dfrac{\sin(l+1/2)\alpha}{\sin \alpha/2}$	$-(2l+1)$	$\dfrac{-\sin(l+1/2)\alpha}{\sin \alpha/2}$
$\vdots$				

The same structure can be seen in $O_h(\equiv O \otimes S_2)$, where it should be borne in mind that there is a correspondence in the notation for various of the operations:

$$S_6 \equiv iC_3, \qquad \sigma_d \equiv iC'_2, \qquad S_4 \equiv C_4, \qquad \sigma_h \equiv iC_2.$$

The symmetry-adapted bases for all such groups, and hence eigenvectors or eigenfunctions, must have specific parity (they belong to g or u irreps).

The other type of direct product group appearing in the crystallographic point group tables is of the form of a rotation group crossed with C_{1h}. C_{1h} contains just the identity and a single reflection, both of which commute with all rotations. The C_{1h} character table is identical to that for S_2; the irreps however are labelled A' and A'', the prime indicating the nature of the bases on reflection. For example $D_{3h} = D_3 \otimes C_{1h}$.

We shall also meet direct product groups of a different form, in consideration of the crystalline periodic problem.

Problems

13.1 Write down the direct product of the two matrices

$$X = \begin{bmatrix} 1 & 0 \\ 1 & 1 \end{bmatrix} \qquad Y = \begin{bmatrix} 1 & 1 & 1 \\ 0 & 0 & 1 \\ 1 & 0 & 1 \end{bmatrix}$$

Note that a normal matrix multiplication XY does not exist. Show that $Y \otimes X \neq X \otimes Y$.

13.2 Consider the naphthalene molecule $C_{10}H_8$, which consists of two benzene-like rings connected along one C–C bond. Consider only the proper rotational operations of the group of the molecule, G. Obtain the symmetry types and degeneracies of the normal modes. Explain how the use of the direct product group $G \otimes S_2$ would simplify evaluation of the eigenvectors. Verify your statement.

14

The symmetry of product functions

14.1 Product functions

There are two major uses of group theory in atomic, molecular and solid-state physics. The first is in the determination of the *symmetry types* and *degeneracies* of the energy levels. The second is the use of the symmetries of the eigenfunctions in order to determine *selection rules*, either for radiation absorption or for the effect of some other perturbation of the system. In order to establish a method for determining selection rules it is necessary to investigate the properties, in particular the symmetry types, of single-particle *product functions*. In so doing we will simultaneously be establishing results that are applicable to the symmetries of many-electron product functions. The distinction between the two types of product function is the following.

Selection rules will involve us with integrations of the form:

$$\mathscr{H}'_{12} = \int_{\text{all space}} \psi_1(\mathbf{r})^* \mathscr{H}' \psi_2(\mathbf{r})\, d\mathbf{r}.$$

Here $\mathbf{r}$ is the coordinate of a single particle (or possibly a single normal-mode coordinate in the quantum theory of molecular vibrations or phonons), and $\mathscr{H}'$ is the Hamiltonian, describing some perturbation to the system. In general $\mathscr{H}'$ is an operator in the coordinate $\mathbf{r}$. We shall be interested in the symmetry of the functions ψ_1^* and $(\mathscr{H}'\psi_2)$ that appear in the integrand; that is, in the symmetry of products of functions *of the same variable*, $\mathbf{r}$.

In contrast, if in a many-particle problem we are able to separate coordinates, then it has already been established that one can write:

$$\mathcal{H} = \sum_i \mathcal{H}_i, \text{ with } \psi = \prod_i \psi_i.$$

Here $\mathcal{H}$ is the many-particle Hamiltonian, ψ the eigenfunction. In the two-particle case, for example,

$$\psi(\mathbf{r}_1,\mathbf{r}_2) = \psi_1(\mathbf{r}_1)\,\psi_2(\mathbf{r}_2),$$

or an antisymmetrized (Slater determinant) combination of functions of this type. The point is that the many-particle function has the form of a product of one-particle functions of *different variables*, $\mathbf{r}_1$, $\mathbf{r}_2$, etc.

Inner direct products

Suppose we have two functions of specific symmetry, that is, each belonging to a specific irreducible representation of a group. Let F^j_k belong to the kth row of the jth irrep. and $F^{j'}_{k'}$ to the k'th row of the j'th irrep. We need not worry whether the variables in the two functions are the same or different; only the symmetry is of concern.

Under a group operation these (symmetry-adapted) functions by definition obey

$$G_r F^j_k = \sum_i F^j_i \Gamma^j_{ik}(G_r), \text{ etc.}$$

Thus for their product function we have

$$\begin{aligned} G_r(F^j_k F^{j'}_{k'}) &= (\sum_i F^j_i \Gamma^j_{ik}(G_r))(\sum_{i'} F^{j'}_{i'} \Gamma^{j'}_{i'k'}(G_r)) \\ &= \sum_{ii'} (F^j_i F^{j'}_{i'})\, \Gamma^j_{ik}(G_r)\, \Gamma^{j'}_{i'k'}(G_r), \end{aligned}$$

$$G_r(F^j_k F^{j'}_{k'}) = \sum_{(ii')} (F^j_i F^{j'}_{i'}) (\Gamma(G_r)^j \otimes \Gamma(G_r)^{j'})_{ii'kk'}.$$

If F^j_k is one of l_j partner functions, Γ^j being of dimension l_j, then there are $(l_j l'_j)$ functions of the form $F^j_i F^{j'}_{i'}$. The above equation demonstrates that these functions transform into linear combinations of each other in accord with the direct product matrix $\Gamma^j \otimes \Gamma^{j'}$.

It is important to note the distinction here between this form of direct product representation, where we have a *single* group and where the particular matrix representing an operation G_r in that group is $\Gamma^j(G_r) \otimes \Gamma^{j'}(G_r)$, and the form of outer direct product groups considered in the previous section. Previously we had *two* distinct groups $\mathcal{A}$, $\mathcal{B}$ and the direct product matrix that represented a product operation $\mathcal{A}_i\mathcal{B}_i$ was $\Gamma^a(\mathcal{A}_i) \otimes \Gamma^b(\mathcal{B}_j)$, a and b referring to irreps of the *different groups*. $\Gamma^a \otimes \Gamma^b$ is known as an outer direct product, $\Gamma^j \otimes \Gamma^{j'}$ is an inner direct product within a group.

There is an additional distinction. $\Gamma^a \otimes \Gamma^b$ was stated to be an *irrep*. of the group $\mathcal{A} \otimes \mathcal{B}$. By contrast $\Gamma^j \otimes \Gamma^{j'}$ (j and j' referring to irreps of the single group) is in general a *reducible* representation of the group G. Suppose for example that

$$\Gamma^j \otimes \Gamma^{j'} = \sum_{j''} a_{j''} \Gamma^{j''}(G_r).$$

Using the reduction formula and the result established in Chapter 13 that the trace of a direct product matrix is the product of the individual traces, we obtain

$$a_{j''} = \frac{1}{g} \sum_r \chi^{j''}(G_r)^* \chi^{j \otimes j'}(G_r) = \frac{1}{g} \sum_r \chi^{j''}(G_r)^* \chi^j(G_r) \chi^{j'}(G_r).$$

From our $(l_j l_{j'})$-set of product functions it is therefore possible to construct linear combinations that are symmetry-adapted functions that belong to various of the irreps of G, as indicated by the values of $a_{j''}$. In this respect note that the order of the direct product is irrelevant as far as the form of the reduction is concerned; that is

$$a_{j''} \text{ for } \Gamma^j \otimes \Gamma^{j'} = a_{j''} \text{ for } \Gamma^{j'} \otimes \Gamma^j.$$

This is to be expected as the order of the original product functions $F^j_k F^{j'}_{k'}$ is of no consequence.

Product functions of the C_{3v} group

(a) ***Two-variable functions***

In order to demonstrate the properties of product functions we now ask, 'What are the properties of functions like $x_1 z_2$, referring to the coordinates of a pair of particles?'

In the C_{3v} group z has A_1 symmetry; (x,y) transform as a pair, with E symmetry. This is true for either particle, z_1 or z_2 both behave like z, etc. Hence a product function of form $x_1 z_2$ belongs to the representation $E \otimes A_1$. This is a trivial direct product, as can be seen from the general relation

$$\chi(A_1 \otimes \Gamma^j) = \chi(\Gamma^j \otimes A_1) = \chi^j \qquad \text{for each } G_r, \text{ because}$$

$$A_1 \otimes \Gamma^j = \Gamma^j \otimes A_1 = \Gamma^j.$$

This is true for any group, providing we substitute for A_1 the totally symmetric irrep. of the group concerned. We will denote this (identity) irrep., the bases of which are functions that transform identically into themselves under all G_r, as Γ^I.

In the case of interest then $x_1 z_2 \in E$. (Note that we shall use the symbol $\in$ to mean 'belongs to' in much of the future work.) By this statement we are to understand that $x_1 z_2$ and a second function, which need not be specified but which is clearly $y_1 z_2$ in this example, transform into linear combinations of each other in accord with the E matrices, under the group operations.

By similar token the product function $x_1x_2 \in E \otimes E$; $E \otimes E$ has characters $\{\chi^E(G_r)\}^2$ and reduces to $A_1 \oplus A_2 \oplus E$ (see Table 4.1). The interpretation now is that x_1x_2 is one of four functions that transform into combinations of each other (x_1x_2, y_1y_2, x_1y_2, y_1x_2) and that from these four functions one linear combination can be constructed of symmetry A_1, one of symmetry A_2 and two that transform as E. It is not too difficult to see what these combinations are. Firstly $(x_1x_2+y_1y_2)$ may be associated with the scalar product $\mathbf{r}_1.\mathbf{r}_2$ of a pair of vectors in the x–y plane; it is a function with the same symmetry as $\mathbf{r}_1.\mathbf{r}_2$. But scalar products are invariant under rotations/reflections, and therefore belong to A_1. Secondly $(x_1y_2-x_2y_1)=(\mathbf{r}_1\wedge\mathbf{r}_2)_z$ has the symmetry of the z-component of an axial vector (the cross-product of any two polar vectors is axial). This is precisely the symmetry of a basis of A_2 (R_z symmetry). Finally, $(x_1x_2-y_1y_2)$ and $(x_1y_2+x_2y_1)$, two functions that are orthogonal to those above, may be seen to transform in accord with E, as will any two orthogonal combinations of these two functions.

Use of axial functions

As an aside here, it is often useful to consider the functions $(x+\mathrm{i}y)$ and $(x-\mathrm{i}y)$ in problems with a preferred z-axis. In the C_{3v} group, for example, these functions have simple transformation properties. Note that $(x\pm\mathrm{i}y)=r\sin\theta\,\mathrm{e}^{\pm\mathrm{i}\phi}$, and for example

$$C_3(\mathrm{e}^{\mathrm{i}\phi},\ \mathrm{e}^{-\mathrm{i}\phi})=(\mathrm{e}^{\mathrm{i}(\phi-2\pi/3)},\ \mathrm{e}^{-\mathrm{i}(\phi-2\pi/3)}),$$

$$\sigma_\beta(\mathrm{e}^{\mathrm{i}\phi},\ \mathrm{e}^{-\mathrm{i}\phi})=(\mathrm{e}^{\mathrm{i}(2\beta-\phi)},\ \mathrm{e}^{-\mathrm{i}(2\beta-\phi)}),$$

where β is the angle between the reflection plane and the x-axis. Hence

$$\Gamma(C_3)=\begin{bmatrix}\mathrm{e}^{-\mathrm{i}2\pi/3} & 0\\ 0 & \mathrm{e}^{\mathrm{i}2\pi/3}\end{bmatrix};\qquad \Gamma(\sigma_\beta)=\begin{bmatrix}0 & \mathrm{e}^{\mathrm{i}2\beta}\\ \mathrm{e}^{-\mathrm{i}2\beta} & 0\end{bmatrix}.$$

The products of such functions also therefore have simple properties:

$$C_3(x_1\pm\mathrm{i}y_1)(x_2\pm\mathrm{i}y_2)=\mathrm{e}^{\mp\mathrm{i}4\pi/3}(x_1\pm\mathrm{i}y_1)(x_2\pm\mathrm{i}y_2).$$

$$C_3(x_1\pm\mathrm{i}y_1)(x_2\mp\mathrm{i}y_2)=(x_1\pm\mathrm{i}y_1)(x_2\mp\mathrm{i}y_2).$$

$$\sigma_\beta(x_1\pm\mathrm{i}y_1)(x_2\pm\mathrm{i}y_2)=\mathrm{e}^{\mp\mathrm{i}4\beta}(x_1\mp\mathrm{i}y_1)(x_2\mp\mathrm{i}y_2).$$

$$\sigma_\beta(x_1\pm\mathrm{i}y_1)(x_2\mp\mathrm{i}y_2)=(x_1\mp\mathrm{i}y_1)(x_2\pm\mathrm{i}y_2).$$

From these results we can determine:

$$(x_1+\mathrm{i}y_1)(x_2-\mathrm{i}y_2)+(x_1-\mathrm{i}y_1)(x_2+\mathrm{i}y_2)\in A_1 \quad \text{as} \quad \Gamma(C_3)=\Gamma(\sigma)=[1].$$

$$(x_1+\mathrm{i}y_1)(x_2-\mathrm{i}y_2)-(x_1-\mathrm{i}y_1)(x_2+\mathrm{i}y_2)\in A_2 \quad \text{as} \quad \Gamma(C_3)=[1],\ \Gamma(\sigma)=[-1].$$

$$(x_1\pm\mathrm{i}y_1)(x_2\pm\mathrm{i}y_2)\in E$$

$$\text{as} \quad \Gamma(C_3) = \begin{bmatrix} e^{-i4\pi/3} & 0 \\ 0 & e^{i4\pi/3} \end{bmatrix}, \quad \Gamma(\sigma) = \begin{bmatrix} 0 & e^{i4\beta} \\ e^{-i4\beta} & 0 \end{bmatrix}.$$

These are precisely the forms of the functions considered above in (a). The advantage of using the complex functions, however, is that it is easier to generalize to more complicated situations—higher-order polynomials for example.

(b) *Functions of a single variable*

By suppressing the suffices on x, etc. in subsection (a) we can consider the symmetries of product functions of a single variable. For example,

$$(x_1x_2+y_1y_2) \to (x^2+y^2) \in A_1;$$

$$(x_1x_2-y_1y_2, x_1y_2+x_2y_1) \to (x^2-y^2, 2xy) \in E;$$

$$(x_1y_2-x_2y_1) \to 0 \qquad \text{as } xy \text{ is indistinguishable from } yx.$$

Thus not all of the irreps in the reduction of $\Gamma^j \otimes \Gamma^{j'}$ can have as their bases, functions of a single variable, whereas they will always have basis functions of two variables. Inspecting the character tables, you will note that the quadratic functions of a single variable, belonging to each irrep. are given, along with the symmetries of the linear functions x, y, z and the axial-vector functions R_x, R_y, R_z.

(c) *Higher-order functions*

In order to see how higher-order functions behave, consider $x_1^3 y_1 z_1 z_2$. By induction from the previous examples, this function belongs to a higher-order direct product representation:

$$x_1^3 y_1 z_1 z_2 \in E^3 \otimes E \otimes A_1 \otimes A_1$$
$$\equiv E^4 \otimes A_1^2 \equiv \Gamma^q, \text{ say.}$$

(E^3 is used as an abbreviation for $E \otimes E \otimes E$). The characters of Γ^q and the reduced form of Γ^q are given below.

	E	$2C_3$	$3\sigma_v$
Γ^q	$2^4.1^2$	$(-1)^4.1^2$	$0^4.1^2$

$$\Gamma^q \Rightarrow 3A_1 \oplus 3A_2 \oplus 10E$$

The interpretation is that $x_1^3 y_1 z_1 z_2$ is one of sixteen functions that transform together. From these sixteen functions, three linear combinations can be constructed that belong to A_1, etc. If we do not distinguish between the particles, then some of these symmetry-adapted basis functions will disappear, corresponding to the fact that different permutations of x^3yz^2, for example, are indistinguishable.

Table 14.1 Summary of the inner direct properties of the C_{3v} group. The arrow indicates reduction.

	E	$2C_3$	$3\sigma_v$		
A_1	1	1	1		z; (x^2+y^2); z^2
A_2	1	1	−1		R_z
E	2	−1	0		(x,y); (R_x,R_y); (x^2-y^2,xy); (xz,yz)
$A_1 \otimes A_1$	1	1	1	$\Rightarrow A_1$	
$\otimes A_2$	1	1	−1	A_2	
$\otimes E$	2	−1	0	E	
$A_2 \otimes A_2$	1	1	1	A_1	
$\otimes E$	2	−1	0	E	
$E \otimes E$	4	1	0	$A_1 \oplus A_2 \oplus E$	

Note in passing that $\Gamma^j \otimes \Gamma^j \Rightarrow A_1$ or ($A_1 \oplus$ other irreps), whereas $\Gamma^j \otimes \Gamma^{j'}$ does not 'contain' A_1 if j and j' differ. This is an important property to which we shall refer again.

14.2 Two-particle states of the spherical-symmetry group

The vector model of the atom

It has been established that the spatial eigenfunctions of a spherically symmetric system belong to the irreps D^{lg} and D^{lu} of the rotation–inversion group $R_3 \otimes S_2$, where the even (g) representation is applicable if l is an even integer, and the u representation if l is odd.

Our problem is: 'If $\psi_{nlm}(\mathbf{r}_1)$ and $\psi_{n'l'm'}(\mathbf{r}_2)$ are eigenfunctions of the one-particle problem, then what are the possible symmetries of two-particle states such as

$$\psi(\mathbf{r}_1,\mathbf{r}_2) = \psi_{nlm}(\mathbf{r}_1)\,\psi_{n'l'm'}(\mathbf{r}_2)?\text{'}$$

Now $\psi_{nlm}(\mathbf{r}_1)$ is one of $(2l+1)$ degenerate states belonging to D^{lp}, where the value of l determines whether we require $p=g$ or u. Similarly $\psi_{n'l'm'}(\mathbf{r}_2)$ is one of $(2l'+1)$ states belonging to $D^{l'p'}$. We shall assume $l \geqslant l'$. The two-particle eigenfunctions therefore belong to the irreps that are contained in the reduced form of $D^{lp} \otimes D^{l'p'}$. The characters of this reducible representation are best formed by remembering that

$$\chi^l(R(\alpha,\xi)) = e^{-il\alpha} + e^{-i(l-1)\alpha} + \cdots + e^{il\alpha}$$

and

$$\chi^l(iR(\alpha,\xi))=(-1)^l\chi^l(R(\alpha,\xi)).$$

The identity and inverse operations themselves are accounted for in such formulae by setting α equal to zero. For the direct product representation:

$$\begin{aligned}\chi^{lp\otimes l'p'}(R)&=(e^{-il\alpha}+e^{-i(l-1)\alpha}+\cdots+e^{il\alpha})(e^{-il'\alpha}+e^{-i(l'-1)\alpha}+\cdots+e^{il'\alpha})\\&=e^{-i(l+l')\alpha}+e^{-i(l+l'-1)\alpha}+\cdots \qquad +e^{i(l+l'-1)\alpha}+e^{i(l+l')\alpha}\\&\qquad +e^{-i(l+l'-1)\alpha}+\cdots \qquad +e^{i(l+l'-1)\alpha}\\&\qquad\qquad +e^{-i(l+l'-2)\alpha}\cdots+e^{i(l+l'-2)\alpha}\\&\qquad\qquad\qquad +e^{-i(l-l')\alpha}+\cdots e^{i(l-l')\alpha};\end{aligned}$$

$$\chi^{lp\otimes l'p'}(iR)=(-1)^{l+l'}\chi^{lp\otimes l'p'}(R).$$

The expansion of the exponential products has been laid out in the above way to demonstrate that

$$\chi^{lp\otimes l'p'}(R)=\chi^{(l+l')p''}(R)+\chi^{(l+l'-1)p''}(R)+\cdots+\chi^{|l-l'|p''}(R),$$
$$\chi^{lp\otimes l'p'}(iR)=\chi^{(l+l')p''}(iR)+\chi^{(l+l'-1)p''}(iR)+\cdots+\chi^{|l-l'|p''}(iR).$$

In addition the choice of p'' as g or u on the right-hand-side is determined by the value of $l+l'$. If this is even then $\chi^{l\otimes l'}(iR)=+\chi^{l\otimes l'}(iR)$ and the even parity case is relevant, and vice versa. As the characters add as above, then the reduction of the direct product matrix representation must be

$$D^{lp}\otimes D^{l'p'}\Rightarrow D^{(l+l')p''}\oplus D^{(l+l'-1)p''}\oplus\cdots\oplus D^{|l-l'|p''}.$$

Note that the parity of *all* the terms on the right-hand side is determined by $(l+l')$ as opposed to the particular l-value of each irrep. This is clarified if we remember that the product of two even functions or two odd functions (which the left-hand-side is associated with) must be an even function; and the product of an even function with an odd function must be odd. The above result is an important one, to which we shall refer in connection with selection rules. For the moment note that it tells us something about the two-particle states. Taking proper regard of the parity of such states it tells us that from states labelled by l and l' we can construct two-particle eigenfunctions that belong to irreps labelled by L, only for L values given by

$$L=l+l',\ l+l'-1,\cdots,|l-l'|.$$

The total number of such states, $\Sigma_L(2L+1)$ is, as required, equal to the total number of orthogonal functions we started with, $[(2l+1)(2l'+1)]$. From these initial states, $[2(l+l')+1]$ linear combinations can be constructed that transform like $D^{l+l'}$, etc.

In the context in which l is associated with the angular momentum of the particle, the above statements describe the *vector model of the atom*. The symmetry arguments presented here represent the justification for this model. One can equally extend the results to cover the many-particle problem.

$$L=(\sum_i l_i),(\sum_i l_i)-1,\ldots,L_{\min},$$

where $L_{\min}$ is the minimum, non-negative combination of $\pm l_i$ values. Finally, it is noted that in dealing with electronic states it is in principle necessary to take account of the spin properties of these states as well as their spatial dependence. Only the latter is considered presently.

Clebsch–Gordon coefficients

Some useful additional comments can be made by considering a particular example. The problem of a pair of particles in $l=1$ (D^{1u}) states exhibits the required properties.

$$D^{1u}\otimes D^{1u}\Rightarrow D^{2g}\oplus D^{1g}\oplus D^{0g}.$$

Now the forms of the one-particle states are, for $i=1,2$,

$$\psi_{n_i 1 m_i}=R_{n_i}(r_i)\begin{cases}\sin\theta_i\,e^{i\phi_i} & (x_i+iy_i),\ m_i=1\\ \cos\theta_i \quad\propto & z_i,\qquad m_i=0\\ \sin\theta_i\,e^{-i\phi_i} & (x_i-iy_i),\ m_i=-1\end{cases}$$

These two triply-degenerate states lead to nine orthogonal two-particle states which must belong to D^{2g}, D^{1g} and D^{0g} respectively. Furthermore the five states belonging to D^{2g} can be put in form such that on rotation by α about the z-axis $R(\alpha,z)\psi=e^{-iM\alpha}\psi$, where $M=-2,\ -1,\ 0,\ 1,\ 2$. The same statement applies to the D^{1g} states, but with $M=-1,\ 0,\ 1$; and to the D^{0g} state, with $M=0$.

Now the basis function

$$f_{m_1m_2}=\psi_{n_1 1 m_1}(\mathbf{r}_1)\,\psi_{n_2 1 m_2}(\mathbf{r}_2)\text{ is proportional to }\exp i(m_1\phi_1+m_2\phi_2),$$

and has the transformation property

$$R(\alpha,z)f_{m_1m_2}=e^{-i(m_1+m_2)\alpha}f_{m_1m_2}.$$

The eigenfunctions labelled by $M\,(=m_1+m_2)=0$ must therefore be linear combinations of functions $f_{m,-m}$, etc. We can therefore write

$$\psi(\mathbf{r}_1,\mathbf{r}_2)=R_{n_1}(r_1)R_{n_2}(r_2)\,Y_M^L(\theta_1,\theta_2,\phi_1,\phi_2),$$

where

$$Y_M^L(\theta_1,\theta_2,\phi_1,\phi_2) = \sum_{m_1} A^L_{m_1,M-m_1} Y^1_{m_1}(\theta_1,\phi_1)\, Y^1_{M-m_1}(\theta_2,\phi_2).$$

Two eigenfunctions are immediately apparent. To obtain $M=\pm 2$ we must have $m_1=m_2=\pm 1$. Thus two of the D^{2g} functions are proportional to $(x_1\pm iy_1)\,(x_2\pm iy_2)$ respectively. Secondly, the D^{0g} functions must be totally symmetric and hence proportional to $\mathbf{r}_1.\mathbf{r}_2$, that is, to

$$[(x_1+iy_1)(x_2-iy_2)+2z_1z_2+(x_1-iy_1)(x_2+iy_2)].$$

The D^{1g} functions can be found by considering the components of the axial vector $\mathbf{r}_1\wedge\mathbf{r}_2$. In this way some of the coefficients $A^L_{m_1,M-m_1}$ can be ascertained. These coefficients are a special case of the Wigner or Clebsch–Gordon coefficients, A, defined through

$$Y_M^{Ll_1l_2} = \sum_{m_1} A^{Ll_1l_2}_{m_1M-m_1} Y^{l_1}_{m_1} Y^{l_2}_{M-m_1}.$$

The coefficients may be found in general by applying the formula for symmetry-adapted basis functions, starting with product functions $Y^{l_1}_{m_1} Y^{l_2}_{m_2}$ as bases.

The angular momentum of one electron may be visualized as a *vector* of magnitude $\{l(l+1)\hbar^2\}^{\frac{1}{2}}$ and with a component in a single chosen direction (defined as the z-direction) of value $m\hbar$. The angular momentum vector lies on a cone defined by these values; the x and y components are otherwise unspecified, in accord with the uncertainty principle. l and m_l are quantized (Figure 14.1a). For two electrons the two corresponding angular momenta are considered to add vectorially (Figure 14.1b). The resultant vector, according to the above symmetry rules, must itself lie on a cone; its magnitude is $\{L(L+1)\hbar^2\}^{\frac{1}{2}}$ and its component is $M\hbar$, since L and M are quantized. As a result of the latter quantization the individual angular momenta are coupled together in that if one were to consider one of the possible orientations of the resultant vector, then the relative orientation of the individual momenta would be fixed (Figure 14.1c).

In Figure 14.1 the combination of two angular momenta (each with $l=1$, $m=1$) to give a resultant with $L=2$, $M=2$, is described. For the $L=2$, $M=0$ state (for example) of the two particles the situation is more complicated as the $L=2, M=0$ cone (Figure 14.2a) can be reached using $m_2=m_1=0$ or $m_2=-m_1=\pm 1$. This is the pictorial representation of the Clebsch–Gordon expression:

$$Y_0^{211} = A^{211}_{00} Y^1_0 Y^1_0 + A^{211}_{1-1} Y^1_1 Y^1_{-1} + A^{211}_{-11} Y^1_{-1} Y^1_1.$$

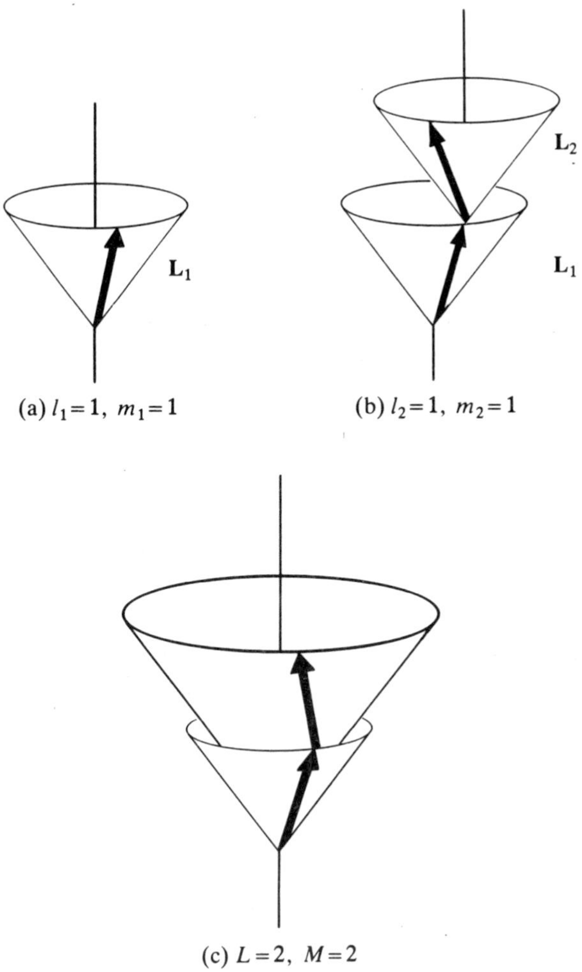

Figure 14.1 (a) Cone of possible vectors ($\mathbf{L}_1$) for the angular momentum of a single electron; (b) cone of vectors obtained by combining a fixed $\mathbf{L}_1$ with arbitrarily oriented $\mathbf{L}_2$. Each $\mathbf{L}_1$ leads to a separate cone. ($L_z = L_{1_z} + L_{2_z}$ is fixed, however); (c) cone of allowed $\mathbf{L}$ vectors, showing one possible combination of $\mathbf{L}_1$ and $\mathbf{L}_2$ that lies on the cone.

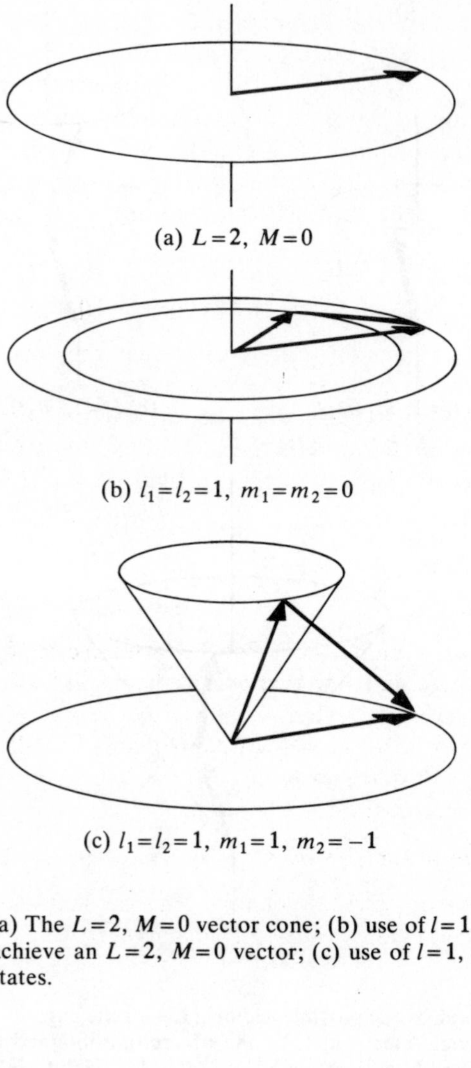

Figure 14.2 (a) The $L=2$, $M=0$ vector cone; (b) use of $l=1$, $m=0$ states to achieve an $L=2$, $M=0$ vector; (c) use of $l=1$, $m_1=1$, $m_2=-1$ states.

14.3 Distinguishable and indistinguishable particles

The D^{1g} two-particle state discussed above is interesting. Three orthogonal functions belonging to D^{1g} must be the axial vector-like components:

$$\psi(\mathbf{r}_1,\mathbf{r}_2)=R_{n_1}(r_1)R_{n_2}(r_2)\frac{1}{r_1r_2}\begin{cases}x_1y_2-y_1x_2\\ y_1z_2-z_1y_2.\\ z_1x_2-x_1z_2\end{cases}$$

If the particles are distinguishable, then x_1y_2 and x_2y_1 lead to distinguishable distribution functions and these states will exist. For indistinguishable particles there are several possibilities. The full, two-particle eigenfunction must be even or odd under particle exchange. Here by full it is intended that the spin contribution is included if relevant.

For zero-spin particles (bosons) the spatial function should be even under particle exchange. It must have the form:

$$\psi=[\psi(\mathbf{r}_1,\mathbf{r}_2)+\psi(\mathbf{r}_2,\mathbf{r}_1)]/\sqrt{2}$$
$$=[R_{n_1}(r_1)R_{n_2}(r_2)-R_{n_2}(r_1)R_{n_1}(r_2)][x_1y_2-x_2y_1]/[r_1r_2\sqrt{2}], \text{ etc.}$$

If the radial states labelled n_1 and n_2 differ, then this function exists; if they are the same, however, the D^{1g} function will not exist.

For spin-$\frac{1}{2}$ functions (fermions) the full eigenfunction must be odd. If the spin contribution is odd, then the spatial function must be even and the above statements still hold. However, if the spin contribution is even, the opposite is true:

$$\psi=[\psi(\mathbf{r}_1,\mathbf{r}_2)-\psi(\mathbf{r}_2,\mathbf{r}_1)]/\sqrt{2}$$
$$=[R_{n_1}(r_1)R_{n_2}(r_2)+R_{n_2}(r_1)R_{n_1}(r_2)][x_1y_2-y_1x_2]/[r_1r_2\sqrt{2}].$$

This function exists even if $n_1=n_2$. For the fermion case, for equal n_1 values, the resulting full eigenfunction is known as the 'triplet' D^{1g} state. The reason is that for two particles the spin contribution can be made even in three ways. In contrast there is only one way of constructing an odd spin function.

The D^{2g} and D^{0g} states have the converse property. The non-vanishing states for $n_1=n_2$ are of the form

$$\psi=[\psi(\mathbf{r}_1,\mathbf{r}_2)+\psi(\mathbf{r}_2,\mathbf{r}_1)]/\sqrt{2}.$$

Therefore boson two-particle functions can exist with these symmetries as can 'singlet' fermion states (of odd spin symmetry).

Problems

***14.1** For the T_d group, obtain the reduced forms of each of the fifteen inner direct product representations ($\Gamma^j\otimes\Gamma^{j'}$). List those reduced representations which contain the symmetric irrep. A_1.

14.2 Consider matrix sets (1) and (2) of Table 3.1. Construct the direct matrix products $\Gamma^j(G_r)\otimes\Gamma^j(G_r)$ for these equivalent irreps, for $G_r\equiv C_3$, σ_b and σ_c. Verify that these four-dimensional matrices satisfy

$$\Gamma(C_3)\Gamma(\sigma_c)=\Gamma(\sigma_b)$$

as required if they are to form part of a representation of the group.

***14.3** Construct a few of the linear combinations of cubic powers of x, y and z that belong to the irreps of the O group.

14.4 (i) For the case of two $l=1$ particles, find the two-particle function

$$Y^{L=0}_{M=0}(\propto \mathbf{r}_1.\mathbf{r}_2)$$

in terms of the single particle spherical harmonics Y^l_m and hence obtain the relative values of

$$A^{011}_{1-1},\ A^{011}_{-11} \text{ and } A^{011}_{00}.$$

(ii) Find orthogonal linear combinations of the components of the axial vector $\mathbf{r}_1 \wedge \mathbf{r}_2$ that can be written in the form

$$r_1 r_2 \sum_m A^{111}_{mM-m} Y^1_m Y^1_{M-m} \text{ for } m=-1,0,1 \text{ and } M=-1,0,1.$$

(iii) Taking $Y^l_m(\theta,\phi)$ to be normalized over all angles

$$\left(\int_0^\pi d\theta \int_0^{2\pi} d\phi \sin\theta\, |Y^l_m|^2 = 1\right),$$

find absolute values for the A-coefficients in part (i) such that Y^L_M is normalized over all angles $(\theta_1,\theta_2,\phi_1,\phi_2)$.

15

Selection rules

15.1 Integrals in quantum mechanics

The quantum-mechanical problems of the hydrogen atom and the harmonic oscillator are soluble exactly. These solutions form a basis for much of the study of electronic motion in atoms, molecules and solids and of vibrational motion in molecules and solids. Nevertheless there are very few other exactly soluble Schrödinger equation problems, so that one has to turn to approximation methods in order to treat almost any real problems.

(i) We discussed one such approximation method in Chapter 12. A complete set of basis functions f_j was introduced and the exact solution to the Hamiltonian problem of interest set up as a linear combination of the f_j. However, in order to make the solution tractable we had to ignore all but a few of the f_j, thereby producing a matrix eigenvalue–equation of finite dimensions. The solution involved the evaluation of integrals of form,

$$\mathscr{H}^f_{ij} = \int f_i^* \mathscr{H} f_j.$$

(Throughout the subsections following this solution a variety of similar integrals of product functions were also introduced.) Although they need to form a complete set for the $\mathscr{H}$-problem, the functions f_j need not have particular symmetries. It is usually the case, though, that one chooses the f_j to be eigenfunctions of some Hamiltonian $\mathscr{H}_0$, thereby having the symmetries of the irreps of the $\mathscr{H}_0$ group. We will show in this section how these symmetries can be exploited in evaluating integrals such as $\mathscr{H}^f_{ij}$.

(ii) Note that if $\mathscr{H}$ can be written as the *sum* of a soluble $\mathscr{H}_0$ and an additional term, $\mathscr{H}'$,

$$\mathcal{H} = \mathcal{H}_0 + \mathcal{H}',$$

then

$$\mathcal{H}^f_{ij} = \int f_i^* (\mathcal{H}_0 + \mathcal{H}') f_j = \epsilon_j \delta_{ij} + \int f_i^* \mathcal{H}' f_j.$$

That is, given that we know the $\mathcal{H}_0$ eigenvalues, ϵ_j, the only remaining problem is to evaluate the $\mathcal{H}'$ matrix elements,

$$(\mathcal{H}')^f_{ij} = \int f_i^* \mathcal{H}' f_j.$$

Definite advantages accrue if these matrix elements are small compared to the energy differences $|\epsilon_j - \epsilon_i|$. One can then expand the solutions to $\mathcal{H}$ to any required order in the $\mathcal{H}'$-elements, with the confidence that higher orders will produce smaller corrections. This is the basis for *time-independent perturbation theory*. $\mathcal{H}'$ is considered to be a perturbation to the original $\mathcal{H}_0$ problem in this situation.

If the $\mathcal{H}'_{ij}$ matrix element is non-zero, then $\mathcal{H}'$ is said to *couple* the original functions (eigenstates of $\mathcal{H}_0$) or to mix them. The coupling leads to eigenfunctions of $\mathcal{H}$ such as

$$\psi_n = a_{in} f_i + a_{jn} f_j + \cdots,$$

where a_{in} and a_{jn} are non-zero.

(iii) $\mathcal{H}'$ matrix elements also arise in the solution of *time-dependent* Hamiltonian problems. The classic example of great interest here is the optical response problem where one is asking for the effect of incident electromagnetic radiation upon some material. Again one divides up the Hamiltonian of the entire system. In the semiclassical theory which will be used here, one concentrates on the material Hamiltonian $\mathcal{H}_0$ (for which a set of solutions is assumed) and the matter-radiation interaction $\mathcal{H}'$. $\mathcal{H}'$ is now time-dependent and instead of considering perturbed stationary states of the system, as in case (ii) above, one considers the $\mathcal{H}_0$ stationary states (which we therefore now describe by ψ_n, E_n) and interprets the effect of $\mathcal{H}'$ as causing transitions between these states. Optical absorption at frequency ω_{nm} must be associated with an allowed transition between states of separation $(E_n - E_m) = \hbar\omega_{nm}$. The point of immediate concern is that such a transition is only allowed provided the states n, m are coupled by the radiation interaction; that is, provided the matrix element $(\mathcal{H}')^{\psi}_{nm}$ is non-zero.

$$(\mathcal{H}')^{\psi}_{nm} = \int \psi_n^* \mathcal{H}' \psi_m.$$

This leads us to the concept of *radiative selection rules*. Not all pairs of states are coupled by a given $\mathcal{H}'$ and so only specific radiative transitions are allowed.

The general point to be made is that there is tremendous physical significance in the determination of whether integrals such as $\mathcal{H}'_{nm}$ or $\mathcal{H}'_{ij}$ are zero or non-zero, regardless of whether we are able to evaluate the magnitude of the non-zero elements. It is precisely to the problem of whether or not $\mathscr{H}'_{ij}$ must be zero from symmetry arguments that group theory gives an almost immediate answer.

Selection rule statement

The group-theoretical statement, which we shall justify below, is the following.

> *Let the irrep. to which the function f_i belongs be denoted Γ^{f_i} and the (generally) reducible representation to which $\mathscr{H}'f_j$ belongs be $\Gamma^{\mathscr{H}'f_j}$.*
> (*i*) *$\int f_i^* \mathscr{H}'f_j$ is zero unless $\Gamma^{\mathscr{H}'f_j}$ contains Γ^{f_i}.*
> (*ii*) *If $\Gamma^{\mathscr{H}'f_j}$ does contain Γ^{f_i}, then the integral may be non-zero but will not necessarily be so.*

When we talk of $\Gamma^{\mathscr{H}'f_j}$ containing Γ^{f_i} we are to understand that $\mathscr{H}'f_j$ is some function, a product of $\mathscr{H}'$ and f_j in the simple case where $\mathscr{H}'$ contains no differential operators. Hence $\mathscr{H}'f_j$ transforms along with some set of functions under the group operations appropriate to $\mathscr{H}_0$. This is the set of functions that form the basis for the (generally) reducible representation in the above statement. On reduction we require

$$\Gamma^{\mathscr{H}'f_j} \Rightarrow a_i \Gamma^{f_i} \oplus \cdots$$

if the integral $(\mathscr{H}')^f_{ij}$ is to be non-zero.

This statement has been given in notation appropriate to the time-independent problem. It is only a matter of reinterpretation to apply the statement to the time-dependent case (the eigenfunctions of $\mathscr{H}_0$ are then written as ψ_i). Equally the statement can be applied to any $\mathscr{H}'^f_{ij}$.

Quite generally for any pair of functions $\mathscr{G}_i$, $\mathscr{G}_j$: $\int \mathscr{G}_i^* \mathscr{G}_j = 0$ unless there is at least one common irrep. in the reduction of the representations $\Gamma^{\mathscr{G}_i}$ and $\Gamma^{\mathscr{G}_j}$. In principle any group could be employed for this reduction. In practice, though, if one knows that one, or both, of the functions are symmetry-adapted functions of the irreps of some particular group (the eigenfunctions of $\mathscr{H}_0$ in the cases cited) then that is the obvious group to choose. Indeed it may be that the $\mathscr{H}_0$ symmetry is the only property of the states that is known.

Proof of the selection rule statement

(i) By definition the functions in our complete sets are orthonormal

$$\int f_i^* f_j = \delta_{ij}.$$

(ii) For the time-independent problem the eigenfunctions of $\mathscr{H}_0$ form a complete set for those of $(\mathscr{H}_0+\mathscr{H}')$ and vice versa. Suppose $f_j=\Sigma_n c_{nj}\psi_n$, where the ψ_n are the $(\mathscr{H}_0+\mathscr{H}')$-eigenfunctions and the c_{nj} are constants:

$$(\mathscr{H}_0+\mathscr{H}')\sum_n c_{nj}\psi_n=\sum_n c_{nj}E_n\psi_n.$$

$$\therefore \mathscr{H}'f_j=\sum_n c_{nj}E_n\psi_n-\epsilon_j f_j=\sum_k \gamma_{kj}f_k,$$

for some set of coefficients γ_{kj}.

(iii) $(\mathscr{H}')^f_{ij}$ is zero unless f_i is contained in this expansion for $\mathscr{H}'f_j$.

(iv) The selection rule is therefore obtained if we show that f_i can be contained in $\mathscr{H}'f_j$ if and only if Γ^{f_i} is contained in $\Gamma^{\mathscr{H}'f_j}$.

(v) Under the group operations f_k transforms into some linear combination of f_k and its partner vectors. Hence $\Sigma_k\gamma_{kj}f_k$ transforms into combinations of all the f_k in the sum, plus all their partner vectors, f_1 to f_N, say:

$$G_r[f_1\ldots f_N]=[f_1\ldots f_N]\ \Gamma^{\mathscr{H}'f_j}(G_r).$$

(vi) $\Gamma^{\mathscr{H}'f_j}(G_r)$ is in blocked form as $f_1,\ldots,f_N$ are symmetry-adapted functions of the group. Hence if f_i is contained in $\Sigma_k\gamma_{kj}f_k$, then Γ^{f_i} must be contained in $\Gamma^{\mathscr{H}'f_j}$.

(vii) The converse statement to (vi) does not follow. Suppose f_i is not contained in $\Sigma\gamma_{kj}f_k$ but that either one of the partner vectors of f_i or an entirely different function but still of Γ^{f_i} symmetry is contained in the sum. Then from (vi) Γ^{f_i} will again appear in $\Gamma^{\mathscr{H}'f_j}$. These are the only ways in which Γ^{f_i} can be in $\Gamma^{\mathscr{H}'f_j}$.

(viii) It follows that:
If Γ^{f_i} is not contained in $\Gamma^{\mathscr{H}'f_j}$ then f_i is certainly not in the sum and $(\mathscr{H}')^f_{ij}$ is zero.
If Γ^{f_i} is contained in $\Gamma^{\mathscr{H}'f_i}$ then either f_i, or one of its partner vectors, or another function of the same symmetry, is in the sum (or several such functions). The integral *may* then be non-zero. It will be so only if f_i is in the sum. Group theory does not give us an elementary method of distinguishing between these cases.

To establish the radiative selection rule, it is sufficient to point out that $\mathscr{H}'$ can be factorized into spatial and time-dependent parts, that we need only consider the former in the integral, and that the above arguments then follow. This completes the proof of the statement of the selection rule theorem. It remains for us to establish the transformation properties of $\mathscr{H}'f_j$ in order to use the theorem. This is not difficult.

(i) If $\mathscr{H}'$ does not contain differential operators, then $\mathscr{H}'f_j$ is simply a product function and

$$\Gamma^{\mathscr{H}'f_j}=\Gamma^{\mathscr{H}'}\otimes\Gamma^{f_j}.$$

$\Gamma^{\mathcal{H}'}$ is in turn simple if $\mathcal{H}'$ belongs to a specific irrep. More generally $\mathcal{H}'$ is a function whose transformation properties can be determined. It can be divided into parts, each of which are basis functions for irreps:

$$\Gamma^{\mathcal{H}'} \equiv \Sigma a_j \Gamma^j.$$

(ii) If $\mathcal{H}'$ is an operator it still has symmetry properties. For example, $\partial/\partial x$ has exactly the same symmetry as x,

$$\frac{\partial}{\partial x} f(x,y,z) \text{ inverts to } -\frac{\partial}{\partial x} f(-x,-y,-z)$$

just as

$$xf(x,y,z) \quad \text{inverts to} \quad -xf(-x,-y,-z)).$$

The irreps of an $\mathcal{H}'$ operator can therefore be established just as in the multiplier example, (i).

Alternative statement of the selection rule

> *If $\mathcal{H}'_{ij}$ is to be non-zero, then the direct product matrix $\Gamma^{f_i^*} \otimes \Gamma^{\mathcal{H}'f_j}$ must contain the completely symmetric representation Γ^I.*

This statement will be justified if we can show that $\Gamma^{f_i^*} \otimes \Gamma^{\mathcal{H}'f_j}$ contains Γ^I if and only if $\Gamma^{\mathcal{H}'f_j}$ contains Γ^{f_i}.

(i) Regardless of the symmetry of f_i the matrix representations that describe the transformation of f_i^* must have elements that are the complex conjugates of those of the f_1 representations.

$$\chi^{f_i^*}(G_r) = \chi^{f_i}(G_r)^*.$$

(ii) The number of times the symmetric irrep. Γ^I appears in the reduction of $\Gamma^{f_i^*} \otimes \Gamma^{f_k}$ is

$$a_I = \frac{1}{g} \sum_r \chi^I(G_r)^* \chi^{f_i}(G_r)^* \chi^{f_k}(G_r).$$

But $\chi^I(G_r)$ is unity for all G_r, and the first orthogonality theorem states that

$$\sum_r \chi^i(G_r)^* \chi^k(G_r) = g\delta_{ik}.$$

Hence

$$a_I = \frac{1}{g} \sum_r \chi^{f_i}(G_r)^* \chi^{f_k}(G_r) = \delta_{ik}.$$

Thus Γ^I appears once, and only once, in the reduction of $\Gamma^{f_i^*} \otimes \Gamma^{f_i}$ and not at all in $\Gamma^{f_i^*} \otimes \Gamma^{f_k}$ if these irreps differ. This is just what we found by inspection for the C_{3v} group in Section 14.1.

(iii) In the present context $a_I = \delta_{ik}$ tells us that $\Gamma^{f_i} \otimes \Gamma^{\mathscr{H}f_j}$ contains Γ^I if and only if $\Gamma^{\mathscr{H}'f_j}$ contains Γ^{f_i}, as required.

Selection rules from a geometric viewpoint

We have now established that for any function $\mathscr{G}$, $\int \mathscr{G}$ is zero unless the symmetric irrep. Γ^I, is contained in the reduced form of $\Gamma^{\mathscr{G}}$, for any group.

This corresponds both to the orthogonality condition and to the matrix element result:

(i) $\int f_i^* f_j = 0$ if Γ^{f_i} and Γ^{f_j} differ, in which case $\Gamma^{f_i} \otimes \Gamma^{f_j}$ does not contain Γ^I. $\int f_i^* f_j$ *may* be non-zero if $\Gamma^{f_i} = \Gamma^{f_j}$, in which case the product does contain Γ^I.

(ii) $\int f_i^* \mathscr{H}' f_j = 0$ unless $\Gamma^{f_i^*} \otimes \Gamma^{\mathscr{H}'} \otimes \Gamma^{f_j}$ contains Γ^I.

Suppose above that on reduction $\Gamma^{\mathscr{G}}$ contains an irrep. of odd (u) parity. This means that $\mathscr{G}$ can be written as a combination of a complete set of functions, one or more of which has odd parity. Consider this part of the sum and divide the integration into regions of positive and negative $\mathbf{r}$. The odd parity contribution to $\mathscr{G}$ changes sign between (x,y,z) and $(-x,-y,-z)$, the two parts of the integral therefore cancel. The same is true if $\Gamma^{\mathscr{G}}$ contains irreps for which the bases change sign on reflection. For any irrep. other than Γ^I similar, if more complicated, arguments can be made. It is only for functions that have the full group symmetry that the integral is non-zero.

The evaluation of $\Gamma^{f_i^*} \otimes \Gamma^{\mathscr{H}'f_j}$

It is usual that the symmetry of $\mathscr{H}'$ will lead us to consider a triple direct product,

$$\Gamma^{f_i^*} \otimes \Gamma^{\mathscr{H}'} \otimes \Gamma^{f_j}.$$

If this contains Γ^I, then $\mathscr{H}'_{ij}$ may be non-zero, otherwise it must be equal to zero.

Because the reduction of direct products is independent of their order, there is a freedom in performing the triple product reduction. If we know the symmetries of f_j and of $\mathscr{H}'$, then the reduction of $\Gamma^{\mathscr{H}'} \otimes \Gamma^{f_j}$ tells us all possible symmetry types of the states to which f_j might couple. However, if we are dealing with a particular pair of states, then $\Gamma^{f_i^*} \otimes \Gamma^{f_j}$ tells us the symmetry types of those perturbations that will couple the states.

It is also useful to remark that if Γ^{l} appears in a direct product (for example if $\Gamma^{\mathscr{H}'}$ or $\Gamma^{f}=\Gamma^{l}$), then the triple product is simplified immediately using the trivial result

$$\Gamma^{j}\otimes\Gamma^{l}=\Gamma^{l}\otimes\Gamma^{j}=\Gamma^{j} \text{ for any } \Gamma^{j}.$$

15.2 Absorption selection rules

Electric-dipole absorption selection rules

The ideas of selection rules are sufficiently important in group theory and quantum mechanics that it is worth our while spending some time on examples. Perhaps the most beautiful results to come out of a symmetry analysis are the selection rules for electric-dipole absorption in free atoms, with which we shall begin.

The Hamiltonian that describes the interaction of radiation with matter may be written:

$$\mathscr{H}'=-e\mathbf{r}.\mathbf{E}\cos(\omega t-\mathbf{q}.\mathbf{r})$$

The coordinate $\mathbf{r}$ refers to the material; $\mathbf{E}$ is the electric vector of the radiation, of frequency ω, propagating with wavevector $\mathbf{q}$. Only a monochromatic beam will be considered presently, although the selection rules are independent of the spectrum of the radiation. The probability, per unit time, that the material system makes a transition from a state ψ_2 to a state ψ_1 may be shown to be proportional to

$$|\int\psi_1^*(-e\mathbf{r}.\mathbf{E}\,\mathrm{e}^{\mathrm{i}\mathbf{q}.\mathbf{r}})\psi_2|^2.$$

The factor $\mathrm{e}^{\mathrm{i}\mathbf{q}.\mathbf{r}}$ in the integral is difficult to handle if the material is localized, as in the case of atoms or molecules. However, the range of the wave functions in such cases (that is the effective size of the atom or molecule) is to a good approximation small by comparison to the radiation wavelength. Typically atomic dimensions are of the order of angstroms(Å) whereas optical transitions occur for radiation wavelengths in excess of 1000Å. Given that $\mathbf{E}$ is the strength of the optical field close to the nuclear position of an atom (or the centre of a molecule), $\mathbf{r}=\mathbf{0}$, then at a distance $\mathbf{r}$ such that $\mathrm{e}^{\mathrm{i}\mathbf{q}.\mathbf{r}}$ is appreciably different from unity $\psi_1^*\psi_2$ will be very close to zero. The factor $\mathrm{e}^{\mathrm{i}\mathbf{q}.\mathbf{r}}$ can therefore be dropped. This is known as the *electric-dipole approximation* because the resulting interaction Hamiltonian is equivalent to the energy of a dipole $(-e\mathbf{r})$ in a uniform (but time-dependent) field, $\mathbf{E}\cos\omega t$. As far as selection rules are concerned we need ask whether or not the matrix element $\int\psi_1^*\mathbf{r}.\mathbf{E}\psi_2$ must be zero. Given that we can define a set of axes for the system, either arbitrarily in the case of a spherically symmetric system, or from the orientation of the material, then the effective $\mathcal{H}'$, for which we are interested in $\Gamma^{\mathcal{H}'}$, is

$$\mathcal{H}' \propto (xE_x + yE_y + zE_z),$$

and the existence of $\mathcal{H}'_{12}$ depends on the existence of one or more of the three matrix elements x_{12}, y_{12}, z_{12}. Thus

$$\Gamma^{\mathcal{H}'} = \Gamma^{(x,y,z)}.$$

For radiation polarized in the chosen z direction, then $\mathcal{H}'$ 'has the symmetry' of z; $\mathcal{H}'$ belongs to the same irrep. as the function z, etc.

Free-atom absorption

For free atoms the energy levels are labelled by the irrep. label l. $\Gamma^l = D^{lg}$ for even l and D^{lu} for odd l. The components x, y, z all belong to the irrep. D^{1u}. The fact that they belong to the same irrep. is a manifestation of the isotropy of the system; there can be no polarization restrictions on the selection rules. (The same will be true of all materials having cubic symmetry.)

The symmetric irrep. in the full rotation–inversion group is labelled D^{0g}. Radiative transitions, within the electric-dipole approximation, will be allowed between states labelled l and l' providing D^{0g} is contained in the direct product

$$\Gamma^{\psi_1^*} \otimes \Gamma^{\mathcal{H}'} \otimes \Gamma^{\psi_2} = \Gamma^{l'} \otimes D^{1u} \otimes \Gamma^{l},$$

or if $\Gamma^{l'}$ is contained in $D^{1u} \otimes \Gamma^l$. It is convenient to separate out the parity argument from that over allowed l values. In our discussion of the vector model of the atom we have already established, for example, that

$$D^1 \otimes D^l \Rightarrow D^{l+1} \oplus D^l \oplus D^{l-1}. \tag{15.1}$$

Also $D^{1u} \otimes \Gamma^l$ must be of opposite parity to Γ^l. (15.2)

The two restrictions on $\Gamma^{l'}$, if it is to be contained in $D^{1u} \otimes \Gamma^l$, are thus

(i) $l' = l+1$, l or $l-1$;
(ii) if l is even l' must be odd, and vice versa.

The second restriction removes from the first the possibility that $l' = l$. In conclusion, electric-dipole absorption for free atoms can occur only between states l and l' such that

$$(l' - l) = \Delta l = \pm 1.$$

There are no polarization restrictions.

Note that if $\Delta l = 1$ is allowed, then $\Delta l = -1$ must also be allowed as the 'direction' of the transition is irrelevant; the direct products do not depend on the order of the functions.

It is also possible to obtain Δm selection rules, but with a slightly different interpretation. In the free atom case there is no preferred direction in the material and we are at liberty to define the $(2l+1)$-degenerate states for a given l as we wish. If we do choose to define a z-direction and states $R_{nl}Y^l_m$ that have specific properties with respect to rotations about the z-axis, then we have states which, although displaying the $R_3 \otimes S_2$ group degeneracy requirements, have the symmetries demanded of the $C_\infty \otimes S_2$ group. $R_{nl}Y^l_m$ belongs to Γ^{lm}, where Γ^{lm} is A_g if l is zero, E_{mg} for even l and E_{mu} for odd l (see Table 11.3). Within this group the radiation Hamiltonian belongs to the irrep. A_u if $\mathbf{E}$ is polarized in the z-direction. It is sensible then to select the z-axis to be in the $\mathbf{E}$-direction when defining the states. The selection rule now requires that A_g be contained in

$$\Gamma^{l'm'} \otimes A_u \otimes \Gamma^{lm}.$$

$A_u \otimes \Gamma^{lm}$ has the same m label as Γ^{lm} but opposite parity. Given that $l' = l \pm 1$ necessarily and therefore $\Gamma^{l'm'}$ must have opposite parity to Γ^{lm}, we now have an additional selection rule.

For plane-polarized radiation, with eigenstates defined with respect to the polarization direction, electric-dipole transitions are allowed only between states for which

$$\boxed{\Delta l = \pm 1, \qquad \Delta m = 0.}$$

Magnetic-dipole transitions

If a given transition is *forbidden* in the electric-dipole approximation it may still be allowed if the radiation interaction is expanded to higher order in the radiation wavevector. The next order of interaction leads to matrix elements of two or three orders of magnitude smaller than typical electric-dipole matrix elements, and is therefore detected only if the electric-dipole transition is forbidden. It is called the magnetic-dipole interaction. Where $\mathscr{H}'_{e\text{-}d} \propto \mathbf{r}.\mathbf{E}$ before, we now have

$$\mathscr{H}'_{m\text{-}d} \propto \mathbf{L}.\mathbf{B},$$

with $\mathbf{L}$ the angular momentum operator and $\mathbf{B}$ the magnetic field component of the radiation.

The symmetry of $\mathscr{H}'_{m\text{-}d}$ is that of L_x, L_y, L_z and these are the components of an axial vector rather than the polar vector for x, y, z. Thus, in the full rotation–inversion group,

$$\Gamma \mathcal{H}'_{m\text{-}d} = D^{1g}.$$

The selection rules fall out immediately, by analogy to the electric-dipole case:

(i) $l' = l+1, l, l-1$;
(ii) the parity of $\Gamma^{l'}$ must be the same as that of Γ^{l}.

Hence, now only $l' = l$ is allowed from restriction (i),

$$\Delta l = 0.$$

In the group $C_\infty \otimes S_2$, for the magnetic field of the radiation polarized in the z direction, $\mathcal{H}'_{m\text{-}d}$ belongs to A_g.

$$\Gamma^{l'm'} \otimes A_g \otimes \Gamma^{lm}$$

contains A_g if $m' = m$ and the initial and final states have the same parity. The later condition is satisfied if $l' = l$, so we have the full selection rules:

$$\boxed{\Delta l = 0, \ \Delta m = 0.}$$

15.3 Selection rules in crystal fields

It is instructive to see how selection rules may change when the environment of an atom is altered. To this end the Ce^{3+}-ion problem, introduced as Problem 10.5, will be investigated. We shall ask what the electric-dipole selection rules are for transitions from the 4f level to the 5s, (a) for the free atom, (b) for the atom at a site of O_h symmetry in a CaF_2 lattice, and (c) when the system is subjected to an additional, static electric field.

(a) For a free atom we know that $\Delta l = \pm 1$ for allowed transitions, whereas $l = 3$ for the 4f states and $l = 0$ for the 5s state. Outside of the crystal environment then the transition is forbidden.
(b) The seven-fold degenerate 4f state must split in the crystal field. Using the characters for operations $R(\alpha, \xi)$ in order to determine the characters of the reducible O_h representation with the seven states as basis one obtains Table 15.1.

Table 15.1

	E	$8C_3$	$6C_4$	$3C_2$	$6C'_2$	i	$8S_6$	$6S_4$	$3\sigma_h$	$6\sigma_d$
Γ_f	7	1	−1	−1	−1	−7	−1	1	1	1

$\Gamma_f \Rightarrow A_{2u} \oplus T_{1u} \oplus T_{2u}$.

At an O_h site the $4f$ state splits into one non-degenerate state, and two triply-degenerate states. The $5s$ state must belong to the non-degenerate, symmetric irrep. A_{1g}, and $\mathscr{H}'_{e\text{-}d}$ belongs to T_{1u} (the irrep for which x, y, z form a basis). Transitions are therefore allowed for those states of Γ symmetry such that the irrep A_{1g} is contained in

$$A_{1g} \otimes T_{1u} \otimes \Gamma.$$

The form of this direct product is such that one obviously makes the product $A_{1g} \otimes T_{1u}$ first ($\Rightarrow T_{1u}$) and then needs only to demand that Γ is equal to T_{1u} for A_{1g} to be contained in the resultant triple product. As one of the Γ_f levels does have T_{1u} symmetry, then a transition is allowed:

$$4f(T_{1u}) \rightarrow 5s(A_{1g})$$

There are no polarization restrictions for this cubic environment.

(c) In the presence of a uniform electric field $\boldsymbol{\varepsilon}$ in the [0,0,1] direction the symmetry of the Ce^{3+} site is reduced to C_{4v}, where the principle axis is that of the uniform field. Reducing Γ_f again, we obtain Table 15.2.

Table 15.2

	E	$2C_4$	C_2	$2\sigma_v$	$2\sigma_d$
Γ_f	7	−1	−1	1	1

$\Gamma_f \Rightarrow A_1 \oplus B_1 \oplus B_2 \oplus 2E$

The level splits into three non-degenerate levels and two doubly-degenerate ones. But if the field is weak the resulting energies must be close to those given by case (b). We should be able to describe the individual splittings of the O_h site levels. To do so, referring to the subgroup problem, Chapter 8, we need to consider the characters of the irreps of the O_h group in the reduced C_{4v} symmetry (Table 15.3).

Table 15.3

	E	$2C_4$	C_2	$2\sigma_v$	$2\sigma_d$
A_{2u}	1	−1	1	−1	1
T_{1u}	3	1	−1	1	1
T_{2u}	3	−1	−1	1	−1

$A_{2u} \Rightarrow B_2,$
$T_{1u} \Rightarrow A_1 \oplus E,$
$T_{2u} \Rightarrow B_1 \oplus E$

As required, these three levels reduce, in total, to precisely the same irreps as Γ_f. The s state belongs to A_1 in the C_{4v} group. To obtain selection rules we now need to note that for the electromagnetic field $\mathbf{E}$ polarized parallel to the principal C_{4v} axis, $\mathcal{H}' \propto z$ and $\Gamma^{\mathcal{H}'} = A_1$. However, for polarization perpendicular to this direction $\mathcal{H}' \propto x$ or y and $\Gamma^{\mathcal{H}'} = E$. There will now be a polarization dependence of the selection rules:

(i) $\mathbf{E} \| \boldsymbol{\varepsilon}$ $\quad A_1 \otimes A_1 \otimes \Gamma$ $\quad$ must contain A_1;
(ii) $\mathbf{E} \perp \boldsymbol{\varepsilon}$ $\quad A_1 \otimes E \otimes \Gamma$ $\quad$ must contain A_1.

Clearly this demands that $\Gamma = A_1$ in case (i), $\Gamma = E$ in case (ii).

Some form of polarization dependence must exist for all groups of symmetry lower than cubic. Figure 15.1 summarizes the Ce^{3+} situation. A few additional points should be made. Firstly the order of the levels in the

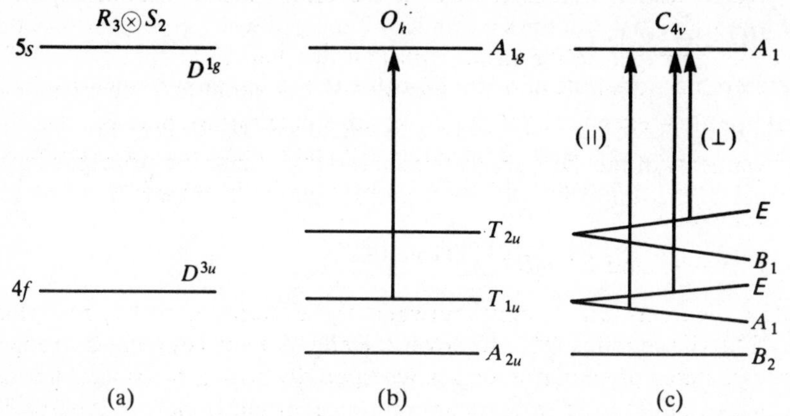

Figure 15.1 Schematic showing the splitting of degeneracies and the change of selection rules as the site symmetry of the Ce^{3+} ion is reduced from $R_3 \oplus S_2$ to O_h to C_{4v}.

reduced symmetry cases can not be specified. Secondly we have assumed that the crystal field splitting is relatively small in comparison to the atomic splitting of the 4*f* to 5*s* states, so that the transitions can be thought of as originating from these levels; we have in mind that the frequencies of all the transitions shown will be similar. Thirdly, whilst we have considered a weak uniform electric field the final selection rules will be independent of its strength. A weak field was considered in order to demonstrate the *grouping* of the levels in diagram (c); again the order within the groups is unspecified. Finally, it is noted that the electromagnetic field is considered to be far weaker yet than the uniform field, in order that we can consider its effect to be a perturbation causing transitions between energy levels established by the time independent Hamiltonian system. From a practical point of view, then, the spectral properties of the Ce^{3+} ion are a strong indication of its environment. In the vicinity of a compensating F^- ion different properties are obtained (Problem 15.1). Once again, spin has not been considered here.

A static perturbation

As an example of a time-independent perturbation coupling various states of an original $\mathcal{H}_0$ problem, we could take the O_h states above and ask how the uniform $\boldsymbol{\varepsilon}$ field mixes them. This would in principle tell us the form of the eigenfunctions in the C_{4v} case, as combinations of the O_h eigenfunctions.

$$\psi_i = f_i + \sum_{j \neq i} [\mathcal{H}'_{ij}/(\epsilon_i - \epsilon_j)] f_j.$$

Here $(\epsilon_i - \epsilon_j)$ is the difference in the energies of the unperturbed levels i and j. The eigenfunction ψ_i is perturbed; f_i and f_j represent unperturbed functions.

It is somewhat more instructive, however, to consider the Stark splitting of a free atom, that is of the $n=1$ and $n=2$ energy levels, from a perturbation viewpoint, comparing the result with that of Chapter 11.

The $\mathcal{H}_0$ problem is now the completely free atom problem. The $n=1$ level is a non-degenerate D^{0g} state (we are still ignoring spin) and the $n=2$ level is made up of a triply degenerate D^{1u} state plus a non-degenerate D^{0g} state. In a uniform electric field $\boldsymbol{\varepsilon}$, in the z-direction, say, the perturbation Hamiltonian is

$$\mathcal{H}' = -e\varepsilon z \text{ and } \Gamma^{\mathcal{H}'} = D^{1u}.$$

Matrix elements of $\mathcal{H}'$ exist between those states f_i and f_j for which $\Gamma^{f_i^*} \otimes \Gamma^{\mathcal{H}'} \otimes \Gamma^{f_j}$ contains D^{0g}. Moreover the effect of any coupling is strongest between states of similar energy and especially strong for states that are degenerate under $\mathcal{H}_0$. The dominant splitting is proportional to the perturbation in the latter case. For the $n=1$, 2 hydrogenic states $\Gamma^{\mathcal{H}'} \otimes \Gamma^{f_j}$ is given by

(i) $n=1$, $D^{1u} \otimes D^{0g} \Rightarrow D^{1u}$
(ii) $n=2$, $D^{1u} \otimes D^{0g} \Rightarrow D^{1u}$
$D^{1u} \otimes D^{1u} \Rightarrow D^{2g} \oplus D^{1g} \oplus D^{0g}$.

These results tell us that D^{0g} states couple only to D^{1u}, whereas D^{1u} states couple to D^{2g} and D^{1g} states (if they exist) as well as to D^{0g}. In particular the $n=2$ states of symmetry D^{0g} and D^{1u}, which exhibited an accidental degeneracy in the H-atom case, are coupled. As a result a strong splitting can occur in the field $\boldsymbol{\varepsilon}$.

Figure 15.2(a) shows the conclusions that can be drawn from group theory. The dashed line indicates that there may be D^{1u} states that do not mix with D^{0g}. Thus whilst we have demonstrated that $\mathcal{H}'_{ij}$ need not be zero for f_i belonging to D^{1u}, and f_j belonging to D^{0g}, there are three D^{1u} states and there is no guarantee that $\mathcal{H}'_{ij}$ will be non-zero for all three. Indeed we know from consideration of the symmetry of the full system $(\mathcal{H}_0 + \mathcal{H}')$ that the $n=2$ level splits into three states in the presence of $\boldsymbol{\varepsilon}$, one doubly-degenerate $|m|=1$ state and a pair of $m=0$ states. The $m=0$ splitting is proportional to the field strength. In the perturbation picture, if we choose to label the original $\mathcal{H}_0$

states as $R_{nl}Y^l_m$, then the $m=0$ states $R_{21}Y^1_0$ and $R_{20}Y^0_0$ mix in the presence of the field ($|m|$ is still a good quantum number) but the $|m|=1$ states are, to a first approximation, unperturbed. They do couple to $|m|=1$ states of $n\neq 2$, however.

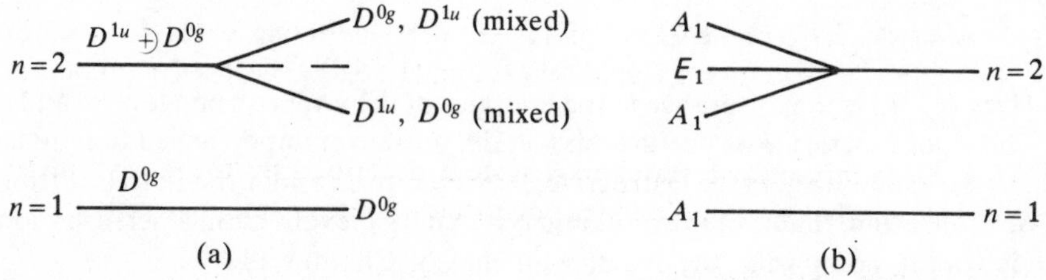

Figure 15.2 Stark splitting of the $n=1$, 2 H-atom levels (a) from the perturbation viewpoint and (b) from consideration of the reduced symmetry.

The Stark spectrum is obtained by considering radiation with **E** polarized parallel and perpendicular to $\boldsymbol{\varepsilon}$ respectively,

(i) $\mathbf{E}\|\boldsymbol{\varepsilon}$, $\mathscr{H}'\propto z$, $\Gamma^{\mathscr{H}'}=A_1$,

(ii) $\mathbf{E}\perp\boldsymbol{\varepsilon}$, $\mathscr{H}'\propto x$ or y, $\Gamma^{\mathscr{H}'}=E_1$.

Hence for transitions from $n=1$ (the A_1 state) to $n=2$,

(i) $\mathbf{E}\|\boldsymbol{\varepsilon}$, $\Gamma^{\mathscr{H}'}\otimes\Gamma^{fj}=A_1\otimes A_1\Rightarrow A_1$,

(ii) $\mathbf{E}\perp\boldsymbol{\varepsilon}$, $\Gamma^{\mathscr{H}'}\otimes\Gamma^{fj}=E_1\otimes A_1\Rightarrow E_1$.

In particular for $\mathbf{E}\|\boldsymbol{\varepsilon}$ the $n=1$ state couples to both of the split A_1, $n=2$ levels. The single absorption line obtained for $\boldsymbol{\varepsilon}=0$, originating from a D^{0g} to D^{1u} allowed transition, will split in the electric field $\boldsymbol{\varepsilon}$ by an amount proportional to $|\boldsymbol{\varepsilon}|$; this is known as the *linear Stark effect*.

Problems

*__15.1__ Obtain the electric-dipole selection rules for $4f$ to $5s$ transitions of Ce^{3+} in the C_{3v} environment that obtains when a nearest-neighbour fluorine atom is ionized. What is the effect of an [0,0,1] uniform electric field on this spectrum? (cf. Problem 10.5).

***15.2** Obtain the magnetic-dipole selection rules for transitions amongst the $n=2$ levels of a free atom in an applied uniform magnetic field (cf. Chapter 11).

15.3 ***Electric quadrupole transitions***
The radiation interaction electric quadrupole term $\mathscr{H}'_{e\text{-}q}$ has the symmetry of the quadratic functions x^2, xy etc. (with the proviso that there is no quadrupole of $(x^2+y^2+z^2)$ symmetry). Determine the allowed quadrupole transitions for an H-atom at a site of T symmetry in a crystal.

***15.4** Determine those states in a system described by the T_d symmetry group that are coupled by a perturbation interaction of symmetry T_2.

15.5 ***The benzene problem***
The one-electron π-orbitals of benzene were discussed in Chapter 12: the order of their energies is

———— B_{2g}

———— E_{2u}

———— E_{1g}

———— A_{1u}

Neglecting electron–electron interactions, in accord with the LCAO model, and noting that each carbon atom contributes one electron to the π-orbitals, a resultant 6-*electron* wave function can be written

$$\psi(\mathbf{r}_1,\mathbf{r}_2,\ldots,\mathbf{r}_6)=\psi_1(\mathbf{r}_1)\psi_2(\mathbf{r}_2)\cdots\psi_6(\mathbf{r}_6)$$

where $\psi_1(\mathbf{r}_1)$ is a one-electron π-orbital for the electron of coordinate $\mathbf{r}_1$.

(i) How would one determine in practice how many electrons each C-atom contributes to the π-orbitals?
(ii) What is the significance of neglecting electron–electron interactions in writing ψ?
(iii) How will this ψ relate to the 42-electron wave functions for the complete benzene molecule?
(iv) If the ground state π-orbital is to be found, what symmetries have the functions ψ_1 to ψ_6 and what principles are used in determining these symmetries?
(v) Use the theory of inner direct products to obtain the symmetries allowed for the 6-electron 'ground state' π-orbital.
(vi) You should have found in (v) that ψ is one of a set of basis functions for a representation of dimension 16. What is the

relationship between the 16 bases functions?
[*Hint*: A representation tells us how functions transform.]

(vii) Why can 6-electron states of different energies be obtained from the same six 1-electron orbitals?
[*Hint*: Re-introducing electron–electron interactions does not change the symmetry of the problem.]

(viii) By raising a *single* electron from an E_{1g} to an E_{2u} symmetry state, one forms some of the excited π-orbitals. What are their symmetries?

(ix) The actual 6-electron ground state has A_{1g} symmetry. If absorption takes place between it and the states in (viii), how many distinct frequencies would be observed and what polarization selection rules are there?

(x) How do the above results compare with the idea of exciting a single electron from E_{1g} to E_{2u} symmetry?

16

Quantization of the normal-mode problem

16.1 Vibrational eigenfunctions

In Part 2 of this text the molecular normal-mode problem was analysed from a classical viewpoint. In Part 3 the quantum-mechanical molecular problem was treated. The Born–Oppenheimer approximation was used to allow independent analyses of the electronic distribution and the nuclear motion. Following the quantum treatment of Chapter 12, the nuclear motion is described by the equation

$$[\sum_n -\frac{\hbar^2}{2M_n}\nabla_n^2 + E_e(\mathbf{R})]X(\mathbf{R}) = E_{\text{mol}}X(\mathbf{R})$$

The nuclear configuration is described by the set of coordinates, $\mathbf{R}$, and $E_e(\mathbf{R})$ is the electronic energy eigenvalue for such a configuration. In contrast the classical problem was treated in the harmonic approximation:

$$[K^d - \omega_i^2 M^d]\begin{bmatrix}\alpha_{1i}\\ \alpha_{2i}\\ \cdot\\ \cdot\\ \cdot\end{bmatrix} = \begin{bmatrix}0\\ 0\\ \cdot\\ \cdot\\ \cdot\end{bmatrix}.$$

The solutions are independent, normal-mode oscillations.

In order to bring the two treatments together, firstly we consider a particular electronic state. $E_e(\mathbf{R})$ takes a minimum value for some effectively equilibrium configuration, which we characterize by $\mathbf{R}_0$.

$$[\sum_n -\frac{\hbar^2}{2M_n}\nabla_n^2 + E_e(\mathbf{R}) - E_e(\mathbf{R}_0)]X(\mathbf{R}) = [E_{\mathrm{mol}} - E_e(\mathbf{R}_0)]X(\mathbf{R}).$$

Thus the energy due to the *motion* of the nuclei satisfies

$$[\sum_n -\frac{\hbar^2}{2M_n}\nabla_n^2 + \Delta V]X = E_{\mathrm{motion}}X,$$

with $\Delta V = E_e(\mathbf{R}) - E_e(\mathbf{R}_0)$ and $E_{\mathrm{mol}} = E_e(\mathbf{R}_0) + E_{\mathrm{motion}}$.

One can now use displacement coordinates for each of the nuclei, and for small displacements the harmonic approximation may be applied; ΔV will then contain only quadratic terms in the displacement coordinates.

Figure 16.1 shows a typical form for the energy E_e of a diatomic molecule, for which the atomic separation R specifies the nuclear configuration potential. The curve is often described by the so-called Morse potential,

$$E_e(R) = E_e(R_0) + D_e[1 - \exp\{-\alpha(R - R_0)\}]^2,$$

where D_e is the molecular dissociation energy.

In the region of R_0, however, the potential approximates to a parabolic well,

$$\Delta V = E_e(R) - E_e(R_0) \approx D_e\alpha^2(R - R_0)^2.$$

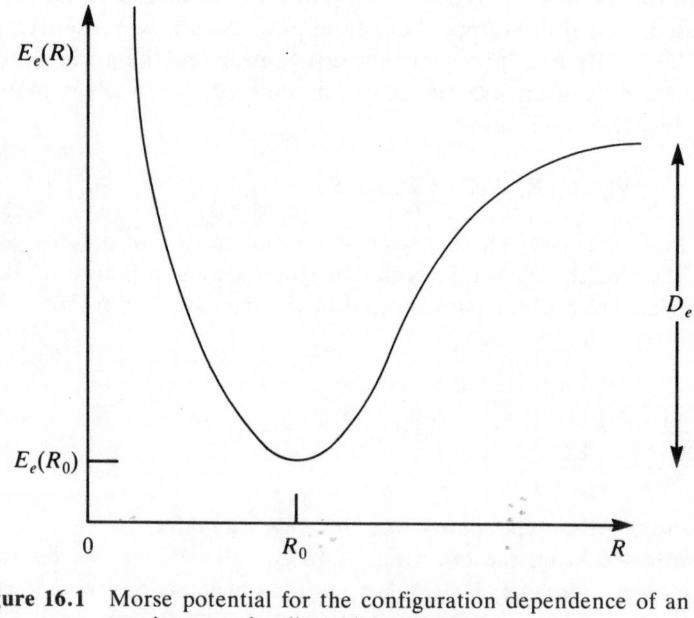

Figure 16.1 Morse potential for the configuration dependence of an electronic state of a diatomic molecule.

Because of the coupled nuclear–nuclear potential energy terms, which are present in $E_e(\mathbf{R})$ above, the displacement coordinates are highly coupled in the nuclear motion equation for polyatomic molecules. Decoupling is achieved, by analogy to the classical case, by converting to the amplitudes of normal-mode eigenvectors, $\mathcal{R}_i$:

$$[\sum_i -\frac{\hbar^2}{2M_i}\nabla_i^2 + \sum_i \tfrac{1}{2}K_i\mathcal{R}_i^2]\boldsymbol{X} = E_{\text{motion}}\boldsymbol{X},$$

Where M_i and K_i are the mass and force matrix-elements $M_{ii}^{\mathcal{D}}$ and $K_{ii}^{\mathcal{D}}$, and are related classically by the eigenfrequencies; $\omega_i^2 = (K_{ii}^{\mathcal{D}}/M_{ii}^{\mathcal{D}})$. Having decoupled the $3N$ nuclear variables the equation is in separable form:

$$\boldsymbol{X} = \prod_{i=1}^{3N} \boldsymbol{X}(\mathcal{R}_i),$$

$$E_{\text{motion}} = \sum_i E_i,$$

where

$$[\frac{-\hbar^2}{2M_i}\frac{\partial^2}{\partial \mathcal{R}_i^2} + \tfrac{1}{2}K_i\mathcal{R}_i^2]\,X(\mathcal{R}_i) = E_i X(\mathcal{R}_i).$$

The translation and rotation motion must be considered separately from the vibrations in the quantum treatment.

For the translations K_i is zero and M_i is equal to the total mass of the molecule, M_T. The molecule translates with momentum $\hbar\mathbf{k}$ like a free particle of mass M_T, with a continuum of allowed energies $\hbar^2k^2/2M_T$. The rotational motion about the centre-of-mass is determined by the principal moments of inertia of the molecule. The allowed energies are quantized with typical spacings of a few wavenumbers (1 wavenumber $\equiv 1\,\text{cm}^{-1} \equiv 0.12\,\text{meV}$†). By contrast vibrational spacings, described below, are typically 10^2–$10^4\,\text{cm}^{-1}$ ($\simeq 0.1$ eV), and electronic level separations exceed $10^4\,\text{cm}^{-1}$ (>1 eV).

For the $3N-6$ vibrational modes ($3N-5$ for linear molecules), each $\boldsymbol{X}(\mathcal{R}_i)$ satisfies the one-dimensional quantum harmonic oscillator equation.

$$\left[\frac{-\hbar^2}{2M_i}\frac{\partial^2}{\partial \mathcal{R}_i^2} + \tfrac{1}{2}M_i\omega_i^2\mathcal{R}_i^2\right] X(\mathcal{R}_i) = E_i X(\mathcal{R}_i).$$

The solutions are well established; the quantized energies form a ladder of equally spaced levels

$$E_i = (n_i + \tfrac{1}{2})\hbar\omega_i,$$

†The wavenumber is defined by $\bar{\nu}$ in the relationship $E = \hbar\omega = hc/\lambda = hc\bar{\nu}$, where $\bar{\nu}$ is determined in cm^{-1} and is given by $\bar{\nu} = E/(6.63\times10^{-34}\times3.00\times10^{10})$, where E is in joules. $1/\bar{\nu}$ is the wavelength (in cm) of the radiation that resonates with the level spacing.

and the eigenfunctions are

$$\psi_{n_i}(R_i) = C_{n_i} H_{n_i}(\alpha_i R_i) \exp(-\alpha_i^2 R_i^2/2).$$

Here

$$\alpha_i = (M_i \omega_i/\hbar)^{1/2},$$

The normalization constants are

$$C_n = (\alpha/2^n n!\ \pi^{1/2})^{1/2},$$

and H_{n_i} is the n_ith Hermite polynomial. Examples of the low-order Hermite polynomials are:

$$H_0(x) = 1; \qquad H_1(x) = 2x; \qquad H_2(x) = (4x^2 - 2).$$

In summary, the form of X for the full nuclear motion is

$$X = TRV,$$

$$E_{\text{motion}} = E_T + E_R + E_V.$$

The vibrational contributions are

$$V = C\{\prod_{i=1}^{3N-6(5)} H_{n_i}(\alpha_i R_i)\ \{\exp(-\sum_i \alpha_i^2 R_i^2/2)\},$$

$$E_V = \sum_i (n_i + \tfrac{1}{2})\hbar\omega_i,$$

where C is the appropriate normalization constant.

The scale of the vibrational energies is such that for diatomic molecules one can think of a ladder of levels lying within the electronic Morse potential, although increasing anharmonicity will cause the spacings of higher levels to differ from $\hbar\omega_i$. The probability of a large nuclear separation increases in the higher levels, and dissociation will eventually occur – typically at n values of twenty or more.

16.2 The symmetry of vibrational states

(a) The ground state

For a polyatomic molecule the vibrational ground state is obtained by setting $n_i = 0$ for each of the $3N-6$ (5) vibrations. The ground state vibrational eigenfunction is therefore

$$V_g = C \exp\left(-\sum_i \alpha_i^2 R_i^2/2\right)$$

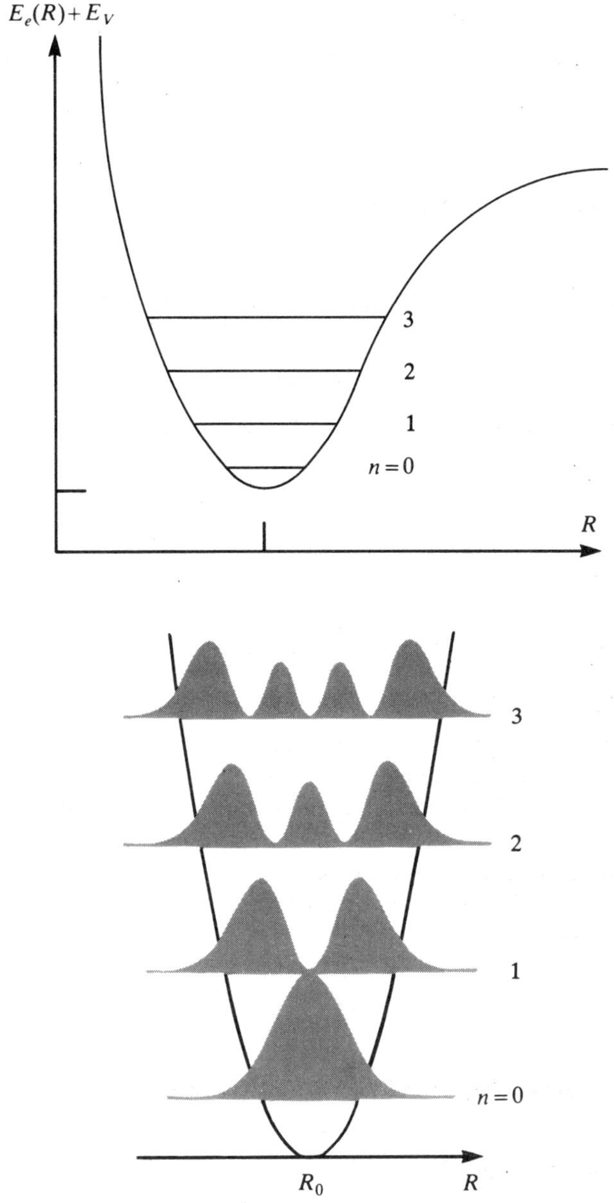

Figure 16.2 (a) Schematic of electonic/vibrational energy levels of a diatomic molecule. The rotational energy separations are considerably smaller; (b) The lower-order harmonic oscillator distribution functions$|\ \psi_{n_i}\ (R_i)|^2$, showing the increased spread at higher energies.

$$= C \exp \left(\frac{-1}{2\hbar} \sum_i M_i \omega_i \mathcal{R}_i^2 \right).$$

The argument of this exponential is invariant under the operations of the group of the molecule. The proof may be achieved by breaking up the sum into a series of sums over degenerate modes. Note that for a set of degenerate eigenvectors $\mathfrak{D}_1$, $\mathfrak{D}_2, \ldots$ that are partner vectors belonging to a specific irrep, both the kinetic and potential energies corresponding to $(\mathcal{R}_1 \mathfrak{D}_1 + \mathcal{R}_2 \mathfrak{D}_2 + \cdots)$ must be invariant under the group operations. Hence the equalities

$$M_{11}^{\mathfrak{D}} = M_{22}^{\mathfrak{D}} = \cdots, K_{11}^{\mathfrak{D}} = K_{22}^{\mathfrak{D}} = \cdots$$

apply, as well as the degeneracy condition, $\omega_1 = \omega_2 = \cdots$.

Thus

$$\sum_i M_i \omega_i \mathcal{R}_i^2 = \sum_j M_j \omega_j \left(\sum_k \mathcal{R}_k^2 \right),$$

where the k-sum is over partner modes, the j-sum is over distinct frequencies. But for any complete set of partner vectors $\sum_k \mathcal{R}_k^2$ is invariant under the group operations (see Problem 9.3). Hence in turn the j-sum above is invariant.

The ground state vibrational eigenfunction is therefore invariant under the molecular group operations: V_g *forms a basis for the symmetric irrep.*

(b) Fundamental excited state

A fundamental excited state of the nuclear motion is defined as one in which one and only one normal mode is in its first excited harmonic level; $n_j = 1$, $n_i = 0$ for all $i \neq j$.

$$V_{f\text{-}j} = C2(\alpha_j \mathcal{R}_j) \exp \left(-\sum_i \alpha_i^2 \mathcal{R}_i^2 / 2 \right).$$

Since it has already been established that the exponential factor is symmetric, the symmetry of V is in general determined by that of the Hermite product. For fundamental levels $V_{f\text{-}j}$ has the symmetry of $\mathcal{R}_j$, which is the irrep. to which $\mathfrak{D}_j$ belongs. The quantum result for fundamental excited states therefore corresponds precisely to the classical symmetry conclusion.

(c) Combination levels

$n_j = 1$, $n_k = 1$, $n_i = 0$ for $i \neq j, k$:

$$V_{c\text{-}j,k} = C4\alpha_j \alpha_k (\mathcal{R}_j \mathcal{R}_k) \exp \left(-\sum_i \alpha_i^2 \mathcal{R}_i^2 / 2 \right).$$

This function has the symmetry of the product $(\mathcal{R}_j\mathcal{R}_k)$. If the eigenvectors $\mathfrak{D}_j$ and $\mathfrak{D}_k$ are non-degenerate (they belong to one-dimensional irreps) then this symmetry is simply of $\Gamma^j \otimes \Gamma^k$. For degenerate modes there are sets of $(\mathcal{R}_j\mathcal{R}_k)$ products and linear combinations that will belong to the specific irreps in the reduction of $\Gamma^j \otimes \Gamma^k$.

(d) Overtone levels

$n_j = 2$, $n_i = 0$ for $i \neq j$:

$$V_{o\text{-}j} = C(4\alpha_j^2\mathcal{R}_j^2 - 2)\exp\left(-\sum_i \alpha_i^2\mathcal{R}_i^2/2\right).$$

Given that $V_{o\text{-}j}$ must have a specific symmetry and that $(-2C \exp)$ is symmetric, then $V_{o\text{-}j}$ must belong to the symmetric irrep. of the group, Γ^I. For non-degenerate modes this result is easily proved from the fact that $\Gamma^j \otimes \Gamma^j$ (to which $\mathcal{R}_j^2$ belongs) is equal to Γ^I.

For the NH_3 molecule, the vibrational modes of which are discussed in Chapter 8, there are six vibrations (two singly-degenerate and two pairs of doubly-degenerate modes).

$$\Gamma^D = \underbrace{A_1 \oplus E}_{\text{translations}} \oplus \underbrace{A_2 \oplus E}_{\text{rotations}} \oplus \underbrace{2A_1 \oplus 2E}_{\text{vibrations}}.$$

Hence the vibrational ground state has A_1 symmetry. The fundamental excited states have A_1, A_1, E, E symmetries. The combination levels have symmetries $A_1 \otimes A_1 = A_1$, $A_1 \otimes E = E$ and $E \otimes E = (A_1 \oplus A_2 \oplus E)$. Overtones of the non-degenerate A_1 levels have A_1 symmetry.

16.3 Absorption selection rules

Having set up the vibrational eigenfunctions as functions of the normal-mode parameters $\mathcal{R}_i$, in order to consider selection rules in the presence of some interaction (such as the matter-radiation interaction) it is necessary to set up the interaction Hamiltonian in terms of the $\mathcal{R}_i$.

In the electric-dipole approximation the radiation interaction with a molecule is

$$\mathcal{H}' = -\mathbf{E} \cdot \mathbf{M},$$

where $\mathbf{M}$ is the electric-dipole moment of the molecule. $\mathbf{M}$ depends on the nuclear configuration of the molecule for any specific electronic state. It may be non-zero in the equilibrium configuration, in which case the molecule is said to have a *permanent* dipole moment. Also $\mathbf{M}$ will in general change under any disturbance from equilibrium:

$$\mathbf{M}(\mathbf{R}) = \mathbf{M}(\mathbf{R}_0) + \Delta\mathbf{M}(\mathbf{R} - \mathbf{R}_0).$$

Such a disturbance might be simplest to picture in terms of individual atomic displacements. However, the amplitudes of the normal-mode displacements can equally well be used. If these are small, and this was the requirement that allowed us to use the harmonic approximation, then a Taylor expansion of **M** can be made:

$$\mathbf{M}(\mathbf{R}) \approx \mathbf{M}_0 + \sum_k \left(\frac{\partial \mathbf{M}}{\partial R_k}\right)_0 R_k.$$

Radiation absorption selection rules between vibrational states are therefore obtained by considering

$$\mathcal{H}'_{ij} = \int V_i^* (-\mathbf{E}).(\mathbf{M}_0 + \sum_k \left(\frac{\partial \mathbf{M}}{\partial R_k}\right)_0 R_k) V_j \, dR_1 \, dR_2 \cdots .$$

(i) As $(-\mathbf{E} \,.\, \mathbf{M}_0)$ is independent of the coordinates R_k, and because different vibrational eigenfunctions must be orthogonal, the first term cannot lead to transitions between different states:

$$\mathcal{H}'_{ij} = -\mathbf{E} \,.\, \mathbf{M}_0 \delta_{ij} + \sum_k (-\mathbf{E}).\left(\frac{\partial \mathbf{M}}{\partial R_k}\right)_0 \int V_i^* R_k V_j \, dR_1 \, dR_2 \cdots .$$

(ii) The kth-transition term in the sum is non-zero only if V_i and V_j are identical in all coordinates other than R_k, again from orthogonality. For $j \neq i$:

$$\mathcal{H}'_{ij} = \sum_k (-\mathbf{E}).\left(\frac{\partial \mathbf{M}}{\partial R_k}\right)_0 C' \int H_{n_{ki}}(\alpha_k R_k) \exp\left(-\frac{\alpha_k^2 R_k^2}{2}\right) R_k$$
$$H_{n_{kj}}(\alpha_k R_k) \exp\left(-\frac{\alpha_k^2 R_k^2}{2}\right) dR_k.$$

C' is included to take care of the normalization coefficients; n_{ki} is the energy level of the kth normal mode in the final state; n_{kj} is its initial energy level.

(iii) From the properties of the Hermite polynomials the only non-zero R_k matrix elements are those for which $n_{ki} = n_{kj} \pm 1$. That is, excitation can only occur from one step of an harmonic ladder to the one immediately above it $(n_{kj} \rightarrow n_{kj} + 1)$. $n_{kj} \rightarrow n_{kj} - 1$ is the de-excitation term and is associated with radiation emission. As only one vibrational mode can be excited at a time, the transition frequencies are the natural (classical) vibrational frequencies, ω_k.

(iv) To see the effect of molecular symmetry on the selection rules, it is more convenient to consider the Cartesian components of the dipole-moment operator,

$$M_x = \sum_i e_i x_i,$$

where e_i is the charge at position x_i of the ith particle, with respect to a fixed reference frame (the point group centre of the molecule). Under any symmetry operation x_i, y_i and z_i transform as polar vector coordinates for each i, so that the vector **M** also transforms as x, y, z.

Hence if the electric-dipole interaction transition matrix elements between states labelled by n_k and $n_k \pm 1$ are to be non-zero, the symmetry condition that must be satisfied is

$$\Gamma^{n_k \pm 1} \otimes \Gamma^{x,y,z} \otimes \Gamma^{n_k} \Rightarrow \Gamma^{l} \oplus \cdots .$$

For transitions between the ground state and a fundamental level ($n_{ki} = 0 \rightarrow n_{kj} = 1$), $\Gamma^{n_k} = \Gamma^{l}$, whilst Γ^{n_k+1} has the symmetry of the normal mode eigenvector, $\Gamma^{\mathcal{D}_k}$. The direct product $\Gamma^{n_k+1} \otimes \Gamma^{n_k}$ will therefore also have the latter symmetry. We have already established that the order of the direct product matrices does not affect the reduced form and that $\Gamma^{j} \otimes \Gamma^{j'} \Rightarrow \Gamma^{l} \oplus \cdots$ only if $j = j'$. Hence the transition is allowed only if $\Gamma^{x,y,z} = \Gamma^{\mathcal{D}_k}$.

In summary, fundamental vibrational absorption between harmonic levels is allowed within the electric-dipole approximation only for modes that have the symmetry of x, y or z. Such modes are said to be infrared active, because the vibrational frequencies lie in the infrared spectral region ($\simeq 1000\,\text{cm}^{-1}$).

Example: NH_3

In Chapter 8 we described the six vibrational modes of ammonia. There are two non-degenerate modes of A_1 symmetry and two pairs of doubly-degenerate E modes (Figure 8.7). Defining the molecular axis through the N-atom as the z axis and the hydrogen plane as the x–y plane, then z has A_1 symmetry whilst (x,y) transform together with E symmetry.

Therefore, for radiation polarized along the (z) axis the two A_1 modes may be excited. For polarization in the (x–y) plane the E modes are absorbing. In either case one expects two observable absorption lines. For a gas of molecules of random orientations there can of course be no preferred polarization selection rules and four lines will be observed.

Problems

16.1 Determine the vibrational frequencies (in radians/s and cm^{-1}) for the following diatomic molecules, given the effective force constants, k:

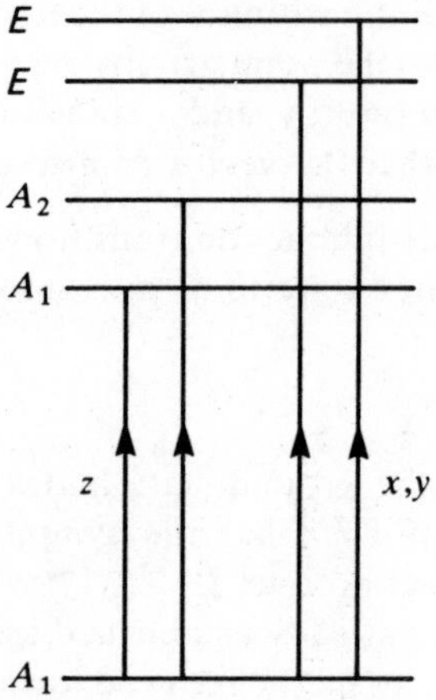

Figure 16.3 Allowed fundamental vibrational transitions of NH_3. Note that the level ordering is not obtained from group theory.

Molecule	H_2	CO	N_2	Na_2	KCl
k ($N m^{-1}$)	573	1902	2297	17	86.

16.2 Determine the symmetry types of the fundamental, combination and overtone levels of diborane, B_2H_6. This molecule has D_{2h} symmetry; the two central H atoms lie equal distances above and below the B–B axis.

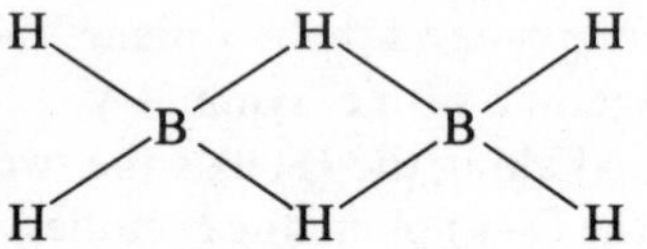

16.3 Determine the infrared active modes of (a) CO_2, (b) H_2O, (c) CH_4 and (d) SF_6.

***16.4** Describe the motion of H_2O in a z-infrared active mode and discuss the direction in which charge is displaced, in order to explain why such a mode may be excited by an oscillating electric field (a force) in the z-direction.

Explain why the A_1 breathing mode of C_6H_6 cannot be infrared active.

16.5 Is it possible to determine whether HCN is a linear or angled molecule from the number of fundamental infrared absorption lines?

***16.6** Figure 16.4 shows three possible configurations for the hydrogen peroxide molecule, H_2O_2. Given that the stable configuration has six observed fundamental vibrational frequencies, determine which configuration is stable.

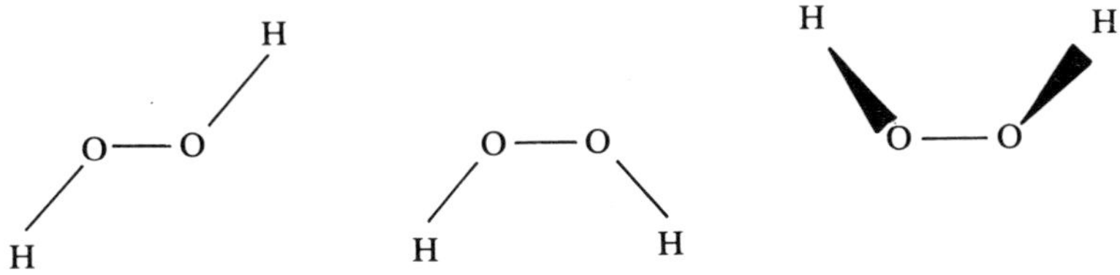

Figure 16.4 Alternative configurations of H_2O_2. In the non-planar version the two hydrogen atoms project, respectively, below and above the plane by equal amounts.

16.7** ***Raman selection rules

At high radiation irradiances it is possible to scatter a photon of energy $\hbar\omega_1$ to one of lower-energy $\hbar\omega_2$ if the excess energy $\hbar(\omega_1-\omega_2)$ matches an excitation energy in the scattering medium. Such a process is known as Stokes–Raman scattering. The selection rule for Raman scattering in which a fundamental vibrational mode is excited is that the excited level must have the symmetry of quadratic functions x^2, y^2, z^2, $xy+yx$, $yz+zy$ or $zx+xz$. A mode of symmetry z^2 will cause scattering from z-polarized ω_1 radiation to z-polarized ω_2 radiation, etc.

Determine the vibrational, Raman selection rules for NH_3.

16.8 Compare the infrared and Raman active modes of H_2O and CO_2.

16.9 ***The exclusion rule***

Explain why the normal-mode vibrations of centrosymmetric molecules may be either absorbing (infrared active) or Raman active but not both. This is known as the exclusion rule.

16.10 Determine the Raman-active mode symmetries for linear molecules.

16.11 ***The benzene problem***

Summarize the infrared and Raman selection rules for vibrational transitions in benzene.

PART FIVE

SYMMETRY AND SOLID-STATE ENERGY LEVELS

17

The crystallographic space groups

17.1 Lattices and crystal point groups

By the end of this final part it is hoped that we will be in a position to label, with some confidence, the energy levels of the valence and conduction bands of semiconductors. The labels will be primarily group theoretical, irreducible representation labels, and will allow us to discuss selection rules between the various states. Before specifically mentioning even semiconductors, though, we need to establish some of the concepts and notations of crystal theory, starting with the classifications of crystal symmetry groups.

Space groups

We have in mind for the moment a perfect, infinite crystal, being a periodic structure that fills all space. The group of all distinct symmetry operations for the crystal is called its space group.

Lattice

Despite the detailed nature of the crystal, as far as its periodicity is concerned it can be simulated by a lattice of points of the same periodicity. Every point in this lattice has exactly the same environment, as far as neighbouring points are concerned. The precise nature of the crystal can be reproduced by associating with each lattice point an identical configuration of atoms.

Lattice translation vectors

A translation of the crystal by a vector **T** that takes one lattice point at **R**, say, to another at **R**+**T** leaves the entire crystal invariant (bearing in mind that we have no crystal boundary). **T** is called a lattice vector.

Primitive translation vectors

Starting at any particular lattice point, then, we can construct three vectors that take us to three chosen nearest neighbour lattice points (such that the defined vectors are not coplanar). If the size of these vectors is minimized and the volume they enclose is minimized also, then they are termed *primitive lattice vectors* $\mathbf{a}_1$, $\mathbf{a}_2$, $\mathbf{a}_3$.

Primitive cell

The primitive vectors define a parallelepiped of volume $\mathbf{a}_1 . \mathbf{a}_2 \wedge \mathbf{a}_3$, which is called the primitive cell. By stacking such cells together one can fill all space, such that the corners of the cells sit at lattice points of the crystal under consideration. Mathematically, for m_1, m_2, m_3 equal to any integers (or zero);

$$\mathbf{T} = m_1\mathbf{a}_1 + m_2\mathbf{a}_2 + m_3\mathbf{a}_3.$$

It should be noted that the primitive cell defined above is not the only cell that stacks in the manner of the crystal. Figure 17.1(a) demonstrates this fact

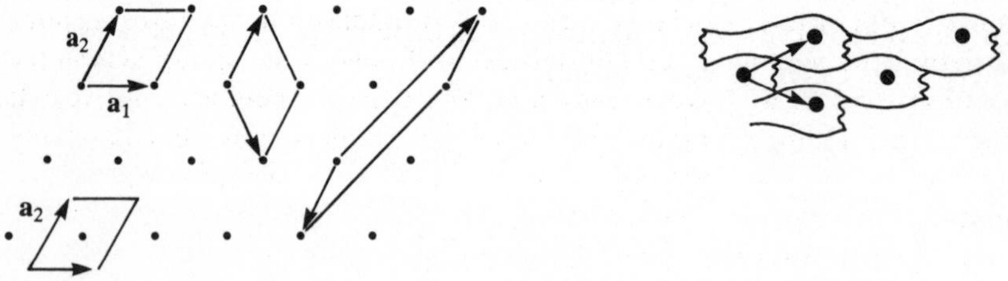

(a) (b)

Figure 17.1 Alternative primitive cells for 2-D lattices.

for the two-dimensional case. The four cells shown all have the same volume, but only those two labelled with $\mathbf{a}_1$ and $\mathbf{a}_2$ contain nearest neighbour translations. Of these two we will choose to define a cell having a lattice point at each corner (the top one) but are not forced to do so. Figure 17.1(b) is included to show that a primitive cell does not have to be a parallelogram (parallelepiped). There are infinitely many shapes that can be stacked to fill space, as exemplified by the attractive diagrams of Escher. For present purposes, though, the primitive cell vectors will be defined as above, the shape of the cell being uniquely specified by the set of parameters.

$$a_1 = |\mathbf{a}_1|,$$

$$\alpha_1 = \cos^{-1} \frac{\mathbf{a}_2 . \mathbf{a}_3}{a_2 a_3}, \text{ etc.,}$$

(Figure 17.2). Different lattice symmetries occur as we let a_1 to α_3 vary over all possibilities.

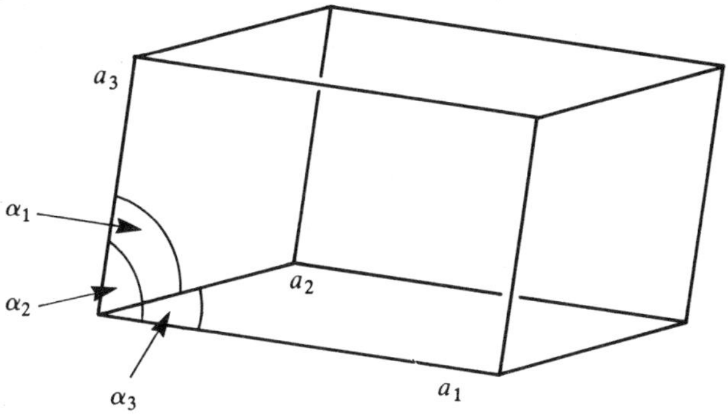

Figure 17.2 Notation for a primitive cell.

Space group operations

The lattices constructed by stacking such cells will have three types of symmetry operation associated with them. In addition to the obvious pure translations by vectors **T**, certain lattices will have point group symmetry operations such as we have already met. For example a cubic primitive cell will generate a lattice for which all the O_h group operations, centred at any lattice point, will leave the crystal invariant. If we associate any structure (more than one atom) with the primitive cell, then it is also possible to obtain symmetry operations that involve point group operations followed by translations by an amount smaller than a primitive vector, i.e. by a vector $\boldsymbol{\tau}$ that lies within the primitive cell. Operations of the latter type are called glide plane or screw axis operations; they will be discussed later in more detail.

The Seitz operator

The Seitz operator, $\{R/\mathbf{t}\}$, notation will be used to describe a general symmetry operation in the complete, space group of a crystal.

$\{R/\mathbf{t}\}$ means:

(i) choose any lattice point as origin;
(ii) perform the point group operation (denoted R) on the crystal;
(iii) then translate the crystal by a vector $\mathbf{t}$.

The effect of $\{R/\mathbf{t}\}$ on a point initially at a position $\mathbf{r}$ is therefore

$$\{R/\mathbf{t}\}\mathbf{r}=R\mathbf{r}+\mathbf{t}.$$

In general $\mathbf{t}$ may be written

$$\mathbf{t}=\mathbf{T}+\boldsymbol{\tau},$$

where $\boldsymbol{\tau}$ is a non-primitive translation (within the primitive cell). The three types of space group operations are then denoted by

$\{E/\mathbf{T}\}$ pure translation, by a lattice vector;

$\{R/\mathbf{O}\}$ pure point group operation;

$\{R'/\boldsymbol{\tau}\}$ glide plane or screw axis operation.

Any combination of these three must also be a space group operation, of course. For example,

$$\{R'/\mathbf{t}\}=\{R'/\mathbf{T}+\boldsymbol{\tau}\}=\{E/\mathbf{T}\}\ \{R'/\boldsymbol{\tau}\}.$$

The translational group

For any crystal the set of all translations $\{E/\mathbf{T}\}$ form a group. It is called the translational group, $\mathcal{T}$, and is fairly clearly a subgroup of the whole space group.

The entire point group

The set of pure point group operations $\{R/\mathbf{O}\}$ and the point group operation portion of the glide/screw operations $\{R'/\mathbf{O}\}$ together form a group. It is called the entire point group of the crystal, but is not a subgroup of the space group because $\{R'/\mathbf{O}\}$ are not in the space group. However, as we shall see, the entire point group is used as a classification scheme for real crystals, primarily because it is characteristic of their macroscopic properties—in particular the three-dimensional macroscopic shape of the (finite) actual crystals. From a macroscopic viewpoint it is not surprising that only the point group portion of the symmetry operations is significant, because it is only the point group operations that can have an observable effect on a real crystal. A translation by an amount less than a lattice vector will never be detectable macroscopically. For those space groups that contain only translations and pure point group operations the entire point group is just $\{R/\mathbf{O}\}$, and in this case it is a subgroup for the system.

The crystallographic point groups

The restriction that the lattice sites of a crystal are periodic has an effect on the number of point groups that are able to display lattice symmetry. For example we shall show that a C_5 point group, which can describe the symmetry of a molecule, does not exist as a lattice point group. The following argument may be used in order to establish to which point groups lattices can belong. In Figure 17.3 concentrate on the atoms labelled A and B. These are meant to represent two nearest neighbour atoms in the lattice, separated by a primitive vector of magnitude a. Suppose that one of the rotation operations in the lattice point group is $R(\alpha, z)$, z corresponding to an axis perpendicular to this paper. Taking atom A as the centre of the point group, there must be a third atom, C, corresponding to the rotation of B by an angle α. Equally we could have taken atom B as the centre. There must be an atom (D) that swings round into position A under the same rotation. Clearly the line CD is parallel to AB. But we have said that the primitive vector in the AB direction has magnitude a. Hence the distance CD must be an integer multiple of a:

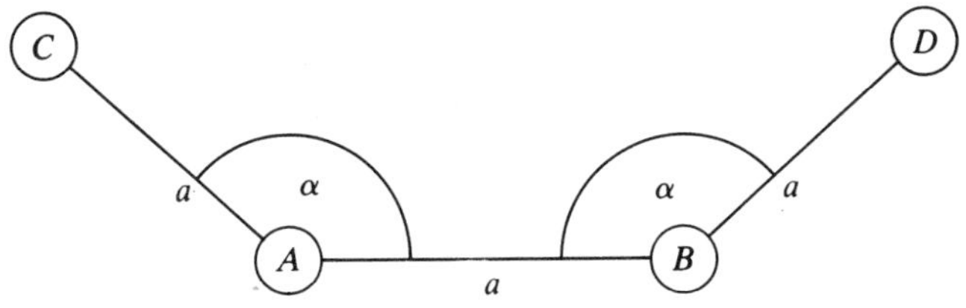

Figure 17.3

$$\mathrm{CD} = a + 2a \cos(\pi - \alpha) = ma,$$

for integer m. Thus

$$\cos \alpha = (1 - m)/2.$$

But cos α must lie between -1 and $+1$, so we have a restriction on m.

$$|(1-m)/2| \leqslant 1.$$

The only integers that obey this relation are $m = -1, 0, 2, 3$. The corresponding values of the angle α are respectively:

$$\alpha = 0, \frac{\pi}{3} \text{ or } \frac{5\pi}{3}, \frac{\pi}{2} \text{ or } \frac{3\pi}{2}, \frac{2\pi}{3} \text{ or } \frac{4\pi}{3}, \pi.$$

Thus the only rotational operations that can appear in crystal lattice point groups are

$$E, 2C_6, 2C_4, 2C_3, C_2.$$

Without proof, there are just 32 point groups that can be constructed without recourse to other rotational operations. These groups constitute the 32 crystallographic point groups, and are tabulated in Appendix A2.

The crystal classes

The entire point groups of all crystals are these same 32 point groups. If the entire point group is G_e, then the crystal is said to belong to the crystal class G_e. (This use of the word class is quite separate from the group-theoretical use of the word.)

The holohedral groups

We have looked at the ideas of crystal lattices from two viewpoints: that of the lattice periodicity—translational symmetry; and that of the point group properties. It is now appropriate to bring the two together by considering (i) the possible primitive cells and (ii) the lattices obtained by stacking such cells, asking what their point group symmetries are.

The symmetry of the primitive cell is defined by five parameters (only the relative values of a_1, a_2, a_3 are important). The most general cell has

$$a_1 \neq a_2 \neq a_3; \qquad \alpha_1 \neq \alpha_2 \neq \alpha_3.$$

This is the prescription for a cell that has no symmetry other than inversion. (Our method of definition of primitive cells requires that they have inversion symmetry.) The point group for this cell, called a triclinic cell, is S_2. As we begin to vary the a_i and α_i, letting two or three of the a_i become equal, for example, or equating the α_i and eventually letting them all become equal to $\pi/2$, then we can generate only seven symmetry types. The corresponding point groups are known as the holohedral groups. We have already noted that they must contain the inversion operation. In addition any axis of three-fold symmetry or greater (n-fold) must be accompanied by n mirror planes containing that axis. (This latter we shall not prove.) In Table 17.1 we list the holohedral groups and the primitive cells of each symmetry.

The crystal systems

When the primitive cells are stacked to form lattices these same seven point groups pertain. However, it does not automatically follow that a primitive cell of a specific symmetry will stack to form a lattice of the same symmetry—it is a consequence of our choice of definition of a cell. We talk of a lattice as belonging to one of the seven crystal systems, each system having a specific point group symmetry (Table 17.1, page 235).

17.2 The Bravais lattices

It is conventional in crystallography to talk of the Bravais lattice of a crystal rather than of the stacking of specific types of primitive cell. The reason is that the Bravais lattices, of which there are 14, are constructed so that they do display the full point group symmetry of the crystal. Some examples are useful in order to illuminate this statement.

Figure 17.4(a) shows a primitive cell with $a_1 \neq a_2 \neq a_3$, $\alpha_1 = \alpha_2 = \pi/2 \neq \alpha_3$. Such a cell has the point group symmetry C_{2h}, the symmetry operations are E, C_{2h}, i, σ_h. Figure (b) shows a second primitive cell, also of C_{2h} symmetry but of half the volume of the former. Finally Figure (c) is included to show that the former of the above C_{2h} *lattices* can be used as a basis from which to generate the latter. Specifically, if atoms are placed centrally on the a_1–a_3 faces of the primitive cell defined in (a), then the lattice it produces on stacking is exactly the same as the lattice produced by stacking cells of shape (b). We have two totally equivalent descriptions of the lattice, one in terms of a primitive cell containing one lattice point (the other seven corners may be thought of as the starting corners of seven adjacent cells) and one in terms of a non-primitive cell, containing two lattice points. This latter is called the 'base-centred unit cell of the monoclinic Bravais lattice'. It is conventional to use the notation a,b,c, α,β,γ for the dimensions and angles of Bravais lattices. In the absence of any centring, the unit cell of a Bravais lattice is identical to a primitive cell.

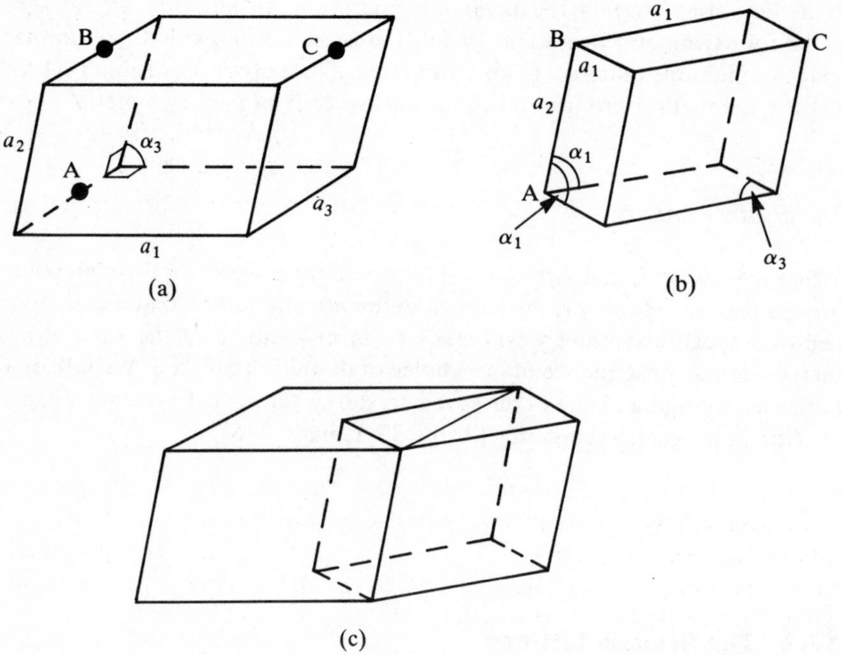

Figure 17.4 Demonstration that a primitive cell with $a_1 = a_3 \neq a_2$, $\alpha_1 = \alpha_2 \neq \alpha_3$ (cell (b)) stacks in the same way as a base-centred monoclinic cell (cell (c)). Cell (a) is monoclinic, AC is a C_2-axis and ABC defines the σ_h plane.

Whilst the two pictures of the lattice were equivalent in that the symmetries of the primitive cell and the Bravais unit cell were the same in the above example, this is not always so. Figure 17.5(a) shows a primitive cell of symmetry D_{3d}.

$$a_1 = a_2 = a_3, \qquad \alpha_1 = \alpha_2 = \alpha_3.$$

In the particular case that the angles are equal to $\pi/3$ the primitive cell symmetry is still D_{3d}, but the cells can easily be shown to stack to produce a lattice of cubic symmetry. Figure 17.5(b) shows a cubic cell that has additional atoms at the centre of each face. Starting from one corner the primitive cell of this lattice is constructed by drawing vectors to the nearest neighbours, the three close-by face-centrings. Completing the cell we see that it is the rhombohedral with $\alpha = \pi/3$. Again, then, we have two equivalent ways of constructing cells. However, the full symmetry of the lattice is now more clearly presented by the face-centred cubic unit cell of the Bravais lattice.

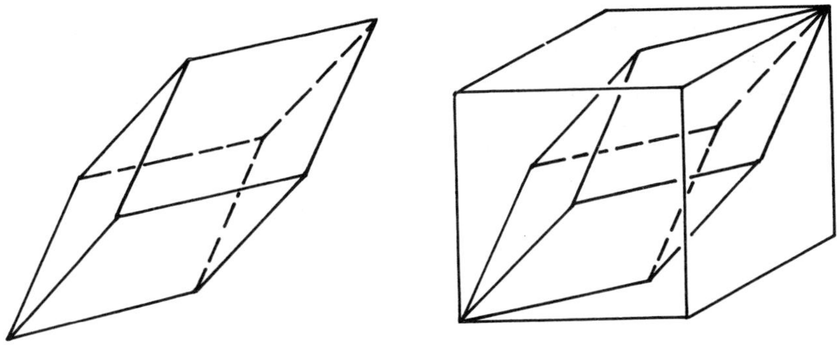

(a) (b)

Figure 17.5 (a) A rhombohedral primitive cell that stacks with cubic symmetry (b).

Again stating the result without proof, there are in total 14 Bravais lattices that can be constructed so as to display the symmetry types of all stacked primitive cells (Table 17.1). Various centrings are used, but in no case is the centring allowed to reduce the symmetry of the lattice. The 14 cells are shown in Figure 17.6.

If there is any doubt as to the difference in symmetry a lattice may have by comparison to its primitive cell it is worth reconsidering the two-dimensional case of Figure 17.1. The primitive cell (Figure 17.7(a)) is described by $a_1 \neq a_2$, $\alpha \neq \pi/2$ and it has a symmetry C_{2h}. By contrast we could consider the lattice to be constructed of body-centred rectangles (Figure 17.7(b)), in which case the full symmetry of the lattice is brought out clearly as D_{2h} $(E, C_2, C'_2, C''_2, i, \sigma_h, \sigma'_v, \sigma''_v)$.

17.3 The crystallographic point groups revisited

We are yet to see how crystals of point group symmetries such as C_{4v}, for which there is no crystal system, can occur. The answer lies in the fact that so far we have basically considered the stacking of only primitive cells. We have not allowed for any structure to be associated with the lattice points so obtained. Clearly, when any structure is placed at each lattice point we run the risk of reducing the symmetry of the crystal. In the extreme case an arbitrary structure will remove all the point group operations and we will be left with only the lattice translational group. Less dramatically a pair of distinct atoms (placed at non-special positions) in the unit cell will remove

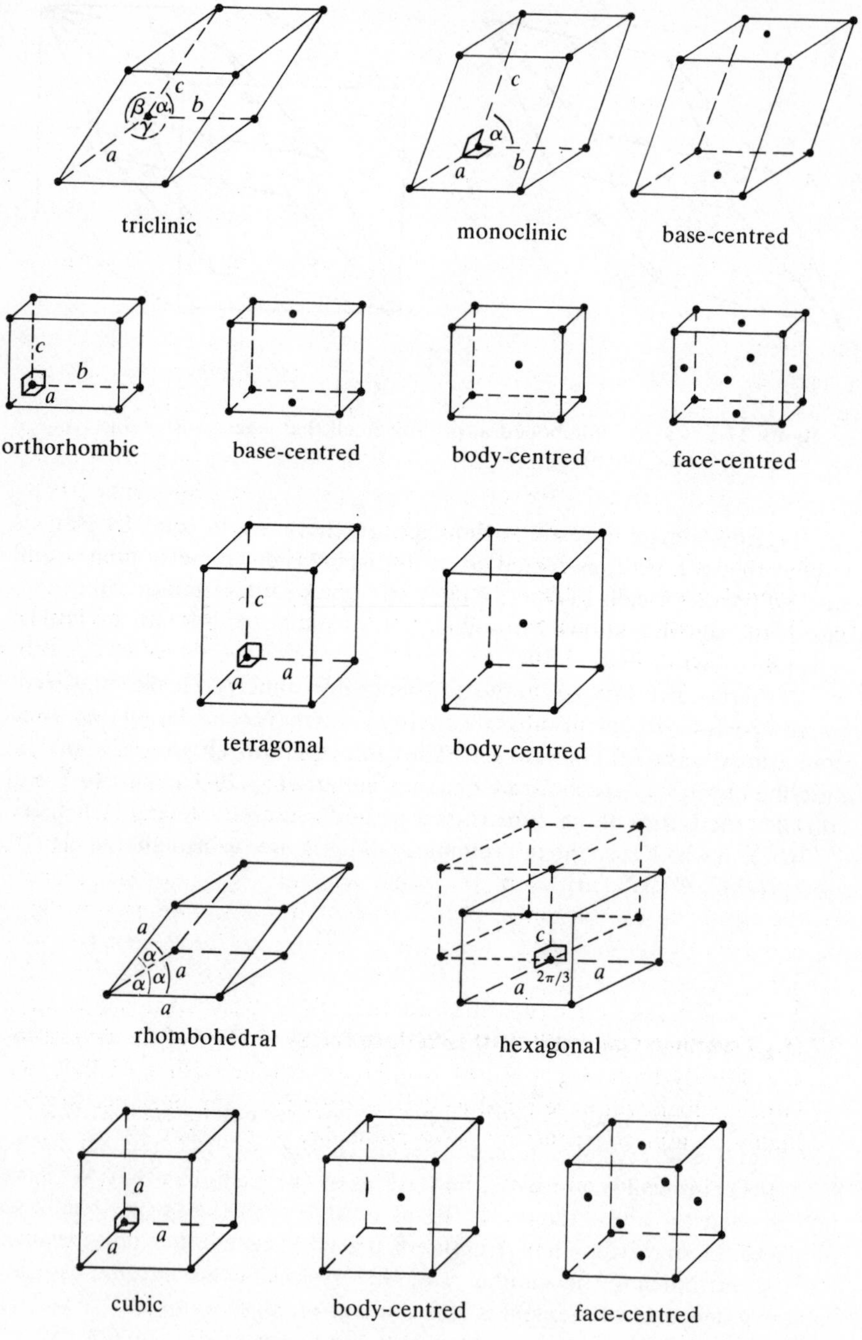

Figure 17.6 The Bravais lattices.

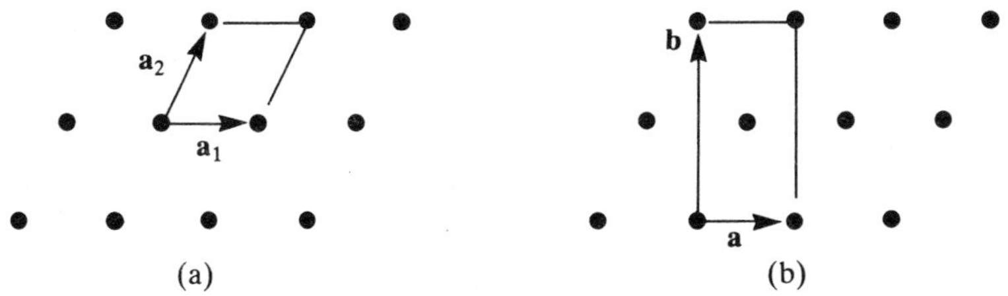

Figure 17.7 (a) Primitive cell and (b) the Bravais unit cell for a 2-D lattice.

the inversion symmetry of the lattice. Figure 17.8 shows how different structures (collections of atoms) placed at the D_{4h} lattice positions, gradually reduce the symmetry of this lattice. There are two possible effects:

(i) The replacement of a single atom on a D_{4h} lattice by the pair of atoms indicated in Figure 17.8(b) changes the overall point group symmetry to C_{4v}. There is no other Bravais lattice of this symmetry (or of symmetry between C_{4v} and D_{4h}) and the resulting crystal will maintain its tetragonal lattice, with the structure  at each lattice point.

(ii) On the further reduction of symmetry, by placing  at the lattice points, Figure 17.8(c), the point group symmetry is reduced to C_{2v}. But now there is a less symmetric lattice, the orthorhombic lattice of D_{2h} symmetry, that can also have its symmetry reduced to C_{2v} by imposing some structure at each lattice point. The appropriate structure will in this case be exactly the same as that we have considered, namely . The practical result of trying to impose such a structure on the D_{4h} lattice is that the forces so introduced *must* cause the lattice itself to change, if only by a very small amount. The lattice symmetry (as opposed to that of the crystal as a whole) must change to D_{2h}, so that we are now dealing with an orthorhombic lattice, plus a structure that reduces the symmetry a little further—to C_{2v}. By similar token, the imposition of the structure  on the D_{4h} lattice (d) will cause the lattice to distort to monoclinic (C_{2h}) symmetry and still further reduce the symmetry of the crystal to C_{1h}. Finally imposing  (e) is only compatible with a triclinic, S_2 lattice, which will have a reduced C_1 symmetry in the presence of this highly asymmetric structure. Two points are to be made then:

(i) The imposition of structure reduces the seven crystal system symmetries to the 32 crystallographic point group symmetries.

(ii) Each of the 32 point group symmetries can be associated with only one lattice. For example, although C_1 is a subgroup of all the lattice point groups, a crystal of C_1 symmetry necessarily has an S_2 lattice because of the forces associated with the structure in each cell.

Table 17.1 summarizes these statements.

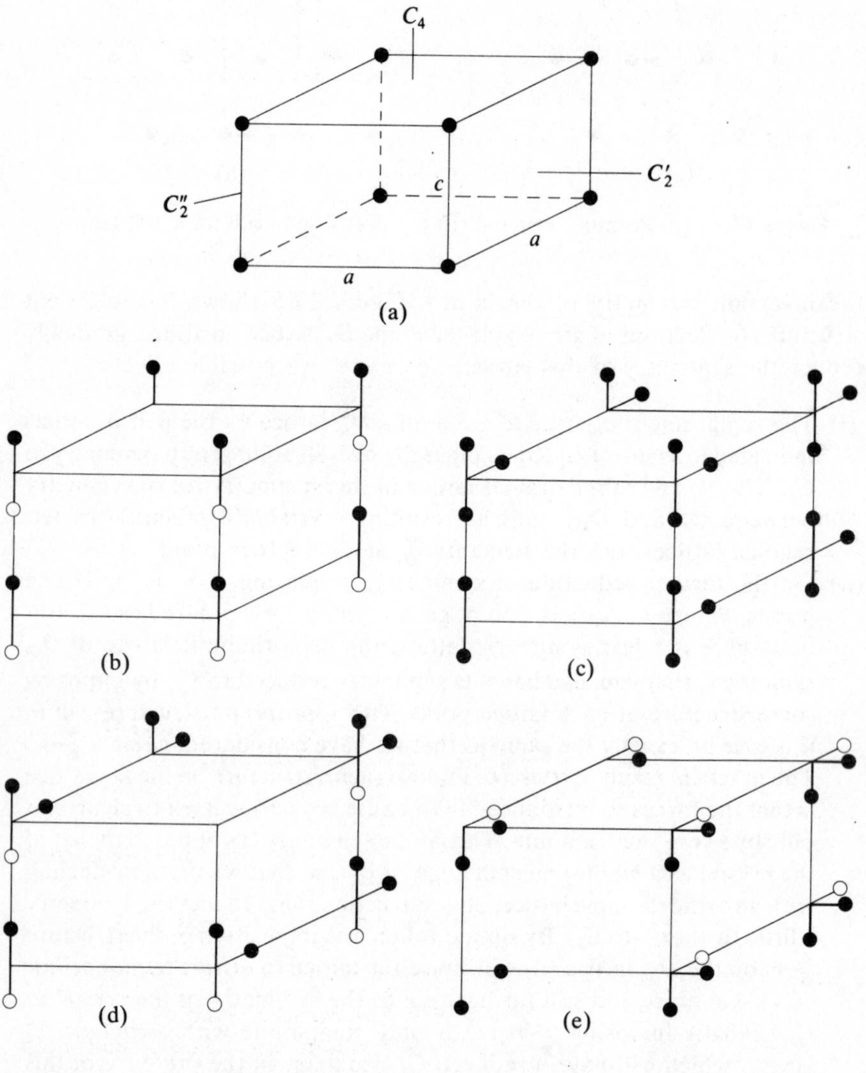

Figure 17.8 The D_{4h} lattice cell (a), indicating typical rotation axes. The symmetry group contains also inversion and reflection planes normal to the above axes. (b)–(e) Reduction of the lattice symmetry by the imposition of structure on each lattice site. (b) C_{4v} (tetragonal); (c) C_{2v} (orthorhombic); (d) C_{1h} (monoclinic); (e) C_1 (triclinic).

Table 17.1 Classification of the crystal symmetry types.

Examples of primitive cell parameters	*Holohedral point group*	*Crystal system*	*Bravais lattices*	*Point groups of Bravais lattices with structure*
$a_1 \neq a_2 \neq a_3$ $\alpha_1 \neq \alpha_2 \neq \alpha_3$	S_2	Triclinic	Simple	C_1
$a_1 \neq a_2 \neq a_3$ $\alpha_1 \neq \alpha_2 = \alpha_3 = \frac{\pi}{2}$	C_{2h}	Monoclinic	Simple, base-centred	C_{1h}, C_2
$a_1 \neq a_2 \neq a_3$ $\alpha_1 = \alpha_2 = \alpha_3 = \frac{\pi}{2}$	D_{2h}	Orthorhombic	Simple, base-centred, body-centred, face-centred	C_{2v}, D_2
$a_1 = a_2 \neq a_3$ $\alpha_1 = \alpha_2 = \alpha_3 = \frac{\pi}{2}$	D_{4h}	Tetragonal	Simple, body-centred	C_4, C_{4v}, C_{4h} S_4, D_4, D_{2d}
$a_1 = a_2 = a_3$ $\alpha_1 = \alpha_2 \equiv \alpha_3 \neq \frac{\pi}{2}$	D_{3d}	Rhombohedral	Simple (R)	C_3, S_6, C_{3v}, D_3
$a_1 = a_2 \neq a_3$ $\alpha_1 = \alpha_2 = \frac{\pi}{2}$ $\alpha_3 = \frac{2\pi}{3}$	D_{6h}	Hexagonal	Simple	C_6, C_{3h}, C_{6h}, C_{6v}, D_6, D_{3h}
$a_1 = a_2 = a_3$ $\alpha_1 = \alpha_2 = \alpha_3 = \frac{\pi}{2}$	O_h	Cubic	Simple, face-centred, body-centred	T, T_h, T_d, O

Note that the centred lattices can also be thought of as primitive lattices with structure at each lattice point (for example with identical atoms at **O** and $(\frac{1}{2}\mathbf{a}+\frac{1}{2}\mathbf{b})$ in the case of the base-centred monoclinic lattice). These, however, are just those structures for which the symmetry is not reduced at all, and for which each 'atom' still has an identical environment.

17.4 Space groups

The symmorphic space groups

Just as structure placed on the simple (primitive) Bravais lattices leads to crystals of reduced symmetry, so will identical structures placed on the centred Bravais lattices lead to the same reductions. In the monoclinic system, for example, there are a total of six different crystal types that can be attained; a simple lattice with C_{2h}, C_{1h} or C_2 symmetry or a base-centred lattice with C_{2h}, C_{1h} or C_2 symmetry. Running through all the systems in Table 17.1, one can generate 61 crystal symmetry types in this way. In addition there are 12 types that are obtained somewhat more subtly. Firstly, on reduction of a hexagonal lattice beyond D_{3h} symmetry, one distorts to a centred rhombohedral lattice rather than a simple one. This gives five extra crystal types. Secondly a further seven types are obtained if careful consideration is made of the orientations of the principal point group axis of the structure compared to that of the lattice. The reader is referred to texts on crystal space groups for further study of these finer points. The resulting crystal symmetry types are known as the 73 symmorphic space groups. Their specification requires the knowledge of (i) the point group (which implies a crystal system) and (ii) the lattice centring.

Non-symmorphic space groups

The symmorphic space groups are not, unfortunately, the only symmetry groups that can be associated with periodic structures. We have yet to consider operations of the type $\{R'/\mathbf{t}\}$, where $\mathbf{t}(=\mathbf{T}+\boldsymbol{\tau})$ is not a lattice translation. Examples of point group operations involving non-primitive lattice translations are discussed below in connection with the diamond structure. This is an important structure for us to consider because it characterizes the elemental semiconductor crystals, germanium and silicon. There are 157 different crystal structures that can be generated using $\{R'/\boldsymbol{\tau}\}$ translations; they are called the non-symmorphic space groups. In total, then, there are 230 space groups.

Space group nomenclature

For a symmorphic space group a sufficient definition is given by a statement of (i) the crystallographic group (the crystal class) and (ii) the centring of the Bravais lattice. For non-symmorphic groups the non-primitive translations associated with each point-group operation are also required. There are three standard terminologies for the 230 space symmetry groups; the Schoenflies, international full symbols and international short symbols. Tables of group properties (Burns and Glazer (1978) Kovalev (1965); see Bibliography, Section (d)) are ordered by crystal class. In the Schoenflies notation the different groups with the same entire point group are simply numbered 1, 2... Thus space group number 225 is labelled O_h^5 in Schoenflies notation, being the fifth tabulated group of the O_h crystal class (out of a total of ten structures having this entire point group).

The international full notation tells us more about the structure. For the O_h^5 group the symbol is

$$F\frac{4}{m}\bar{3}\frac{2}{m}.$$

The F tells us that the lattice is face-centred. The rest of the symbol tells us the point group symmetry, $4/m$ indicates a four-fold axis with a perpendicular mirror plane, $\bar{3}$ indicates a three-fold axis with additional S_6 operations. The full symbol expresses the face-centred cubic group. The international short symbol $Fm3m$ contains the same information in abbreviated (and therefore less clear) form.

17.5 Example: the diamond structure

The diamond structure may be thought of as a pair of interpenetrating, face-centred cubic lattices, or as a single face-centred lattice with a structure of two carbon atoms per primitive cell. If one of the two atoms is set at (0,0,0), then the second lies at $(\frac{1}{4},\frac{1}{4},\frac{1}{4})$, on the scale of the Bravais lattice (see Figure 17.9, caption). There are then eight atoms per unit cell of the Bravais lattice. The positions of any atoms may be written:

$$\begin{aligned}\mathbf{A}=&(n_1,n_2,n_3),\quad (n_1+\tfrac{1}{4},n_2+\tfrac{1}{4},n_3+\tfrac{1}{4}),\\ &(n_1+\tfrac{1}{2},n_2+\tfrac{1}{2},n_3),(n_1+\tfrac{3}{4},n_2+\tfrac{3}{4},n_3+\tfrac{1}{4}),\\ &(n_1+\tfrac{1}{2},n_2,n_3+\tfrac{1}{2}),(n_1+\tfrac{3}{4},n_2+\tfrac{1}{4},n_3+\tfrac{3}{4}),\\ &(n_1,n_2+\tfrac{1}{2},n_3+\tfrac{1}{2}),(n_1+\tfrac{1}{4},n_2+\tfrac{3}{4},n_3+\tfrac{3}{4}),\end{aligned}$$

where n_1, n_2, n_3 take any integer values, positive, negative or zero; see Figure 17.9. Because there is no atom at the body centre of the unit cell, or at $(\frac{3}{4},\frac{3}{4},\frac{3}{4})$, it is clear that the two atoms in the primitive cell do not have identical positions (we are not simply dealing with a new lattice). Taking (0,0,0) as our point origin, the lattice has T_d symmetry, with group elements E, $8C_3$, $3C_2$, $6S_4$, $6\sigma_d$. Although the face-centred cubic lattice has O_h symmetry, the presence of structure at each point has, apparently, reduced this symmetry. For example, under a C_4 rotation an atom at $(\frac{1}{4},\frac{1}{4},\frac{1}{4})$ swings around to $(-\frac{1}{4},\frac{1}{4},\frac{1}{4})$, a combination that does not appear in any of the **A** points. However, by inspection of other centres in the unit cell, different symmetry operations can be found.

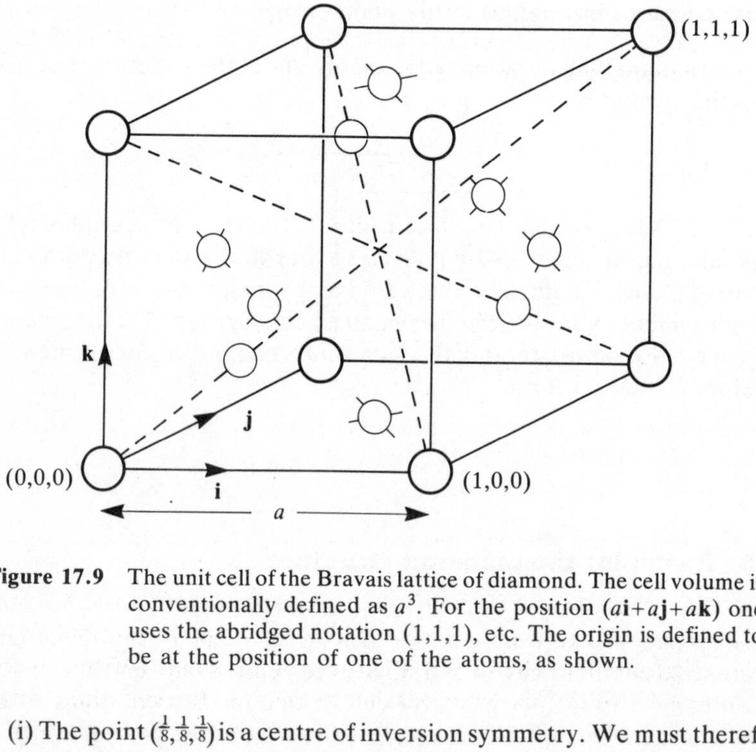

Figure 17.9 The unit cell of the Bravais lattice of diamond. The cell volume is conventionally defined as a^3. For the position $(a\mathbf{i}+a\mathbf{j}+a\mathbf{k})$ one uses the abridged notation (1,1,1), etc. The origin is defined to be at the position of one of the atoms, as shown.

(i) The point $(\frac{1}{8},\frac{1}{8},\frac{1}{8})$ is a centre of inversion symmetry. We must therefore expect to include the operation i in some way in the diamond space group. The combined operation $\{i/\langle\frac{1}{4},\frac{1}{4},\frac{1}{4}\rangle\}$, meaning inversion about (0,0,0) followed by a non-primitive translation by $\langle\frac{1}{4},\frac{1}{4},\frac{1}{4}\rangle^{\dagger}$ has exactly the same effect as inversion about $(\frac{1}{8},\frac{1}{8},\frac{1}{8})$ and is used to describe inversion symmetry in the diamond space group.

†Note that we have used $\langle\frac{1}{4},\frac{1}{4},\frac{1}{4}\rangle$ to denote a translation by $a(\mathbf{i}/4+\mathbf{j}/4+\mathbf{k}/4)$ etc.

(ii) *Glide planes*. The plane parallel to the j–k plane but passing through $(\frac{1}{8},\frac{1}{8},\frac{1}{8})$ (see Figure 17.10) is a *glide plane*. Known as the diamond glide, the lattice can be transformed into itself by reflection in the plane followed by a translation of $\langle 0,\frac{1}{4},\frac{1}{4}\rangle$. Again there is an alternative description of such an operation, in terms of $\{R'/\boldsymbol{\tau}\}$; namely a reflection in the j–k plane, followed by a translation by $\langle\frac{1}{4},\frac{1}{4},\frac{1}{4}\rangle$.

In the glide plane description the atom originally at $(n_1+\frac{1}{4},n_2+\frac{1}{4},n_3+\frac{1}{4})$, for example, reflects to $(-n_1,n_2+\frac{1}{4},n_3+\frac{1}{4})$, then translates to $(-n_1,n_2+\frac{1}{2},n_3+\frac{1}{2})$, which is a face centre of the original lattice. The second description of the operation reflects the atom to $(-n_1-\frac{1}{4},n_2+\frac{1}{4},n_3+\frac{1}{4})$, then translates it to $(-n_1,n_2+\frac{1}{2},n_3+\frac{1}{2})$, with the same effect. The term glide plane is used for any operation involving a non-primitive translation *in the plane* of reflection and can thus only be used for the first description of the operation. In general a glide plane must involve a non-primitive translation $\boldsymbol{\tau}_g$ such that $2\boldsymbol{\tau}_g$ is a lattice vector, as performing the operation a second time has the effect of cancelling the reflection.

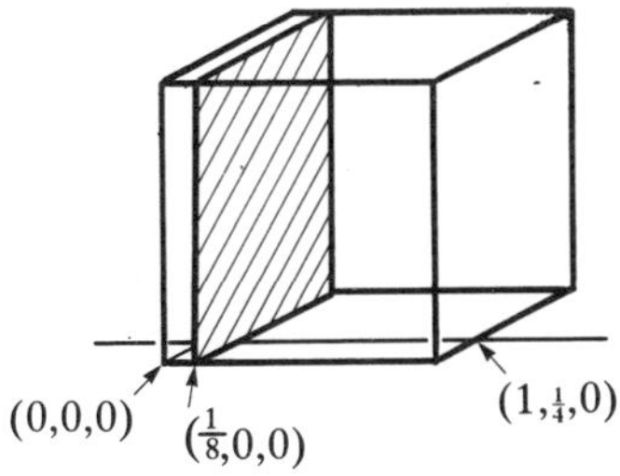

Figure 17.10 Schematic of a diamond glide plane and a C_4 screw axis perpendicular to it.

(iii) *Screw axes*. The final type of non-primitive space group operation is associated with a *screw axis*. In diamond the axis indicated in Figure 17.10, which again does not pass through (0,0,0), has the property that a C_4 rotation about it, followed by a translation by $\langle\frac{1}{4},0,0\rangle$, transforms the lattice into itself. Such an operation is called a screw axis operation and is defined wherever the non-primitive translation is parallel to the axis. In general the rotation angle must be a rational fraction of 2π, such that a series of screw-axis operations effects a pure translation. In the diamond example a rotation by C_4 about the **i** axis, followed by the translation $\langle\frac{1}{4},\frac{1}{4},\frac{1}{4}\rangle$, which we note is not parallel to the axis, effects the same result as the screw-axis description. In summary the full space group of the diamond structure consists of the lattice translations $\{E/\mathbf{T}\}$ combined with the 48 operations:

E, $8C_3$, $3C_2$, $6S_4$, $6\sigma_d$ in the form $\{R/\mathbf{O}\}$

and

$$i,\ 8S_6,\ 3\sigma_v,\ 6C_4,\ 6C'_2 \qquad \text{in the form } \{R'/\boldsymbol{\tau}\}$$

where $\boldsymbol{\tau}=a(\mathbf{i}+\mathbf{j}+\mathbf{k})/4$ for all R'.

The *entire point group* of the crystal is therefore the O_h symmetry group. The space group is labelled O_h^7 (number 227) or, in international notation,

$$F\frac{4_1}{d}\bar{3}\frac{2}{m}.$$

The subscript on the 4 indicates the presence of 4-fold screw axes, and the d indicates the diamond-glide planes perpendicular to such axes.

17.6 Magnetic groups

In all of the symmetry groups considered so far, whether for solids, molecules or atoms, the potential function has been used to define the symmetry. In turn the potential has been considered to be due to the electrostatic Coulomb interaction between the charged particles in the system. Group operations that leave the geometric structure of the material invariant do not change the charge distribution (the time-averaged charge density). However, in ferromagnetic and antiferromagnetic crystals the time-averaged distribution of the current density, which gives rise to magnetic interactions, may be reversed under some operations that do not change the charge density. Furthermore a completely new operation (M) may be introduced, which reverses the current density without effecting any spatial change in the material geometry. The combination of one of those spatial operations which reverse the current with the operation M is a symmetry operation of the material group. By the introduction of the M operation one obtains an increase in the total number of space groups to 1651, involving 122 different point groups. Those which must be used for magnetic materials are also known as the black-and-white groups, because the operation M has the simple property of changing a current or magnetic moment to one of the opposite sign (black-to-white or vice versa). By colouring parts of a lattice black or white one obtains the geometric coloured structures that have the magnetic space group symmetries.

Problems

***17.1** Verify the equality $\{R'/\mathbf{t}\}=\{E/\mathbf{T}\}\{R'/\boldsymbol{\tau}\}$, where $\mathbf{t}=\mathbf{T}+\boldsymbol{\tau}$.

***17.2** Determine the number of nearest neighbour atoms and the nearest neighbour distance for (a) the simple primitive cubic lattice, (b) a face-centred cubic lattice and (c) a body-centred cubic lattice.

17.3 Determine the symmetry of a simple orthorhombic lattice containing a pair of atoms of the structure  at each lattice site. Will the lattice distort due to the forces involved in the pair structure? Repeat for the structure  .

17.4 Obtain the symmetry group operations for the two planar objects shown in Figure 17.11, including spatial operations G_r, the black-to-white reversal operation M, and combinations MG_r.

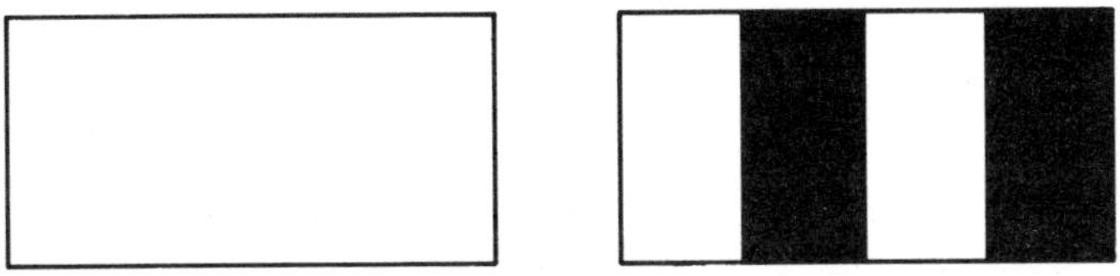

Figure 17.11

17.5 Face-centred and base-centred tetragonal lattices do not appear in the list of Bravais lattices (Table 17.1). Show that the former is in effect present because it is equivalent to the body-centred tetragonal lattice, and find an equivalent to the base-centred lattice.

18

The symmetry of crystal states

The usefulness of group theory in quantum mechanics is manifested by the fact that eigenfunctions (or eigenvectors) of a symmetric system must relate to the symmetry. In our group-theoretical language, they belong to the irreducible representations of the symmetry group in question. This statement is just as relevant to the solid-state case as it was in our consideration of atoms and molecules. The crystal space groups, however, are rather more complicated than the atomic or molecular point groups. It is the purpose of this chapter to break down the properties of the space group and show that the eigenfunctions of a crystal system can be thought of on the one hand as belonging to the irreps of the translational group of the crystal, and on the other hand as belonging (at the same time) to a particular point group. The operations in this point group will be discussed; having found them we will be in a position to label the electronic states (for example) by the conventional solid-state crystal wavevector, as well as by a point group irrep. label. From the latter labelling, selection rules can be derived as in the atomic and molecular cases.

18.1 Subgroups of the space group

There is an important group-theoretical result that we shall need to have in mind when separating the translational and point group symmetries of a crystal. It is a result that we have in effect already used but have never stated explicitly. We know that the eigenfunctions of any system must be symmetry-adapted functions belonging to the irreps of the group of that system. We also know that when the symmetry is lowered and the group reduces to one of its subgroups, then these same symmetry-adapted functions form bases for representations of the subgroup, and that these representations may or may not be irreducible.

$$\Gamma^j(\text{group}) = \sum_{j'} a_{j'} \Gamma^{j'}(\text{subgroup}).$$

Given a set of basis functions (partner functions) belonging to Γ^j of the group, it is always possible however to make linear combinations of these functions alone, that form bases for those particular subgroup irreps $\Gamma^{j'}$ in the sum $\Sigma a_{j'}\Gamma^{j'}$. These orthogonal linear combinations must of course still be a basis for Γ^j (they still transform amongst themselves only). One can conclude therefore that

(i) symmetry-adapted functions belonging to the irreps of a group can be *chosen* such that they also belong to irreps of any one subgroup of that group;
(ii) eigenfunctions (eigenvectors) for any symmetric problem can be *chosen* such that they are bases for the irreps of any one subgroup of the symmetry of the problem.

We have in fact used precisely the latter result in our discussion of the eigenfunctions of the free atom system. Consideration of the full rotation group told us that eigenfunctions were bases for the $(2l+1)$-fold degenerate irreps of $R_3 \otimes S_2$. Given a set of $(2l+1)$ degenerate eigenfunctions, we are at liberty to construct *any* $(2l+1)$ orthogonal combinations of these functions for use in some problem. The most popular choice is the set ψ_{nlm}, each member being chosen to be a function of specific m-quantum number; that is the set is *chosen* to be a symmetry-adapted basis of the axial group $C_{\infty v}$, a subgroup of $R_3 \otimes S_2$. The p-orbitals used in certain molecular physics problems provide an alternative choice, as far as the $l=1$ states are concerned. They happen to be symmetry-adapted functions of any of the point group subgroups of $R_3 \otimes S_2$, providing the point group axes/axis are defined in the $x,y/z$ directions.

For the present, though, we are concerned with the space group problem. It is sensible to concentrate on symmorphic space groups and to reserve comments on non-symmorphic crystals until we have completed the treatment of the simpler case. We will need to consider not one but two subgroups of the full space group, $\mathcal{S}$.

The operations in a symmorphic space group, $\mathcal{S}$, may be written

$$\{R/\mathbf{T}\} = \{E/\mathbf{T}\}\{R/\mathbf{O}\},$$

where $\{R/\mathbf{O}\}$ is an operation in the crystal point group, $\mathcal{P}$. Suppose we consider any subgroup of this point group, $\mathcal{P}'$, with operations $\{R'/\mathbf{O}\}$. Then a subgroup of $\mathcal{S}$ can be constructed by combining these operations with the translational group operations, to form $\{R'/\mathbf{T}\}$. We can check that these do form a group. A product of two such operations has the effect

$$\{R'_2/\mathbf{T}_2\}\{R'_1/\mathbf{T}_1\}\mathbf{r}=\{R'_2/\mathbf{T}_2\}(R'_1\mathbf{r}+\mathbf{T}_1)=(R'_2R'_1)\mathbf{r}+(R'_2\mathbf{T}_1+\mathbf{T}_2)$$
$$=\{R'_3/\mathbf{T}_3\}\mathbf{r},\text{ say.}$$

$\{R'_3/\mathbf{T}_3\}$ defined with $R'_3 \equiv R'_2R'_1$ and $\mathbf{T}_3 \equiv R'_2\mathbf{T}_1+\mathbf{T}_2$ is another member of the subgroup, so that closure is obeyed. Also the inverse of $\{R'_1/\mathbf{T}_1\}$ is $\{(R'_1)^{-1}/-(R'_1)^{-1}\mathbf{T}_1\}$, which is again in the subgroup. We shall call this subgroup $\mathcal{S}'$. A second subgroup, not only of $\mathcal{S}$ but also of $\mathcal{S}'$, is the translational group itself. This hierarchy of subgroups is expressed by Table 18.1.

Table 18.1

Space group →	*space sub-group* →	*translational group*
$\mathcal{S}$	$\mathcal{S}'$	$\mathcal{T}$
$\{R/\mathbf{T}\}$	$\{R'/\mathbf{T}\}$	$\{E/\mathbf{T}\}$
$F^{\mathcal{S}} \in \Gamma^{\mathcal{S}}$	$F^{\mathcal{S}'} \in \Gamma^{\mathcal{S}'}$	$F^{\mathcal{T}} \in \Gamma^{\mathcal{T}}$

The eigenfunctions of the full space group problem in general form bases for the irreps $\Gamma^{\mathcal{S}}$. Either these same eigenfunctions, or suitable combinations of the degenerate eigenfunctions, form bases for the irreps $\Gamma^{\mathcal{S}'}$ of the space subgroup $\mathcal{S}'$. In turn these, or suitable combinations of partner functions, form bases for the irreps $\Gamma^{\mathcal{T}}$.

It is appropriate to consider first the symmetry-adapted bases for the translational group before looking at the space subgroup problem. In Section 18.4 we will discuss particular choices of $\mathcal{S}'$ for which the symmetry-adapted functions can be chosen such as to be the bases for the irreps of both the translation group and for a point group—'the group of the wavevector **k**'.

18.2 The translational group $\mathcal{T}$

Cyclic boundary conditions

Most efficient use of the translational symmetry of a crystal is made if we slightly modify our model for a real crystal. So far a crystal structure has been assumed to be an infinite, perfect periodic structure. In practice of course a real crystal is of finite dimensions and strictly speaking does not have any translational symmetry. However, typically a crystal sample contains perhaps 10^{22} or more atoms. By far the majority of these atoms lies within the interior of the crystal, relatively far from the surface. Thus provided one is interested in the properties of the bulk of the material the near-surface atoms should be of little consequence. In turn the mathematical model one uses to describe the crystal should not be too sensitive to the boundary conditions imposed at the surface. The new picture of a crystal we wish to introduce here involves *cyclic boundary conditions*. In one dimension cyclic boundary conditions imposed on a translational problem can be visualized by taking the finite-periodic structure and bending it around to join one end with the other (Figure 18.1(a)). Translation by the full finite extent of the

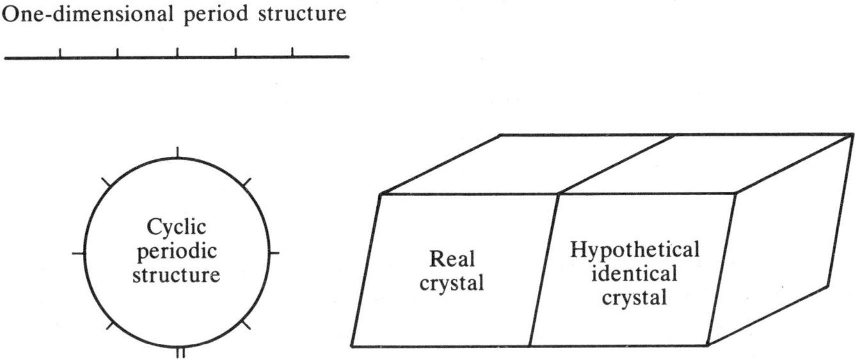

(a) (b)

Figure 18.1 Visualization of cyclic boundary conditions in one and three dimensions.

structure therefore brings it to an identical position. In three dimensions this mathematical condition can be visualized by considering crystals stacked throughout all three-dimensional space such that the real crystal is reproduced in every respect on a full translation. This cyclic boundary condition is known in solid-state theory as the Born–von Karman periodic boundary condition.

The operations of translation by **T** in the (cyclic) translational group $\mathcal{T}$ may be written as

$$E;T_1,T_1^2\ldots T_1^{N_1-1};\ T_2,T_2^2\ldots,T_2^{N_2-1};T_3,T_3^2\ldots T_3^{N_3-1};T_1T_2,\ldots.$$

Here T_1 denotes a translation by the primitive cell vector $\mathbf{a}_1$, etc. The real crystal is supposed to consist of a total of $N_1N_2N_3$ unit cells ($=N$) such that its dimension in the $\mathbf{a}_1$ direction is $N_1|\mathbf{a}_1|$ etc. The series of $\mathbf{a}_1$ translations ceases at $T_1^{N_1-1}$ because of the cyclic condition

$$T_1^{N_1} \equiv E.$$

Now any two translations commute, so their order is irrelevant. In particular any two translations in different directions commute. Hence $\mathcal{T}$ can be considered as the direct product of three groups, one for each primitive vector direction,

$$\mathcal{T} = \mathcal{T}_1 \otimes \mathcal{T}_2 \otimes \mathcal{T}_3.$$

The properties of $\mathcal{T}$ can therefore be obtained by considering just one of these groups. In particular we wish to find the irreps and their symmetry-adapted functions. This requires us to build up the character table of a cyclic group.

The cyclic group $\mathcal{T}_1$

The group $\mathcal{T}_1$ consists of the N_1 elements E, T_1, $T_1^2, \ldots, T_1^{N_1-1}$.

(i) Because all these elements commute we have

$$G_s^{-1}\, G_r G_s = G_s^{-1}\, (G_s G_r) = G_r.$$

Hence each element G_r is in a class of its own. There are thus N_1 classes in the group and N_1 irreps, all of which must be one-dimensional.

(ii) Because $T_1^2 \equiv T_1 T_1$, matrix representations of these operations must obey $\Gamma(T_1^2) = \Gamma(T_1)\Gamma(T_1)$. But the characters of the irreducible representations are just the irreps themselves for one-dimensional matrix irreps. Hence

$$\chi^{j_1}(T_1^2) = [\chi^{j_1}(T_1)]^2.$$

(iii) Extending the above statement to power N_1 and remembering that $T_1^{N_1}$ is the identity and therefore has a unit matrix irrep.,

$$\chi^{j_1}(T_1^{N_1}) = [\chi^{j_1}(T_1)]^{N_1} = 1.$$

Hence the characters of the irreps for the operation T_1 of an N_1-fold cyclic group are the N_1 roots of unity. With j_1 an integer we can write

$$\chi^{j_1}(T_1) = \exp\{-2\pi i j_1/N_1\} = \omega^{j_1}, \text{ say.}$$

For the time being we will choose the roots given by $j_1 = 0$ to $N_1 - 1$ but note that any N_1 consecutive j_1-values could be used. Table 18.2 shows the resulting cyclic group characters.

In general the character for the operation $T_1^{m_1}$ for the j_1th irrep is

$$\chi^{j_1}(T_1^{m_1}) = \omega^{m_1 j_1} = \exp\{-2\pi i \frac{m_1 j_1}{N_1}\}$$

Table 18.2 Character table for the cyclic group of order N_1.

	E	T_1	T_1^2	$\cdots$	$T_1^{N_1-1}$	
Γ^0	1	1	1	$\cdots$	1	
Γ^1	1	ω	ω^2		ω^{N_1-1}	
Γ^2	1	$(\omega)^2$	$(\omega^2)^2$		$(\omega^2)^{N_1-1}$	$\omega=\exp(\frac{-2\pi i}{N_1})$
.	.	.	.		.	
.	.	.	.		.	
.	.	.	.		.	
Γ^{N_1-1}	1	ω^{N_1-1}	$(\omega^{N_1-1})^2$		$(\omega^{N_1-1})^{N_1-1}$	

The **k** *label*

The irreps of the three-dimensional cyclic group $\mathcal{T}$, obtained using our direct-product group results, must be

$$\Gamma^{\mathcal{T}}=\Gamma^{j_1}\otimes\Gamma^{j_2}\otimes\Gamma^{j_3},$$

all of which are one-dimensional.

For a particular translation

$$\chi^{(j_1\otimes j_2\otimes j_3)}(\mathbf{T})=\exp\left\{-2\pi i\left(\frac{m_1j_1}{N_1}+\frac{m_2j_2}{N_2}+\frac{m_3j_3}{N_3}\right)\right\},$$

where

$$\mathbf{T}\equiv m_1\mathbf{a}_1+m_2\mathbf{a}_2+m_3\mathbf{a}_3,$$

and the symmetry-adapted functions belonging to such an irrep. must transform into a constant (χ) times themselves. j_1, j_2, j_3 label the irreps of the group $\mathcal{T}$. A simpler notation is obtained however if we define a vector from the values of these j integers.

We define a vector **k** such that

$$\chi^{j_1\otimes j_2\otimes j_3}=\exp\,(-i\mathbf{k}.\mathbf{T}).$$

In turn this demands that

$$\mathbf{k}.(m_1\mathbf{a}_1+m_2\mathbf{a}_2+m_3\mathbf{a}_3)=2\pi\;\frac{m_1j_1}{N_1}+\frac{m_2j_2}{N_2}+\frac{m_3j_3}{N_3}\quad.$$

Now consider **k** to have components $j_i\mathbf{b}_i/N_i$. These $\mathbf{b}_i$ vectors must obey the relation

$$\mathbf{b}_i.\mathbf{a}_j=2\pi\delta_{ij}.$$

They will do so if

$$\mathbf{b}_1 = 2\pi \frac{\mathbf{a}_2 \wedge \mathbf{a}_3}{\mathbf{a}_1 . (\mathbf{a}_2 \wedge \mathbf{a}_3)} \text{ etc.}$$

The irreps of $\mathcal{T}$ can now be labelled $\Gamma^{\mathbf{k}}$, where **k** can take the specific values obtained for $j_1 = 0$ to $N_1 - 1$, etc.

$$\mathbf{k} = \frac{j_1}{N_1}\mathbf{b}_1 + \frac{j_2}{N_2}\mathbf{b}_2 + \frac{j_3}{N_3}\mathbf{b}_3.$$

The symmetry-adapted functions $F^{\mathbf{k}}(\mathbf{r})$ obey the transformation

$$\{E/\mathbf{T}\}F^{\mathbf{k}}(\mathbf{r}) = F^{\mathbf{k}}(\mathbf{r}) \exp(-i\mathbf{k}.\mathbf{T}).$$

The vector **k** and its components $\mathbf{b}_1, \mathbf{b}_2, \mathbf{b}_3$ have the following interpretation.

(i) ***k**-space.* As the primitive translations $\mathbf{a}_i$ are non-coplanar, then the vectors $\mathbf{b}_j$ will also be non-coplanar. They can be thought of as the primitive vectors for the 'space of **k**' and a lattice can be constructed in this space by analogy to the real-space lattice. The lattice points can be denoted (for integer n_i) as

$$\mathbf{K} = n_1\mathbf{b}_1 + n_2\mathbf{b}_2 + n_3\mathbf{b}_3.$$

(ii) A primitive cell of this **k** space will contain within it $N_1N_2N_3$ points of vector positions $(j_1\mathbf{b}_1/N_1 + j_2/N_2\mathbf{b}_2 + j_3/N_3\mathbf{b}_3)$. These N discrete **k** values are the labels of the irreps of $\mathcal{T}$.

(iii) Just as in the real lattice we can define cells with the same property as the primitive cell in a number of ways. If it is to contain N distinct **k** values, however, then no two positions in the cell should be separated by a vector greater than or equal to one of the **K** vectors. If this restriction is not obeyed, then both $\mathbf{k}_1$ and $\mathbf{k}_1 + \mathbf{K}$ will be in the cell for some $\mathbf{k}_1$, **K**. In this case the characters for the two irreps associated with these **k** values are

$$\exp[-i\mathbf{k}_1.\mathbf{T}] \quad \text{and} \quad \exp[-i(\mathbf{k}_1 + \mathbf{K}).\mathbf{T}].$$

But $\mathbf{k}.\mathbf{T} = m_1n_1\mathbf{b}_1.\mathbf{a}_1 + m_2n_2\mathbf{b}_2.\mathbf{a}_2 + m_3n_3\mathbf{b}_3.\mathbf{a}_3$

$$= \text{integer} \times 2\pi.$$

Hence both $\mathbf{k}_1$ and $\mathbf{k}_1 + \mathbf{K}$ label the same irrep.; they are indistinguishable from a group-theoretical viewpoint.

(iv) *The proximity cell.* Rather than use the points in the primitive cell of **k**-space to label the irreps of $\mathcal{T}$ it is conventional to define a symmetric cell by the following method. The lattice of **K** points is constructed. Starting at one of these points one draws lines to the nearest neighbour points and constructs planes normal to each of these lines, passing through their central points. Taking the next nearest neighbour **K** points, the same procedure is carried out. The set of planes so constructed will enclose symmetrically a region around the origin of volume identical to that of the primitive cell, and containing no points of separation in excess of any **K**. Only the first few neighbours of the origin will be required to define completely this region of **k**-space, which we will call the *proximity cell.* Proximity cells may be stacked so as to fill all space, as required.

(v) The N discrete **k** labels are contained within and on a portion of the surface of the proximity cell. There are however certain positions on the surface that are separated by one of the vectors **K**. Such positions are equivalent regarding the irrep. they label and so must only be included once. As a result only one half of the surface of the zone is significant, and only one quarter of the corner points.

(vi) *Reciprocal space and the first Brillouin zone.* Our group-theory considerations of translational symmetry have thus led us to picture a space of **k** values with a periodic lattice of **K** positions. In solid-state physics such a lattice was introduced in crystallography as the *reciprocal lattice.* The proximity cell we have defined is known as the *first Brillouin zone* of reciprocal space. What is strictly new in our methodology is the interpretation of the discrete positions defined in this Brillouin zone as purely the *labels* of irreps to which the eigenfunctions of the full problem can be chosen to belong. As yet we put no other physical interpretation on **k**.

Example: The Brillouin zone for the diamond lattice

Pursuing our example of the diamond structure in order to demonstrate the construction of a Brillouin zone, remember that the lattice structure is face-centred cubic. In order to obtain the reciprocal lattice vectors, however, we need the primitive translations of the spatial lattice $\mathbf{a}_1$, $\mathbf{a}_2$, $\mathbf{a}_3$. Figure 17.9 showed the lattice, with the dimension of the cubic Bravais unit cell defined as a. The primitive cell is constructed by vectors from the origin to the nearest face centres, for example†

$$\mathbf{a}_1=\frac{a}{2}(\hat{\mathbf{j}}+\hat{\mathbf{k}}) \qquad \mathbf{a}_2=\frac{a}{2}(\hat{\mathbf{k}}+\hat{\mathbf{i}}) \qquad \mathbf{a}_3=\frac{a}{2}(\hat{\mathbf{i}}+\hat{\mathbf{j}}).$$

The volume of the primitive cell is $a^3/4$. Hence the reciprocal-lattice primitive vectors are

$$\mathbf{b}_1=2\pi\frac{\mathbf{a}_2\wedge\mathbf{a}_3}{\mathbf{a}_1.\mathbf{a}_2\wedge\mathbf{a}_3}=\frac{2\pi}{a}(-\hat{\mathbf{i}}+\hat{\mathbf{j}}+\hat{\mathbf{k}}).$$

†Carets are introduced here to avoid confusion between the unit vector in the **k** direction (one of three orthogonal directions) and the newly defined **k** reciprocal-space vector.

Similarly

$$\mathbf{b}_2=\frac{2\pi}{a}(-\hat{\mathbf{j}}+\hat{\mathbf{k}}+\hat{\mathbf{i}}),\qquad \mathbf{b}_3=\frac{2\pi}{a}(-\hat{\mathbf{k}}+\hat{\mathbf{i}}+\hat{\mathbf{j}}).$$

These vectors define a primitive cell that stacks to form a body-centred cubic reciprocal lattice (Figure 18.2(a)). Starting from one reciprocal lattice point, the nearest neighbours lie at the eight symmetrically placed positions

$$\mathbf{K}_1=\frac{2\pi}{a}(\pm\hat{\mathbf{i}}\pm\hat{\mathbf{j}}\pm\hat{\mathbf{k}}),$$

and the next nearest neighbours lie at the six positions

$$\mathbf{K}_2=\frac{2\pi}{a}(\pm 2\hat{\mathbf{i}}),\qquad \frac{2\pi}{a}(\pm 2\hat{\mathbf{j}}),\qquad \frac{2\pi}{a}(\pm 2\hat{\mathbf{k}}).$$

Planes constructed normally to the midpoints of these vectors are sufficient to define the first Brillouin zone of the system. It is a truncated rhombohedral in shape, as shown in Figure 18.2(b).

Special points on the Brillouin zone are labelled. For example X, L correspond to the positions in $\mathbf{k}$ space given by

$$X\equiv\frac{\mathbf{b}_2+\mathbf{b}_3}{2}\equiv\frac{2\pi}{a}\hat{\mathbf{i}}\equiv(1,0,0),$$

$$L\equiv\frac{\mathbf{b}_1+\mathbf{b}_2+\mathbf{b}_3}{2}\equiv\frac{2\pi}{a}(\hat{\mathbf{i}}+\hat{\mathbf{j}}+\hat{\mathbf{k}})\equiv(1,1,1).$$

In $\mathbf{k}$ space the 'volume' of the Brillouin zone is

$$\mathbf{b}_1.\mathbf{b}_2\wedge\mathbf{b}_3=4(2\pi/a)^3=\Omega_K.$$

Within this volume there are N distinct $\mathbf{k}$ labels, each of which can therefore be considered to occupy a 'volume'

$$\Delta_K=\Omega_K/N=(2\pi)^3/V.$$

This latter is a general result for all symmetry types. V, the crystal volume, is equal to $N\Omega$, where Ω is the real space primitive cell volume. Ω in turn is equal to $(2\pi)^3/\Omega_K$ for all symmetries.

18.3 Bloch functions, and the symmetry of energy bands

Bloch's theorem

We have established that the eigenfunctions of a problem of space group symmetry $\mathcal{S}$ can be chosen such that they are symmetry-adapted functions of

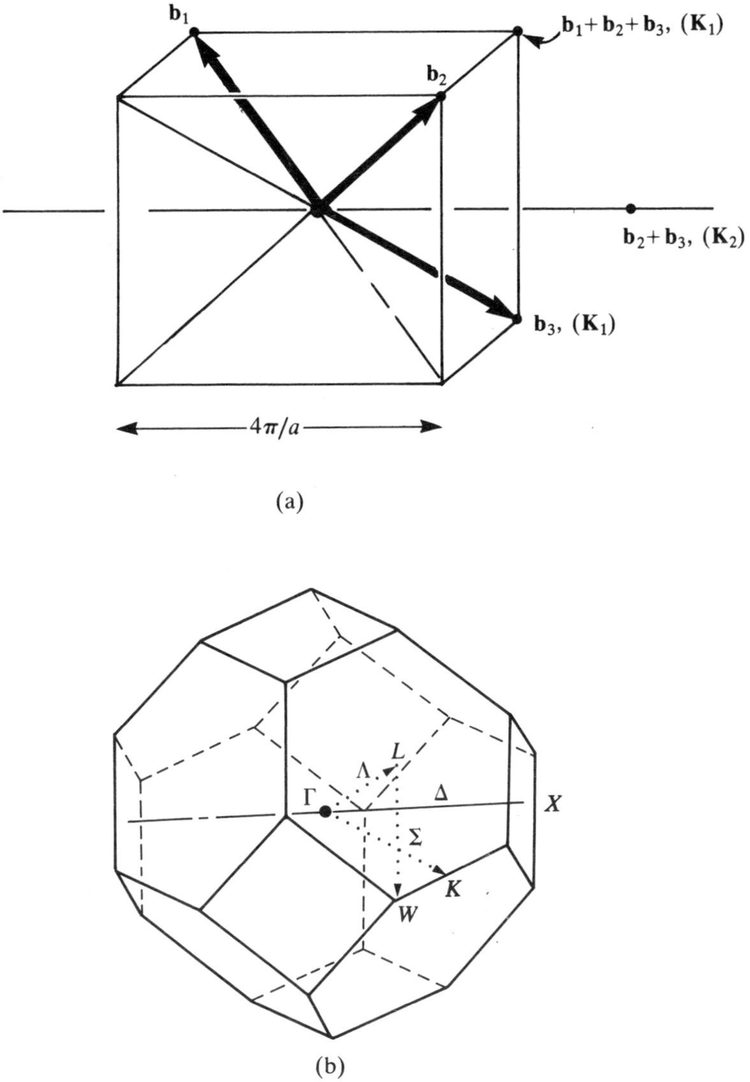

Figure 18.2 (a) The primitive lattice vectors of the reciprocal lattice of a face-centred cubic crystal. Showing nearest ($\mathbf{K}_1$) and next nearest neighbour ($\mathbf{K}_2$) lattice positions. The reciprocal lattice is body-centred cubic; (b) The Brillouin zone for the above lattice.

the irreps of $\mathcal{T}$. These we can now label by $\mathbf{k}$ and we will be able to find some set of symmetry-adapted functions belonging to $\Gamma^{\mathbf{k}}$ for any given problem; denote them by $F^{\mathbf{k}}_{n'}(\mathbf{r})$. The eigenfunctions must be specific linear combinations of these functions,

$$\psi_{n\mathbf{k}}(\mathbf{r}) = \sum_{n'} b_{nn'} F^{\mathbf{k}}_{n'}(\mathbf{r}).$$

Now the effect of a translation on $\psi_{n\mathbf{k}}(\mathbf{r})$ is given (i) by the value of function that moves to position $\mathbf{r}$ under the translation and (ii) by the symmetry of $\psi_{n\mathbf{k}}(\mathbf{r})$, exhibited by the fact that it belongs to $\Gamma^{\mathbf{k}}$:

(i) $\{E/\mathbf{T}\}\ \psi_{n\mathbf{k}}(\mathbf{r}) = \psi_{n\mathbf{k}}(\{E/\mathbf{T}\}^{-1}\mathbf{r})$

$$= \psi_{n\mathbf{k}}(\mathbf{r}-\mathbf{T}).$$

(ii) $\{E/\mathbf{T}\}\ \psi_{n\mathbf{k}}(\mathbf{r}) = \psi_{n\mathbf{k}}(\mathbf{r})\, e^{-i\mathbf{k}.\mathbf{T}}$.

Combining these results, and multiplying by $e^{-i\mathbf{k}.(\mathbf{r}-\mathbf{T})}$, we obtain

$$\psi_{n\mathbf{k}}(\mathbf{r}-\mathbf{T})\, e^{-i\mathbf{k}.(\mathbf{r}-\mathbf{T})} = \psi_{n\mathbf{k}}(\mathbf{r})\, e^{-i\mathbf{k}.\mathbf{r}} = U_{n\mathbf{k}}(\mathbf{r}), \qquad \text{say.}$$

Hence the eigenfunctions of the system have the form

$$\boxed{\psi_{n\mathbf{k}}(\mathbf{r}) = U_{n\mathbf{k}}(\mathbf{r})\, e^{i\mathbf{k}.\mathbf{r}}}\ ,$$

where $U_{n\mathbf{k}}(\mathbf{r})$ has the periodicity of the lattice. This is Bloch's theorem. We shall call the function $U_{n\mathbf{k}}(\mathbf{r})$ the Bloch function. **k** must lie in the first Brillouin zone.

Energy band symmetries

The form of $\psi_{n\mathbf{k}}(\mathbf{r})$ enables us to make a significant statement concerning the energies of the eigenstates. The energy levels of the system can clearly be labelled by $n,\mathbf{k}$ in the same way as the eigenfunctions:

$$\mathscr{H}\psi_{n\mathbf{k}}(\mathbf{r}) = E_{n\mathbf{k}}\psi_{n\mathbf{k}}(\mathbf{r}).$$

That is, to each point in **k** space we can allot a set of energy levels, one for each n value. Eventually we will have in mind that n labels a band of energies $E_n(\mathbf{k})$, where E_n is treated as a function of **k** even though only discrete **k** values have any physical meaning. Any operation of $\mathscr{S}$, by definition, leaves the Hamiltonian $\mathscr{H}$ invariant and therefore does not change the energy of the system. An eigenfunction $\psi_{n\mathbf{k}}$, however, may be altered, either by a change of phase or by a transformation to some linear combination of functions degenerate with $\psi_{n\mathbf{k}}$. This is after all what we mean when we say that $\psi_{n\mathbf{k}}$ belongs to an irrep. of the group $\mathscr{S}$.

Consider such a transformation:

$$\begin{aligned}\{R/\mathbf{T}\}\ \psi_{n\mathbf{k}}(\mathbf{r}) &= \psi_{n\mathbf{k}}(\{R/\mathbf{T}\}^{-1}\mathbf{r})\\ &= \psi_{n\mathbf{k}}(R^{-1}\mathbf{r}-R^{-1}\mathbf{T})\\ &= U_{n\mathbf{k}}(R^{-1}\mathbf{r}-R^{-1}\mathbf{T})\exp\{-i\mathbf{k}.(R^{-1}\mathbf{r}-R^{-1}\mathbf{T})\}.\end{aligned}$$

Now $\mathbf{k}.R^{-1}\mathbf{r}$ is a scalar and is therefore unaltered by any point group transformation of the coordinate system. For example it must be equal to $R\mathbf{k}.R(R^{-1}\mathbf{r})$, which is equal to $(R\mathbf{k}).\mathbf{r}$. Also $R^{-1}\mathbf{T}$ is some lattice transformation $\mathbf{T}'$. Hence

$$\{R/\mathbf{T}\}\ \psi_{n\mathbf{k}}(\mathbf{r}) = [U_{n\mathbf{k}}(R^{-1}\mathbf{r}-\mathbf{T}')\,\mathrm{e}^{+\mathrm{i}\mathbf{k}.\mathbf{T}'}]\,\mathrm{e}^{-\mathrm{i}R\mathbf{k}.\mathbf{r}}.$$

The function in square brackets has the translational periodicity of the lattice. Thus on the right-hand side we have a function of the form $U(\mathbf{r})\,\mathrm{e}^{-\mathrm{i}R\mathbf{k}.\mathbf{r}}$, which is the form demanded of $\psi_{n'(R\mathbf{k})}(\mathbf{r})$ for some level labelled by n'. The effect of $\{R/\mathbf{T}\}$ on $\psi_{n\mathbf{k}}(\mathbf{r})$ is therefore to produce another eigenfunction, degenerate with $\psi_{n\mathbf{k}}(\mathbf{r})$ but associated with the **k** space position $R\mathbf{k}$. The energy levels of the system must thus display the symmetry

$$E_n(\mathbf{k}) = E_{n'}(R\mathbf{k}).$$

That is,

energy bands, as functions of **k**, *have the symmetry of the point group of the crystal lattice.*

The star of k

The set of **k** values given by performing all the point group operations $\{R/\mathbf{O}\}$ on **k** is known as the *star of* **k**. For a general point in the Brillouin zone the total number of such values will equal the order of the point group, ρ. At **k** positions of higher symmetry there are fewer values in the star of **k** and consequently fewer states of energy equal to $E_n(\mathbf{k})$.

18.4 The point group symmetry of eigenfunctions

In Section 18.2 we considered the translational properties of the crystal space group. The point group properties can also give us some important results. It is therefore appropriate now to return to the ideas of Section 18.1 concerning the association of the eigenfunctions of $\mathcal{S}$ and the symmetry-adapted functions of some space subgroup $\mathcal{S}'$. For the present the arguments are still restricted to symmorphic space groups.

Consider a degenerate set of eigenfunctions of $\mathcal{S}$,

$$\psi_{n_1\mathbf{k}_1}(\mathbf{r}),\ \psi_{n_2\mathbf{k}_2}(\mathbf{r}),\ldots\,.$$

This set of functions forms a basis for some irrep $\Gamma^{\mathscr{S}}$ of $\mathscr{S}$. That is, under any of the space group operations,

$$\{R/\mathbf{T}\}\ \psi_{n_i\mathbf{k}_i}=\sum_j \psi_{n_j\mathbf{k}_j}\ \Gamma^{\mathscr{S}}_{ij}\ (\{R/\mathbf{T}\}).$$

Also the eigenfunctions may have been chosen to be bases for the irreps of the space subgroup $\mathscr{S}'$.

$$\{R'/\mathbf{T}\}\ \psi'_{n_l\mathbf{k}_l}=\sum_m \psi'_{n_m\mathbf{k}_m}\ \Gamma^{\mathscr{S}'}_{ml}\ (\{R'/\mathbf{T}\}).$$

The sum over m may be smaller or equal to that over j, depending respectively on whether $\Gamma^{\mathscr{S}}$ reduces in the subgroup or remains an irrep. If we choose the subgroup $\mathscr{S}'$ such that the R' operations are those point group operations in $\mathscr{P}$ which transform some particular $\mathbf{k}$ into itself, then

$$\{R_{\mathbf{k}}/\mathbf{T}\}\ \psi_{n_l\mathbf{k}}=\sum_m \psi_{n_m\mathbf{k}}\ \Gamma^{\mathscr{S}_\mathbf{k}}_{ml}\ (\{R_{\mathbf{k}}/\mathbf{T}\}).$$

The group of **k**

The $\{R_{\mathbf{k}}/\mathbf{T}\}$ operations constitute what is known as the space group of **k**, $\mathscr{S}_{\mathbf{k}}$. The operations $\{R_{\mathbf{k}}/\mathbf{O}\}$ form the point group of **k**, $\mathscr{P}_{\mathbf{k}}$. For a position within the Brillouin zone the point group of **k** is straightforward. All of the operations $\{R/\mathbf{O}\}$ in $\mathscr{S}$ transform **k** to some point also within the zone. The identity, and possibly certain other operations, leave **k** completely unchanged. For a **k** position on the zone boundary, however, certain operations transform **k** to another boundary point separated from the first by a reciprocal lattice vector **K**. This corresponds to precisely the same irrep. as **k** and should therefore be treated as indistinguishable from it. Such operations must therefore be included in the group of **k**.

If we now use the relation $\{R_{\mathbf{k}}/\mathbf{T}\}=\{E/\mathbf{T}\}\ \{R_{\mathbf{k}}/\mathbf{O}\}$ and the translational property of $\psi_{n_l\mathbf{k}}$, then the operations in the space group of **k** give

$$\begin{aligned}\{R_{\mathbf{k}}/\mathbf{T}\}\ \psi_{n_l\mathbf{k}}&=\{E/\mathbf{T}\}\ \{R_{\mathbf{k}}/\mathbf{O}\}\ \psi_{n_l\mathbf{k}}\\&=\{E/\mathbf{T}\}\sum_m \psi_{n_m\mathbf{k}}\ \Gamma^{\mathscr{S}_\mathbf{k}}_{ml}\ (\{R_{\mathbf{k}}/\mathbf{O}\})\\&=\sum_m \psi_{n_m\mathbf{k}}\ \mathrm{e}^{-i\mathbf{k}.\mathbf{T}}\ \Gamma^{\mathscr{S}_\mathbf{k}}_{ml}\ (\{R_{\mathbf{k}}/\mathbf{O}\}).\end{aligned}$$

Comparing this with the previous result, then

$$\Gamma^{\mathscr{S}_\mathbf{k}}\ (\{R_{\mathbf{k}}/\mathbf{T}\})=\mathrm{e}^{-i\mathbf{k}.\mathbf{T}}\ \Gamma^{\mathscr{S}_\mathbf{k}}\ (\{R_{\mathbf{k}}/\mathbf{O}\}).$$

Simultaneous translation group and point group labelling

The reason for undertaking this argument is that we now wish to demonstrate that the eigenfunctions $\psi_{n\mathbf{k}}$ not only form bases for the irreps of $\mathcal{S}_{\mathbf{k}}$ but also belong to the irreps of the crystallographic point group $\mathcal{P}_{\mathbf{k}}$. The latter is a group of finite order, whose properties we know how to handle. It is clear that the $\psi_{n\mathbf{k}}$ must form bases of representations of some sort for $\mathcal{P}_{\mathbf{k}}$ as these are simply a particular choice of the operations of $\mathcal{S}_{\mathbf{k}}$ for all of which the $\psi_{n\mathbf{k}}$ transform in sets. However, we need to establish that these representations are irreducible. Now the $\Gamma^{\mathcal{S}_{\mathbf{k}}}$ matrix representation we know to be irreducible with respect to our chosen $\psi_{n\mathbf{k}}$. This means that at least one (if not many) of the matrices $\Gamma^{\mathcal{S}_{\mathbf{k}}}\{R_{\mathbf{k}}/\mathbf{T}\}$ is of a form that is not block diagonal. We have now established that the matrices $\Gamma^{\mathcal{S}_{\mathbf{k}}}\{R_{\mathbf{k}}/\mathbf{T}\}$, for different $\mathbf{T}$ values, are directly proportional to each other. Hence if one of the $\Gamma^{\mathcal{S}_{\mathbf{k}}}\{R_{\mathbf{k}}/\mathbf{T}\}$ is not block diagonal, then equally one of the $\Gamma^{\mathcal{S}_{\mathbf{k}}}\{R_{\mathbf{k}}/\mathbf{O}\}$ is not. Therefore the matrix set $\Gamma^{\mathcal{S}_{\mathbf{k}}}\{R_{\mathbf{k}}/\mathbf{O}\}$ is an irreducible set of matrices that represents the group of $\mathbf{k}$, $\mathcal{P}_{\mathbf{k}}$, and the eigenfunctions $\psi_{n\mathbf{k}}$ from a basis for this irrep.:

$$\{R_{\mathbf{k}}/\mathbf{O}\}\,\psi_{n_l\mathbf{k}} = \sum_m \psi_{n_m\mathbf{k}}\,\Gamma^{\mathcal{S}_{\mathbf{k}}}_{ml}(\{R_{\mathbf{k}}/\mathbf{O}\}).$$

We will therefore be able to label the eigenfunctions not only by $\mathbf{k}$ (the irreps of one subgroup of $\mathcal{S}$) but also by the point group symmetry labels of $\mathcal{P}_{\mathbf{k}}$ (a completely different subgroup of $\mathcal{S}$). The hierarchy of subgroups is shown in Figure 18.3. Just as in the atomic and molecular cases, we will be able to use the symmetry properties of the various irreps in order to establish selection rules for radiative transitions and in order to determine which states mix under various constant perturbations.

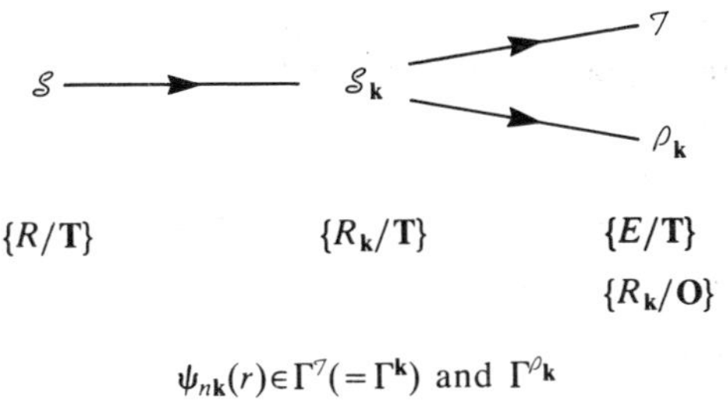

Fig. 18.3

The group of k *in the non-symmorphic case*

Despite the fact that the above arguments have been restricted to cover only symmorphic groups it is possible to associate the eigenfunctions for non-symmorphic systems with point groups. However, because on the one hand the entire point group of the crystal [$\{R/\mathbf{O}\}$ and $\{R'/\mathbf{O}\}$ operations] is not a subgroup of $\mathcal{S}$, and on the other hand the operations $\{R/\mathbf{O}\}$ plus $\{R'/\boldsymbol{\tau}\}$ do not form a group, it is not a trivial matter to extend the results of symmorphic space groups and we will not attempt to present the extension. The following results summarize the situation for non-symmorphic groups.

(i) The symmetry of $E_n(\mathbf{k})$ holds good for all operations in the entire point group:

$$E_n(\mathbf{k}) = E_{n'}(R\mathbf{k}) = E_{n''}(R'\mathbf{k}).$$

(ii) The eigenfunctions for non-symmorphic space groups belong to a group that contains the elements $\{R_\mathbf{k}/\mathbf{O}\}$ and $\{R'_\mathbf{k}/\boldsymbol{\tau}\}$. $R_\mathbf{k}$ and $R'_\mathbf{k}$ must leave **k** invariant. Although these *operations* do not form a group it can be established that their *matrix representations*, with Bloch functions as bases, do obey closure.

(iii) For **k** values within the Brillouin zone the characters for this effective group are of value $\exp(i\mathbf{k}.\boldsymbol{\tau})$ times those of the point group obtained from $\{R_\mathbf{k}/\mathbf{O}\}$ plus $\{R'_\mathbf{k}/\mathbf{O}\}$—*the entire point group of* **k**. In particular, for $\mathbf{k}=0$ the character table is precisely that of the entire point group of **k**.

(iv) For **k** values on the zone boundary it is necessary to determine the group character table by more complicated methods. Nevertheless, the eigenfunctions of $\mathcal{H}$ still belong to irreps of some group and can be labelled accordingly.

Problems

***18.1** Consider the primitive hexagonal lattice. Determine the reciprocal lattice vectors $\mathbf{b}_1$, $\mathbf{b}_2$, $\mathbf{b}_3$ for such a structure. Hence show that the first Brillouin zone is also of hexagonal symmetry. Determine the magnitude of the smallest Brillouin zone boundary vectors (i) in the plane of the hexagon (the $\mathbf{b}_1$–$\mathbf{b}_2$ plane), and (ii) normal to the $\mathbf{b}_1$–$\mathbf{b}_2$ plane (in the $\mathbf{b}_3$ direction). Verify that the Brillouin zone volume Ω_k is $(2\pi)^3/\Omega$, where Ω is the real-space primitive cell volume.

19

Semiconductor energy bands

19.1 Semiconductor lattices

The idea of examining 230 space group crystal structures is an awesome thought. Fortunately, though, the majority of semiconductor materials crystalize into one of just four structures: diamond (O_h^7), zinc blende or sphalerite (T_d^2), wurtzite (C_6^4), and rocksalt (O_h^5). The reason must lie in the nature of the bonding between the atoms in the various elemental or compound materials. The outermost atomic electrons, which form the bonds, must be free enough to contribute to conductivity once excited across a particularly small energy gap, and yet be concentrated in regions between the nuclei in such a manner as to reduce the total lattice energy by comparison to that for a collection of free atoms.

Figure 19.1(a) is a schematic of the electronic energy levels of an atom. The outermost occupied state lies approximately 10 eV (of the order of 1 Rydberg) below the ionization limit, in the nth atomic energy levels, say. The $(n-1)$ levels will lie several Rydbergs below this region. Table 19.1 shows those atoms that most commonly appear in semiconductor compounds.

Table 19.1 Elements commonly found in semiconductor compounds.

I	II	III	IV	V	VI	VII
	Be	B	C	N	O	F
	Mg	Al	Si	P	S	Cl
Cu	Zn	Ga	Ge	As	Se	Br
Ag	Cd	In	Sn	Sb	Te	I
	Hg	Tl	Pb	Bi		

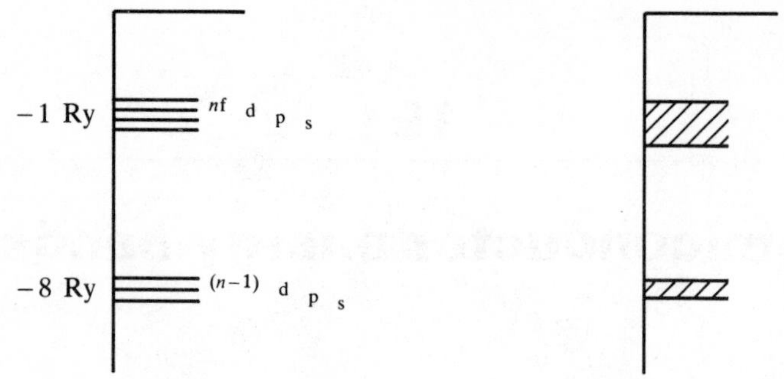

(a) (b)

Figure 19.1 (a) Schematic of electronic levels in atoms and (b) bands in the solid state.

The elemental semiconductors are the crystalline states of the group IV atoms silicon and germanium, having four electrons per atom outside of closed shells. In the free atom these electrons would occupy two ns-states (one of each spin) and two of the six np-states. As vast numbers of atoms are brought together to form a solid crystal the atomic states must broaden to form bands. They cannot remain as precisely degenerate, identical states, or the Pauli exclusion principle would be violated. The core electrons $(n-1)$, $(n-2),\ldots$ have atomic distributions that are sufficiently smaller than those for the nth level to be relatively undisturbed by the presence of adjacent atoms; they form very narrow energy bands. There are two extreme pictures however for the broadening that does occur for the nth-level states. On the one hand we can think of the atoms releasing the four outer electrons so that they can move freely throughout the crystal. This is the thinking behind the so-called empty lattice model for the semiconductor nth band, a model that gives a surprisingly good basis for the real situation. In the opposite extreme these electrons can be thought of as being shared between the ns and np atomic states in such a manner as to form bonds between adjacent atoms, similar to the formation of molecular bonding orbitals. The structure of the elemental semiconductor materials is governed by the physical nature of the bonds that can be formed in this way. The material is stable because whilst it takes a few electron volts of energy to promote atomic s-electrons to p-like orbits in order to complete the bonds, there is a more substantial reduction in the overall energy of the agglomeration of atoms because the interaction energy between adjacent atoms is lowered. The natural bond formed from four electrons per atom is the sp^3 hybridized bond, as in the molecule CH_4. This leads to a tetragonal coordination between nearest neighbour atoms (Figure 19.2) and commonly to a face-centred-cubic lattice structure with interpenetrating lattices displaced by $(\frac{1}{4},\frac{1}{4},\frac{1}{4})$.

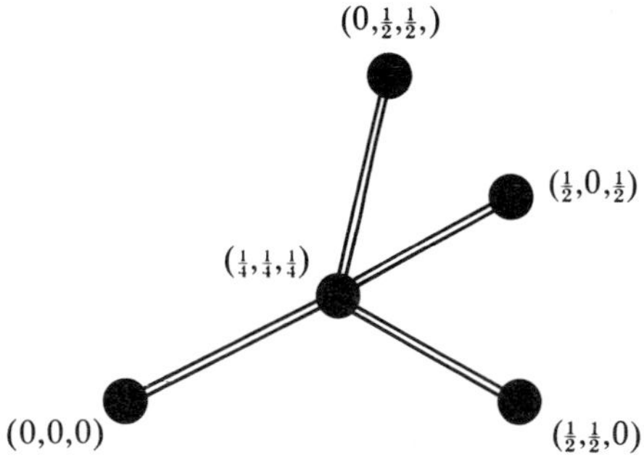

Figure 19.2 Tetrahedral bonding.

This sharing of electrons between adjacent atoms and therefore, by a passing-on process, with all the atoms in the crystal, is known as covalent bonding. The electrons concentrate along the bonds within any one unit cell yet are free enough to move throughout the crystal. The same basic form of bonding is to be expected of all compounds of group III and group V elements—indeed for most $A^{N} B^{VIII-N}$ compounds. However, as one tends to $A^{I} B^{VII}$ compounds, it may become energetically more favourable for the outer electron on atom A to become localized in the vicinity of atom B. If this happens, then in effect each atom has closed subshells of electrons. No other state can be then formed of energy close to that of the outermost subshell of either the A^{+} or B^{-} atoms, and the compound is an insulator, stabilized by the ionic interaction between adjacent atoms. The natural configuration in this instance is the rocksalt (NaCl) structure. The III–V and II–IV compounds fall between the two camps. There is a measure of covalency, all eight outer electrons being shared between adjacent (dissimilar) atoms, and yet the ionicity this would create is avoided to some extent as the eight electrons occupy states that are 'somewhat' more localized on the group V or VI anions. These compounds again have tetrahedral coordination, but because the atoms are dissimilar the interpenetrating face-centred-cubic lattice of the III–V and most II–VI compounds has only T_d point group symmetry, whereas the elemental structures Si and Ge have O_h entire point group symmetry as discussed in Chapter 18. The exceptions are CdS and CdSe (and one phase of ZnS), which form hexagonal C_6^4 (wurtzite) structures. Figure 19.3 is included to demonstrate that this change in structure is a feature of the next nearest neighbour configuration; the tetrahedral coordination is maintained.

Table 19.2 summarizes the structures for a large number of binary compound semiconductors. Alloys of these materials on the whole have a similar Bravais lattice structure to their parent compounds but, just as the

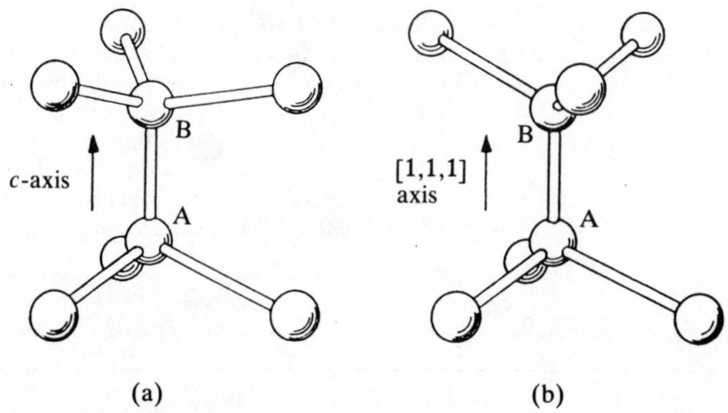

Figure 19.3 Orientations of adjacent atomic tetrahedra in wurtzite (a) and zincblende (b). The A and B refer to the two different species of atom.

space group changes on going from Ge to InSb, so it will change again for ternary compounds, etc. As shown in the table the binary IV–VI compounds PbS, PbSe, PbTe and SnTe have a different bonding structure altogether to the $A^N B^{VIII-N}$ materials, having a face-centred-cubic lattice of A atoms and a similar interpenetrating B lattice displaced by ($\frac{1}{2}$,0,0), as does NaCl. We will return briefly to the bonding properties of semiconductors when the empty lattice model has been investigated.

19.2 The free-electron model

One piece of experimental evidence leads to the use of a simple model for the outermost occupied energy band and the unoccupied bands of semiconductors. (These are termed the valence and conduction bands respectively.) The evidence comes from the soft X-ray emission of these materials. After electron impact (which expels core electrons) the outer electrons fall to the narrow core levels. It is found that for many semiconductors the range of photon energies over which the soft X-ray emission is strong is of the order of 15–30 eV. This is in turn very close to the value of the Fermi energy of a gas of free electrons that has the density of the valence electrons of a semiconductor. One is therefore led to consider a model where the $(n-1)$ states and below consist of narrow bands but the n states coalesce into a valence band of width around 20 eV. For the core electrons the state eigenfunctions are dominated by the localized potential due to the particular atoms. For the valence electrons, however, one can almost ignore the periodic variation in

Table 19.2 Characteristic fundamental band gaps (in eV) for common elementary and binary compound semiconductors. i indicates an indirect gap, capital letters indicate the crystal structure(s): D,diamond$-O_h^7$; Z,zincblende$-T_d^2$; W,wurtzite$-C_{6v}^4$; R,rocksalt$-O_h^5$; O,orthorhombic$-D_{2h}$; Rh,rhombohedral$-D_4$; T,trigonal$-D_3^4$; OR,orthorhombic distorted rocksalt; M,monoclinic. For details of these and further compound and alloy semiconductors see Landolt and Bornstein (see Bibliography, Section (g)).

(a) Tetrahedrally bonded materials

IV	C	Si	Ge	α–Sn
C	5.5i,D			
Si	2.6i,Z/W	1.2i,D		
Ge			0.74i,D	
α–Sn				0.09,D

III–V	N	P	As	Sb
B	3.8,W	2.0i,Z	1.5i,Z	
Al	5.9,W	2.5,Z	2.2,Z	1.7,Z
Ga	3.5,W	2.4i,Z	1.5,Z	0.81,Z
In	2.4,W	1.4,Z	0.41,Z	0.24,Z

II–VI	O	S	Se	Te
Zn	3.4,W	3.6,Z/W	2.8,Z/W	2.4,Z
Cd	1.3i,R	2.5,Z/W	1.8,Z/W	1.6,Z
Hg	2.2,O/Rh	2.3,T	–.06,Z	–.3,Z

I–VII	F	Cl	Br	I
Cu		3.4,Z	3.1,Z	3.1,Z
Ag	2.8i,R	3.2i,R	2.7i,R	3.0,W

(b) Non-tetrahedral bonded materials

IV–VI compounds

IV–VI	O	S	Se	Te
Ge		1.7,OR	1.1,OR	0.15,R
Sn		1.1,OR	0.9,OR	2.1,R
Pb	2.0,i	0.29,R	0.15,R	0.19,R

Group VI elements

VI	S	Se	Te
	3.6,O	1.9i,T 2.5,M	0.33,T

Group V elements

V	P	As	Sb	Bi
	.33,O	.17,Rh	.10	.015

the magnitude of the electrostatic potential. These electrons can almost be considered to be 'free', restricted only by the boundaries of the crystal itself, but experiencing a uniform potential within these boundaries (Figure 19.4).

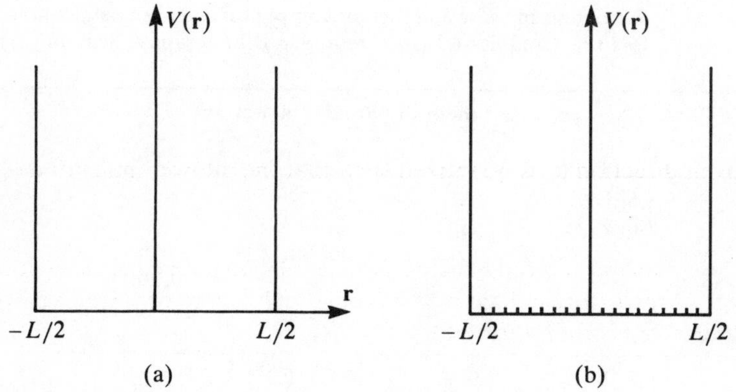

Figure 19.4 (a) Free-electron model of a semiconductor and (b) the empty lattice model.

If we ignore the lattice of nuclei and core electrons altogether, and ignore the interactions between the valence electrons other than their contribution to the uniform background potential, then the one-electron problem to be solved reduces to the free-electron problem,

$$\frac{-\hbar^2}{2m}\nabla^2\psi(\mathbf{r}) = E\psi(\mathbf{r}).$$

The solutions are plane-wave eigenfunctions, and 'parabolic' energies.

$$\psi_{\mathbf{k}'}(r) = C \exp(\mathrm{i}\mathbf{k}'.\mathbf{r}), \qquad E_{\mathbf{k}'} = \frac{\hbar^2}{2m}|\mathbf{k}'|^2.$$

C is a normalization constant, and $\mathbf{k}'$ can take *any* value. (Hereafter in this section we will ignore normalization constants.) We use the notation $\mathbf{k}'$ rather than $\mathbf{k}$ here as the latter symbol is reserved for the $\mathbf{k}$ labels within the first Brillouin zone for a crystal lattice.

The simplest boundary condition that can be imposed is to set $\psi_{\mathbf{k}'}(\mathbf{r})$ to zero at the crystal surfaces; that is, to assume that the potential rises to infinity, thereby restricting all the electrons to the material. The resulting quantization is almost but not quite the same as obtained by imposition of periodic boundary conditions. For consistency we shall therefore employ the latter here. Thus we demand

$$\psi_{\mathbf{k}'}(\mathbf{r}+a_1\mathbf{L}_1+a_2\mathbf{L}_2+a_3\mathbf{L}_3)=\psi_{\mathbf{k}'}(\mathbf{r}).$$

Here a_i are positive or negative integers, or zero, and the $\mathbf{L}_i$ are crystal edges. This condition is satisfied for all a_i providing the eigenfunction is repeated in each adjacent crystal volume, that is for

$$C \exp[\mathrm{i}\mathbf{k}'.\mathbf{r}]=C \exp[\mathrm{i}\mathbf{k}'.(\mathbf{r}+\mathbf{L}_1)], \text{etc.}$$

Thus

$$\mathbf{k}'.\mathbf{L}_i=2\pi n_i \qquad \text{for} \quad n_i=0,\pm 1,\pm 2,\ldots$$

For a given direction $\mathbf{k}'$ is quantized such that the allowed magnitudes are

$$k'=2\pi n/L.$$

The $\mathbf{k}'$ states can be pictured as being associated with points in $\mathbf{k}'$ space; each allowed $\mathbf{k}'$ value is surrounded by a 'volume' in $\mathbf{k}'$ space of size $(2\pi)^3/V$, in which there is no other allowed $\mathbf{k}'$.

Fermi energy

If electrons are allotted to these states, starting at the lowest and building up to the Fermi level at energy $E_F=\hbar^2(k'_F)^2/2m$, then k'_F is obtained by requiring that the volume in $\mathbf{k}'$ space enclosed by the 'surface' at $\mathbf{k}'_F$ is equal to the product of the number of electrons to be accommodated, N_F, and the volume per $\mathbf{k}'$ state:

$$N_F(2\pi)^3/V=2\times\frac{4}{3}\pi k'^3_F.$$

At this stage we need to comment on the electron spin. For the time we will assume that each $\mathbf{k}'$ state can accommodate two electrons of opposite spins. This is the origin of the factor 2 on the right of the above equation. On the scale of the energy bands at which we are looking presently this is a reasonable approximation. For $A^{\mathrm{N}}B^{\mathrm{VIII-N}}$ materials we hope to accommodate four electrons per atom on average, and in both the face-centred-cubic diamond and sphalerite structures there are two atoms in each primitive cell. Thus $N_F=8N$, where the crystal consists of N primitive cells. Finally, then,

$$E_F=\frac{\hbar^2}{2m}\left(3\pi^2\frac{8N}{V}\right)^{2/3}.$$

For a primitive cell volume V/N typical of semiconductors, this Fermi energy has a value close to the 20 eV value of the soft X-ray emission range.

19.3 The empty-lattice model

The next stage of sophistication in our valence-conduction band model is the inclusion of the lattice periodicity, without introducing any change of magnitude for the electrostatic potential across the primitive cell. Whilst the *Hamiltonian* has discrete solutions for $\mathbf{k}'$ values of unlimited magnitude, we now wish to *label* eigenfunctions by $\mathbf{k}$ values that lie within the first Brillouin zone. Hence replace $\mathbf{k}'$ by $\mathbf{k}+\mathbf{K}_n$, where $\mathbf{K}_n$ is some reciprocal lattice vector. We have

$$\psi_{n\mathbf{k}} = C \exp \mathrm{i}(\mathbf{k}+\mathbf{K}_n).\mathbf{r},$$
$$E_n(\mathbf{k}) = \hbar^2|\mathbf{k}+\mathbf{K}_n|^2/2m$$

and

$$\mathbf{K}_n = n_1\mathbf{b}_1 + n_2\mathbf{b}_2 + n_3\mathbf{b}_3.$$

k-*state volume*

The volume associated with each $\mathbf{k}$ label is Ω_k/N, where Ω_k is the primitive cell or Brillouin zone volume:

$$\Omega_K = \mathbf{b}_1.\mathbf{b}_2 \wedge \mathbf{b}_3 = (2\pi)^3/\Omega,$$

Ω being the primitive cell volume in real space. (For the face-centred-cubic structure $\Omega = a^3/4$.) As $N\Omega = V$, each $\mathbf{k}$ label is associated with a region $(2\pi)^3/V$ regardless of the structure of the lattice—as per the free-electron model.

Example (a): The one-dimensional model

We are interested now to plot $E_n(\mathbf{k})$ as a function of $\mathbf{k}$; the energy band structure, for various $\mathbf{K}_n$ vectors. Firstly we consider the one-dimensional case. In one dimension, for a 'lattice' of period a the 'reciprocal lattice' has period $b = 2\pi/a$ and the 'Brillouin zone' extends from $-b/2$ to $+b/2$. We will employ the convention of labelling the Brillouin zone boundaries $(-1,0,0)$ and $(1,0,0)$; our scaling for k is thus the unit π/a in this example—different scalings occur for different lattice structures. The reciprocal lattice vectors in one dimension are

$$K_n = n\frac{2\pi}{a}, \qquad n = 0, \pm 1, \pm 2, \ldots,$$

and the energies of the k states become

$$E_n(k)=\frac{\hbar^2}{2m}\left(k+\frac{2\pi n}{a}\right)^2.$$

It is useful to introduce dimensionless parameters at this stage:

$$\kappa=ka/\pi \qquad \text{and} \qquad \epsilon=E\Big/\left[\frac{\hbar^2}{2m}\left(\frac{\pi}{a}\right)^2\right].$$

κ ranges from -1 to $+1$ across the Brillouin zone. Figure 19.5 shows $\epsilon_n(\kappa)$

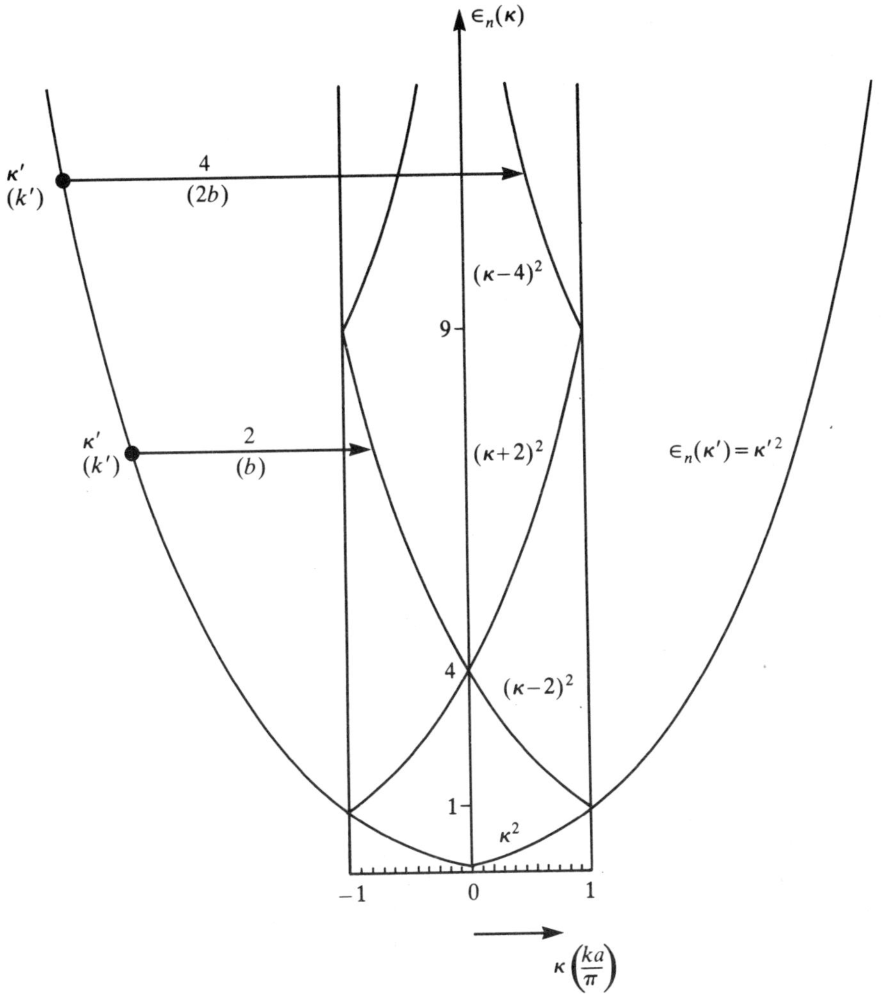

Figure 19.5 Energy levels in the first Brillouin zone in the one-dimensional free-electron model.

plotted for various n values:

$$\epsilon_n(\kappa) = E_n(k)\frac{2m}{\hbar^2}\left(\frac{a}{\pi}\right)^2 = (\kappa + 2n)^2.$$

The free-electron parabola $\epsilon(\kappa) = \kappa^2$ is included to demonstrate that one is merely translating the state at some κ' (k') value outside of the zone by an integer number of reciprocal vectors (nb) in order to bring it into the zone at $\kappa' + 2n$, ($k' + nb$). The markings along the κ axis within the zone are to remind us that only specific κ-values have any physical significance. $\kappa = \pm j/2N$ label the irreps of the translation group of the system. Each point vertically above one of these markers and lying on one of the energy curves corresponds to one of the allowed states of the system. Typically there are 10^7 markers in each direction across the Brillouin zone for a real semiconductor.

It is of no physical significance in this one-dimensional model, but if we hypothesize that we wish to allot $8N$ electrons into these free-electron bands, then $2N$ are accommodated in the lowest parabola ($\epsilon \leqslant 1$), $2N$ more up to $\epsilon = 4$, and the remaining $4N$ up to $\epsilon = 16$. This is the position of the Fermi energy. All higher states are empty at low temperature and form the conduction band.

Example (b): Face-centred-cubic structures

In three dimensions the translation of states at $\mathbf{k} + \mathbf{K}_n$ into the Brillouin zone is far less easy to visualize than in one dimension. In addition, we obtain different results for different $\mathbf{k}$ directions; we will concentrate on the k_x direction [100].

Figure 19.6 is included as an aid to visualizing the process of translating by $\mathbf{K}_n$ vectors. It is a diagram of the planes that determine six of the faces of the Brillouin zone of a face-centred-cubic lattice (Chapter 18). The shaded area indicates that part of the face that is at the zone boundary when the zone has been constructed in full.

Various reciprocal lattice points are shown as solid circles. The cube may be thought of as being in a space of $\mathbf{k}'$ values which label states of the Hamiltonian, their energy depending only on the magnitude of $\mathbf{k}'$—its distance from the centre of the cube. On Figure 19.6(a) the four wavy lines extend a distance $2\pi/a$ out from the four lattice vectors†

†Note that we introduce $\hat{\mathbf{i}}, \hat{\mathbf{j}}, \hat{\mathbf{k}}$ here to indicate a set of orthogonal unit vectors, where $\mathbf{i}$, $\mathbf{j}$, $\mathbf{k}$ have been used previously. The carets are superscribed in this section to avoid any possible confusion between the irrep. label $\mathbf{k}$ and the unit vector in the $\hat{\mathbf{k}}$ direction.

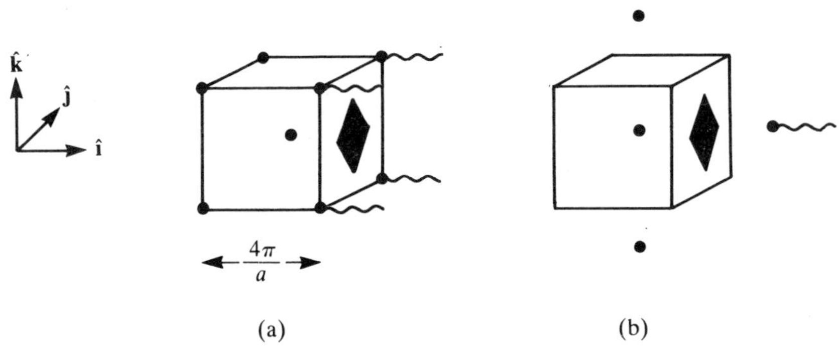

Figure 19.6

$$\mathbf{K}_n = \frac{2\pi}{a}(\hat{\mathbf{i}} \pm \hat{\mathbf{j}} \pm \hat{\mathbf{k}}).$$

The energy at points along each of these lines is

$$E(\mathbf{k}') = E_n(\mathbf{k}) = \frac{\hbar^2}{2m}|\mathbf{k}+\mathbf{K}_n|^2 = \frac{\hbar^2}{2m}\left[\left(k_x + \frac{2\pi}{a}\right)^2 + \left(\pm\frac{2\pi}{a}\right)^2 + \left(\pm\frac{2\pi}{a}\right)^2\right].$$

But on subtraction of the appropriate $\mathbf{K}_n$ each wavy line translates to the same $\mathbf{k}$ line, namely 0 to $2\pi/a$ along the k_x direction. Thus we have a fourfold degenerate set of states of the above energies. Similarly the wavy line in Figure 19.6(b) translates by $\mathbf{K}_n = (2\pi/a)(2\hat{\mathbf{i}})$ back to this same $\mathbf{k}$ line, and has the associated energy

$$E_n(\mathbf{k}) = \frac{\hbar^2}{2m}|\mathbf{k} + \frac{2\pi}{a}2\hat{\mathbf{i}}|^2.$$

There is only one set of states that translates back in this way; this energy level is therefore nondegenerate (barring spin effects).

Changing to the dimensionless parameters, for the face-centred-cubic lattice we define a vector $\boldsymbol{\kappa}$ and energy $\epsilon_n(\boldsymbol{\kappa})$ such that κ_x (the $\hat{\mathbf{i}}$-component of $\boldsymbol{\kappa}$) is unity at the Brillouin zone boundary in the $\hat{\mathbf{i}}$ direction. Thus

$$\boldsymbol{\kappa} = \mathbf{k}\Big/\left(\frac{2\pi}{a}\right), \qquad \epsilon = E\Big/\left[\frac{\hbar^2}{2m}\left(\frac{2\pi}{a}\right)^2\right].$$

Hence if

$$\mathbf{k}' = \mathbf{k} + \frac{2\pi}{a}(p_x\hat{\mathbf{i}} + p_y\hat{\mathbf{j}} + p_z\hat{\mathbf{k}}) = \mathbf{k} + \mathbf{K}_n,$$

then

$$\epsilon_n(\boldsymbol{\kappa}) = (\kappa_x + p_x)^2 + (\kappa_y + p_y)^2 + (\kappa_z + p_z)^2.$$

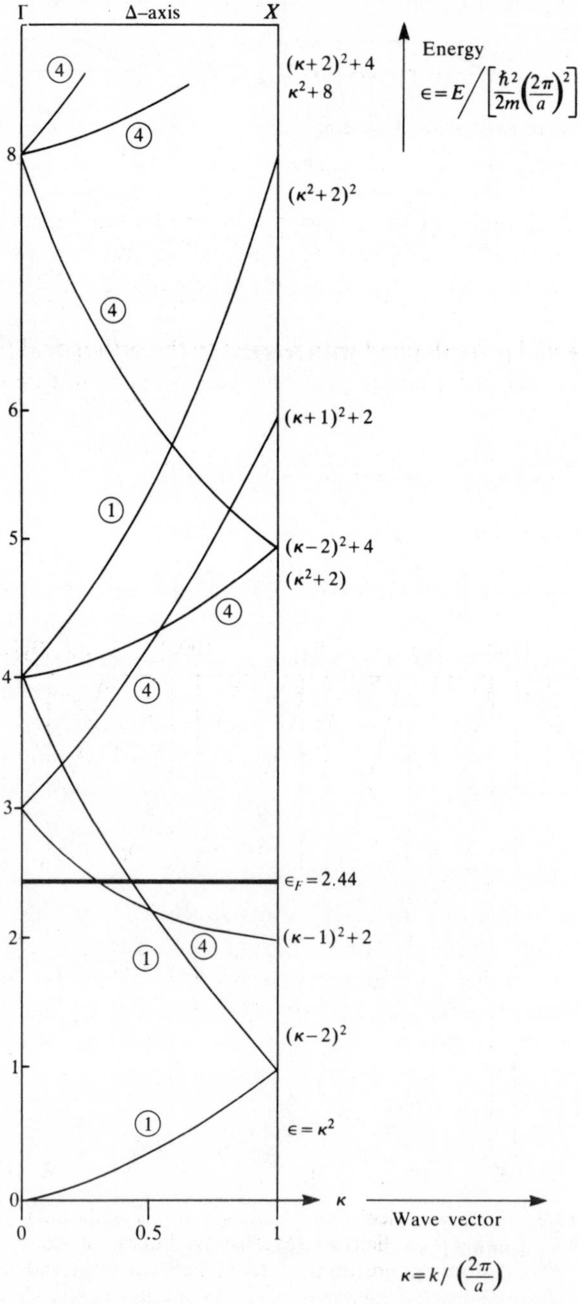

Figure 19.7 Empty lattice model for the face-centred-cubic band structure. $\epsilon_F = 2.44$ for $8N$ states. State degeneracies are circled.

The four degenerate states referred to above have energy, for **k** in the k_x-direction,

$$\epsilon_n(\kappa)=(\kappa_x+1)^2+(\pm 1)^2+(\pm 1)^2=(\kappa_x+1)^2+2;$$

the non-degenerate state has energy

$$\epsilon_n(\kappa)=(\kappa_x+2)^2.$$

Figure 19.7 summarizes the empty lattice model band structure in the k_x direction, obtained by considering the small $\mathbf{K}_n$ vectors. Only positive κ_x values are considered, as the structure is symmetric.

Note that κ and $\mathbf{p}$ are defined with respect to the orthogonal $(\hat{\mathbf{i}},\hat{\mathbf{j}},\hat{\mathbf{k}})$ vectors, whereas the reciprocal lattice primitive vectors need not be orthogonal to each other.

$$\mathbf{K}_n=\frac{2\pi}{a}(p_x\hat{\mathbf{i}}+p_y\hat{\mathbf{j}}+p_z\hat{\mathbf{k}})\equiv(n_1\mathbf{b}_1+n_2\mathbf{b}_2+n_3\mathbf{b}_3).$$

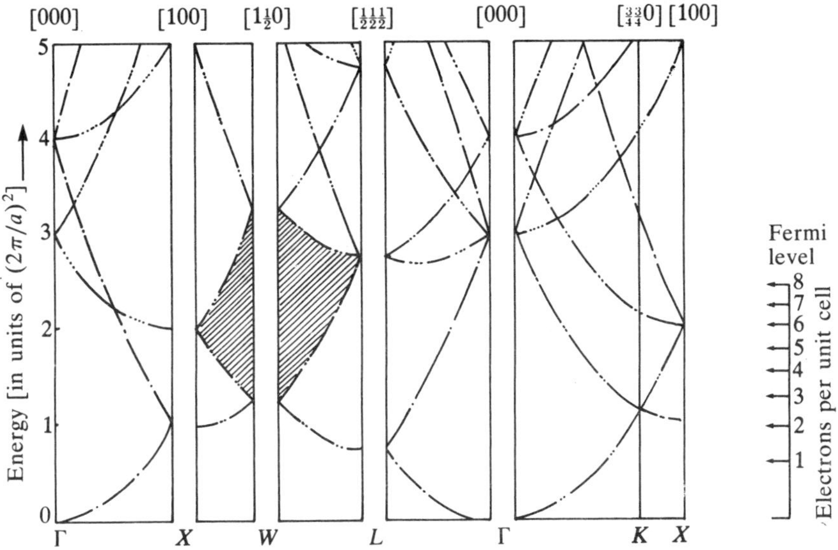

Figure 19.8 'Empty lattice' band structure for crystals having the translational periodicity of a face-centred-cubic lattice. The heavy lines denote the positions of the highest valence band and the lowest conduction band in diamond-type and zinc-blende-type crystals, as well as in crystals having the rock salt or the antifluorite structure (*after* Herman, see Bibliography, Section (f)).

The particular Brillouin zone point (1,0,0) is called the X point; the [100] axis in Figure 19.7 is called the Δ axis. Figure 19.8 depicts the bands in various other special directions of the Brillouin zone; these are defined in Figure 18.2.

Also shown in these figures is the position of the Fermi energy. For face-centred-cubic lattices N/V, the primitive cell volume, has been defined as $a^3/4$. Hence

$$E_F = \frac{\hbar^2}{2m}\left(\frac{2\pi}{a}\right)^2\left(\frac{12}{\pi}\right)^{2/3},$$

$$\epsilon_F = \left(\frac{12}{\pi}\right)^{2/3} = 2.44.$$

Because we are discussing the three-dimensional case, and because of the overlapping of the free-electron bands even in any given direction, we cannot add up the number of states of a given **k** value lying below E_F and expect to obtain eight for each **k** as in the one-dimensional case. Now, for certain **k** values, of which (0,0,0) is a good example, there are less than eight states, whilst at others, (1,1,1), there are more than eight.

Table 19.3 summarizes the $\mathbf{K}_n$ vectors corresponding to each band that appeared in Figure 19.7. The corresponding eigenfunctions, exp i $(\mathbf{k}+\mathbf{K}_n).\mathbf{r}$, and the state degeneracies, are also tabulated.

19.4 Nearly-free electrons

In the empty lattice model we have used only the translational symmetry of the material. We now want to account for the fact that the eigenfunctions must belong to the point group of **k** and reintroduce the finite magnitude of the lattice periodic potential, in order to break up the continuum band structure of the empty lattice energies into recognizable bands. It is significant that the irreps of the crystallographic point groups to which the eigenfunctions of the crystal Hamiltonian should belong are at most three-dimensional. Therefore the energy levels cannot be more than three-fold degenerate. In the empty lattice model several of the bands exceeded this degeneracy, and in particular at $\kappa=0, \epsilon=3$ there was an eight-fold degeneracy. These are accidental degeneracies due to the simplicity of the model, and should be split in reality. We must turn to the point-group symmetry of the structure to determine the form of the splitting. It is sensible to treat for a symmorphic crystal and we will select the most common semiconductor structure, sphalerite, with the T_d^2 space group, having in mind the material InSb for example.

Table 19.3 Empty lattice model eigenfunctions and energy, for points on the [1,0,0], Δ-axis.

$\mathbf{k}$	κ	$\mathbf{K}_n$	$\epsilon_n(\kappa)$	$\psi_n(\kappa)$	degeneracy
$\frac{2\pi}{a}(0,0,0)$	0	$\frac{2\pi}{a}(0,0,0)$	0	1	1
		$(\pm 1,\pm 1,\pm 1)$	3	$\exp \frac{i2\pi}{a}(\pm x \pm y \pm z)$	8
		$(\pm 2,0,0)$ $(0,\pm 2,0)$ $(0,0,\pm 2)$	4	$\exp \frac{i4\pi}{a}(\pm x)$, $\exp \frac{i4\pi}{a}(\pm y)$, $\exp \frac{i4\pi}{a}(\pm z)$	6
$\frac{2\pi}{a}(1,0,0)$	$\hat{\mathbf{i}}$	$(0,0,0)$	1	$\exp \frac{i2\pi}{a} x$	2
		$(-2,0,0)$	1	$\exp \frac{i2\pi}{a}(-x)$	
		$(-1,\pm 1,\pm 1)$	2	$\exp \frac{i2\pi}{a}(\pm y \pm z)$	4
$\frac{2\pi}{a}(\kappa,0,0)$	$\kappa\hat{\mathbf{i}}$	$(0,0,0)$	κ^2	$\exp \frac{i2\pi}{a}\kappa x$	1
		$(-2,0,0)$	$(\kappa-2)^2$	$\exp \frac{i2\pi}{a}\{(\kappa-2)x\}$	1
		$(-1,\pm 1,\pm 1)$	$(\kappa-1)^2+2$	$\exp \frac{i2\pi}{a}\{(\kappa-1)x \pm y \pm z\}$	4
		$(1,\pm 1,\pm 1)$	$(\kappa+1)^2+2$	$\exp \frac{i2\pi}{a}\{(\kappa+1)x \pm y \pm z\}$	4
		$(0,\pm 2,0)$ $(0,0,\pm 2)$	κ^2+4	$\exp \frac{i2\pi}{a}(\kappa x \pm 2y)$, $\exp \frac{i2\pi}{a}(\kappa x \pm 2z)$	4
		$(2,0,0)$	$(\kappa+2)^2$	$\exp \frac{i2\pi}{a}\{(\kappa+2)x\}$	1

Example: The point group of **k** *for a lattice of tetrahedral symmetry*

The entire point group for sphalerite is T_d; there are no non-primitive translations. The point group of **k**, $\rho_{\mathbf{k}}$, is that subgroup of the T_d operations which, when operating on **k**, either leave it invariant or change it by a reciprocal lattice vector. Thus for **k** values in the [1,0,0] direction there are three associated point groups:

(i) At $\mathbf{k}=0$ all of the T_d operations leave **k** precisely at zero.
(ii) At $\mathbf{k}=(2\pi/a)(1,0,0)$, with reference to Figure 19.9, the operations E, C_2, 2σ leave k at the X point. The operations $2S_4$ and $2C'_2$ transform X to $-X$; these points are separated by the **K**-vector $(2\pi/a)(2,0,0)$. The group of **k** is thus the D_{2d} group, of order eight.
(iii) At a general k_x point only the four operations E, C_2, 2σ form the group of **k** as $2S_4$ and $2C'_2$ transform k_x to $-k_x$, which is now less than a **K**-vector away. This group is C_{2v}.

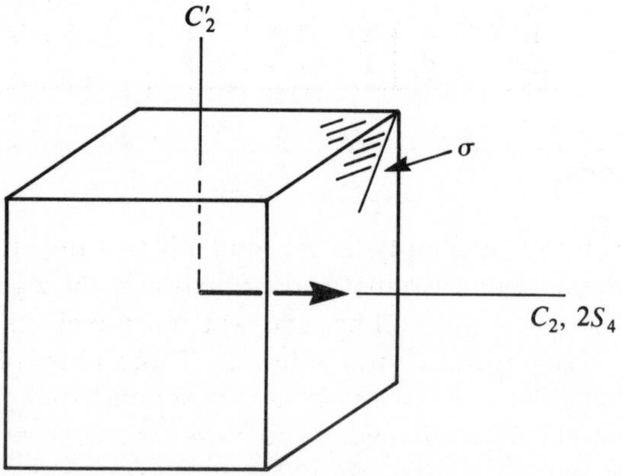

Figure 19.9 Operations in the group of **k** at X.

The characters of these groups are given in Table 19.4. Along with the familiar chemical notation we include the solid-state notation for the irreps. This notation is not unique, but two forms are in common use, following Bouckaert *et al.* (1963) and Koster (1957) (see Bibliography, Section (d)). Note that the use of Γ in Table 19.4(a) refers strictly to the $\mathbf{k}=0$, Γ-point. Generally if Γ appears with subscripts one is dealing with an irrep of the point group at $\mathbf{k}=0$ for a solid. Previously in this text we have used Γ with superscripts to denote irreps and without them to denote reducible representations of a group.

Table 19.4 Character tables for the groups of **k** along the k_x-axis, for the sphalerite symmetry groups.

		T_d	E	$8C_3$	$6S_4$	$3C_2$	$6\sigma_d$	
Γ_1	Γ_1	A_1	1	1	1	1	1	
Γ_2	Γ_2	A_2	1	1	−1	1	−1	(a)
Γ_{12}	Γ_3	E	2	−1	0	2	0	$\mathbf{k}=0$
Γ_{25}	Γ_5	T_1	3	0	1	−1	−1	
Γ_{15}	Γ_4	T_2	3	0	−1	−1	1	

	D_{2d}	E	$2S_4$	C_2	$2C_2'$	$2\sigma_d$	
X_1	A_1	1	1	1	1	1	
X_4	A_2	1	1	1	−1	−1	(b)
X_2	B_1	1	−1	1	1	−1	X point
X_3	B_2	1	−1	1	−1	1	
X_5	E	2	0	−2	0	0	

	C_{2v}	E	C_2	σ_v	σ'_v	
Δ_1	A_1	1	1	1	1	
Δ_2	A_2	1	1	−1	−1	(c)
Δ_3	B_1	1	−1	1	−1	Δ axis
Δ_4	B_2	1	−1	−1	1	

Basis functions

The significance of the empty lattice model is that the energies and eigenfunctions produced give us a first approximation to the eigenstates at each **k**. These functions form a useful basis for the more realistic situation, where they will be mixed by the crystal potential. Those states of closest energies are most strongly mixed and so our next best approximation is to use each *degenerate* set of empty lattice eigenfunctions as a basis for improved functions. The improved functions must be symmetry-adapted functions belonging to the irreps of $\mathscr{P}_{\mathbf{k}}$. Our next task is therefore to take each degenerate set in turn, find the characters for its matrix representation, and reduce this representation. It will prove useful to have available a table that shows how x, y, z transform under a typical member of each class of the point groups, in order to obtain the characters (Table 19.5).

Consider first the Γ-point. At $\epsilon=0$ there is just one empty lattice function, $f=$constant. This clearly transforms into itself under all T_d operations and forms a basis for the symmetric irrep. Γ_1 (A_1). At $\epsilon=3$, however, there are eight states,

$$f \propto \exp\left[\mathrm{i}\frac{2\pi}{a}(\pm x \pm y \pm z)\right]$$

(see Table 19.3). Using Table 19.5 these functions will transform amongst

Table 19.5 Transformation properties of x, y and z under the symmetry operation of the groups of **k** (Table 19.4).

T_d	E	C_3	S_4	C_2	σ_d
x	x	y	$-y$	$-x$	y
y	y	z	x	$-y$	x
z	z	x	$-z$	z	z

D_{2d}	E	S_4	C_2	C'_2	σ_d
x	x	$-x$	x	$-x$	x
y	y	$-z$	$-y$	$-y$	z
z	z	y	$-z$	z	y

C_{2v}	E	C_2	σ_v	σ'_v
x	x	x	x	x
y	y	$-y$	z	$-z$
z	z	$-z$	y	$-y$

Table 19.6

	E	$8C_3$	$6S_4$	$3C_2$	$6\sigma_d$
$\Gamma(\epsilon=3)$	8	2	0	0	4

themselves under the symmetry operations of T_d and the characters of their 8-dimensional reducible representation are given in Table 19.6. For example, $(+x+y+z)$ transforms to itself under C_3, as does $(-x-y-z)$. All other functions transform into different ones: for example, $C_3(+x+y-z) = (-x+y+z)$, etc. Under the σ_d operation used in Table 19.5 the four functions $(+x+y\pm z)$, $(-x-y\pm z)$ are invariant. Reducing Γ we obtain

$$\Gamma(\epsilon=3) \Rightarrow 2\Gamma_1 \oplus 2\Gamma_{15} \quad \text{or} \quad 2A_1 \oplus 2T_2.$$

This has the following interpretation. From the eight basis functions we can construct two symmetry-adapted combinations belonging to Γ_1 and six belonging to Γ_{15}. Combinations of the symmetry-adapted functions belonging to specific rows to these two irreps are the eigenfunctions for our new model. The hitherto eightfold-degenerate level must split into two singly and two triply-degenerate levels, of unspecified ordering.

The same procedure can be carried out for higher Γ-point energy levels, and for other **k** points. At $X(\epsilon=1)$ there are two degenerate states,

$$f \propto \exp\,[\mathrm{i}(2\pi/a)(\pm x)],$$

and at $\epsilon=2$ four states, $f \propto \exp\left[\mathrm{i}\frac{2\pi}{a}(-x\pm y\pm z)\right]$.

The characters for their representations and their reduced forms are given in Table 19.7.

Table 19.7

	E	$2S_4$	C_2	$2C'_2$	$2\sigma_d$		
$\Gamma(\epsilon=1)$	2	0	2	0	2	$\Rightarrow X_1 \oplus X_3$	or $A_1 \oplus B_2$
$\Gamma(\epsilon=2)$	4	0	0	0	2	$\Rightarrow X_1 \oplus X_3 \oplus X_5$	or $A_1 \oplus B_2 \oplus E$

The other states that will prove to be relevant in our discussion of the band structure are those along the Δ-axis lying below $\epsilon=3$ (Table 19.8).

Table 19.8

		E	C_2	σ_v	σ'_v	
$f \propto \exp \frac{i2\pi}{a} \kappa_x x.$	$\Gamma(0<\epsilon<1)$	1	1	1	1	$\Rightarrow \Delta_1$ or A_1
$f \propto \exp \frac{i2\pi}{a} (\kappa_x - 2).$	$\Gamma(1<\epsilon<4)$	1	1	1	1	$\Rightarrow \Delta_1$ or A_1
$f \propto \exp \frac{i2\pi}{a} [(\kappa_x - 1)x \pm y \pm z].$	$\Gamma(2<\epsilon<3)$	4	0	2	2	$\Rightarrow 2\Delta_1 \oplus \Delta_3 \oplus \Delta_4$ or $2A_1 \oplus B_1 \oplus B_2$

Splitting of degeneracies

As a demonstration that the accidental degeneracies will be split in a finite crystal field we will consider the $\boldsymbol{X}$ point at energy $\epsilon=1$. In the empty lattice model the degenerate eigenfunctions at this position are $\psi = \exp[i(\pi/a)(\pm x)]$. As the field is reintroduced these two functions, to a first approximation, form a basis from which eigenfunctions at $\boldsymbol{X}$ and of energies close to $\epsilon=1$ can be constructed. These eigenfunctions must have X_1 and X_3 symmetry respectively—they are the symmetry-adapted combinations of $\exp[i(\pi/a)(\pm x)]$. It is easy to see that such combinations are the standing waves:

$$\psi^{X_1} \propto \cos(\pi x/a), \qquad X_1 \text{ symmetry}$$

$$\psi^{X_3} \propto \sin(\pi x/a), \qquad X_3 \text{ symmetry.}$$

Now introduce a model-potential that has (i) T_d symmetry and (ii) lattice periodicity:

$$V(r) = V_0 + V_1 \left(\cos\frac{2\pi x}{a} + \cos\frac{2\pi y}{a} + \cos\frac{2\pi z}{a}\right)$$

is a suitable simple model potential (having somewhat more than the required T_d symmetry). The functions ψ^{X_1}, ψ^{X_3} are indeed eigenfunctions of the Hamiltonian and their energy eigenvalues differ by the amount

$$\Delta E = \left[\frac{\int (\psi^{X_1})^* V(\mathbf{r}) \psi^{X_1}}{\int (\psi^{X_1})^* \psi^{X_1}} - \frac{\int (\psi^{X_3})^* V(\mathbf{r}) \psi^{X_3}}{\int (\psi^{X_3})^* \psi^{X_3}} \right]$$

$$= \left[\frac{V_1 \int_0^a \cos^2 (\pi x/a) \cos (2\pi x/a)}{\int_0^a \cos^2 (\pi x/a)} - \frac{V_1 \int_0^a \sin^2 (\pi x/a) \cos (2\pi x/a)}{\int_0^a \sin^2 (\pi x/a)} \right]$$

$$= V_1.$$

As expected, the finite size of the periodic potential serves to open up an energy gap between states that were degenerate in the empty lattice limit but have different **k** point group symmetries.

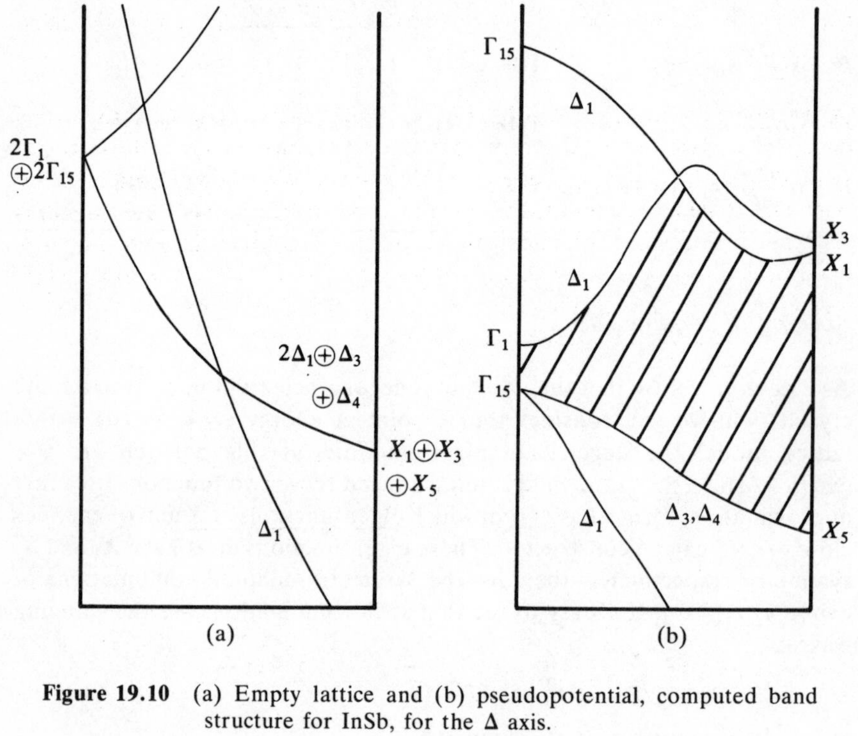

Figure 19.10 (a) Empty lattice and (b) pseudopotential, computed band structure for InSb, for the Δ axis.

Figure 19.10(a) shows the empty lattice bands with the point group labelling. By contrast Figure 19.10(b) shows a pseudopotential computation for the InSb band structure. What we should like to be able to do without having to resort to computation is to take the empty lattice results and see if we could anticipate any of the orderings of the states that derive from accidentally degenerate groupings. It turns out that there are several clues as to the ordering.

(i) ***Compatibility relations***

The problem of moving in **k**-space from a point of high symmetry (e.g. Γ) to an adjacent position of lower symmetry ($k \approx 0$, along the Δ axis) is directly analogous to the subgroup problem for symmetries in real space. Thus if we consider a Γ_{15} state in the sphalerite structure there are three degenerate eigenfunctions.

Extremely close to $\mathbf{k}=0$ the eigenfunctions and their energies hardly change, but now we know that the states have the symmetries of the Δ irrep. Hence there are three combinations of the original eigenfunctions that are bases for certain irreps of Δ. These irreps could be found by taking the three functions, obtaining the characters for their reducible matrix representation of the Δ group and reducing it. This is precisely the subgroup problem and we can use the characters of Γ_{15} in order to solve it directly from Table 19.4:

$$\Gamma_{15} \Rightarrow \Delta_1 \oplus \Delta_3 \oplus \Delta_4.$$

When we move away from the Γ_{15} point, in the k_x direction, the triply degenerate level must therefore split into three non-degenerate states of symmetries $\Delta_1, \Delta_3, \Delta_4$ respectively. Again the order of the splitting is uncertain. The same arguments follow for other symmetry points. A set of *compatibility relations* can be constructed. In order to form the entire energy band ($k_x=0$ to $k_x=\pi/a$ for example) only certain irreps of the lower symmetry are compatible with certain irreps at the higher symmetry points, as given by the subgroup reductions.

Table 19.9 Compatibility relations for the $\Gamma - X$ axis of a T_d system.

Γ_1	Γ_2	Γ_{12}	Γ_{25}	Γ_{15}
Δ_1	Δ_2	$\Delta_1 \oplus \Delta_2$	$\Delta_2 \oplus \Delta_3 \oplus \Delta_4$	$\Delta_1 \oplus \Delta_3 \oplus \Delta_4$
X_1	X_4	X_2	X_3	X_5
Δ_1	Δ_2	Δ_2	Δ_1	$\Delta_3 \oplus \Delta_4$

Table 19.9 summarizes the $\Gamma - X$ symmetry restrictions for sphalerites. The Γ_{15} states split into Δ_1, Δ_3 and Δ_4 but must join X_1 (or X_3) and X_5 states when they meet the X point. (If we treat the combination $\Delta_3 \oplus \Delta_4$ as a single state Δ_5, we can see now the origin of the subscript labelling of the Γ point irreps. This labelling tells us how the states will split: $\Gamma_{15} \Rightarrow \Delta_1 \oplus \Delta_5$, etc.) Note that the symmetries imposed by consideration of the empty lattice eigenfunctions are consistent with the compatibility relations, as they must be.

(ii) ***Time-reversal symmetry***

We know that the symmetry of $E(\mathbf{k})$ is the same as that of the crystal point group. Hence for centrosymmetric crystals (with an inversion centre) the energy bands at $(-\mathbf{k})$ are the same as at $\mathbf{k}$. This result may however be generalized to *all* crystals, by consideration of the time reversal symmetry of the Hamiltonian problem. It is usual for the Hamiltonian operator to be real (rather than just Hermitian, as required of an operator representing an observable). This is certainly the case for all Hamiltonians we have considered so far. Consequently if ψ is a solution of the time-independent equation

$$\mathscr{H}\psi = E\psi,$$

then, taking the complex conjugate of both sides,

$$\mathscr{H}\psi^* = E\psi^*$$

and ψ^* is another (or the same) eigenfunction, degenerate with ψ. As far as Bloch functions are concerned the complex conjugate of $U_{n\mathbf{k}}(r)\, e^{i\mathbf{k}.\mathbf{r}}$ is $U^*_{n\mathbf{k}}(r)\, e^{-i\mathbf{k}.\mathbf{r}}$. Whatever the precise nature of U^*_n it still has lattice periodicity and we can therefore re-label it as some $U_{n'}$. Further, as the associated exponential arguement is $(-i\mathbf{k}.\mathbf{r})$ it is appropriate to use the relabelling $U^*_{n\mathbf{k}}(\mathbf{r}) \rightarrow U_{n'(-\mathbf{k})}(\mathbf{r})$. The point is that for any state at $\mathbf{k}$ there must be a degenerate state at $-\mathbf{k}$, regardless of the spatial symmetry of $V(\mathbf{r})$.

$$E_{n'}(-\mathbf{k}) = E_n(\mathbf{k}) \qquad \text{for some } n'.$$

Such a result may be thought of as being a consequence of time reversal symmetry. This is because the function $\psi^*(t)$ evolving in the positive time direction traces the past history of $\psi(t)$. It is the time reversed function, obeying the same equation as $\psi(-t)$:

$$\text{if} \qquad \mathscr{H}\psi(t) = i\hbar\, \partial\psi/\partial t,$$

$$\text{then} \qquad \mathscr{H}\psi(-t) = i\hbar\, \partial\psi(-t)/\partial(-t) = -i\hbar\, \partial\psi(-t)/\partial t,$$

$$\text{and} \qquad \mathscr{H}\psi^*(t) = [i\hbar\, \partial\psi(t)/\partial t]^* = -i\hbar\, \partial\psi^*(t)/\partial t.$$

This time-reversal symmetry can lead to additional degeneracies of the energy bands. For the present we shall simply site the case of significance, that of the Δ_3 and Δ_4 bands. Time reversal sends a Δ_3 state at k_x to a degenerate state at $-k_x$ having the same symmetry. The spatial operation S_4 also reverses the sign of k_x. The resulting state must again have the same energy as the original one (as S_4 is in the crystal space group) but can now be shown to have Δ_4 symmetry and is therefore distinct from the time reversal state. Consequently Δ_3 and Δ_4 states must be degenerate at all k_x, not just for $k_x \rightarrow 0$ or $2\pi/a$, as is indicated by the compatibility results.

(iii) ***Bonding considerations***

A final, extremely useful indication of the ordering some of the energy bands is obtained by considering the opposite extreme from the empty lattice approximation, namely the tight binding approximation. Just as in the molecular case, in the immediate vicinity of a nucleus the electronic distribution functions must closely approximate the atomic functions. It is therefore worthwhile constructing linear combinations of such functions and determining their symmetry properties. If $f_j(\mathbf{r}-\mathbf{R}_i)$ is an atomic function centred at a nucleus ($\mathbf{R}_i$) in the primitive cell denoted by T, construct a function of form

$$\psi_{n\mathbf{k}}(\mathbf{r})=\mathrm{e}^{\mathrm{i}\mathbf{k}.\mathbf{r}}\sum_{j,\tau,i} a_{nij}f_j(\mathbf{r}-\mathbf{R}_i),$$

where the sum is over the orthonormal set of atomic functions, over all the primitive cells, and if necessary over each nucleus in the cell. $\psi_{n\mathbf{k}}(\mathbf{r})$ satisfies the requirements of a crystal electronic eigenfunction and, if one of the atomic orbital states dominates, closely represents an atomic orbital near each nucleus. From energy considerations one would choose to construct these LCAO states from the *ns* and *np* atomic orbitals in order to describe the valence and lower conduction states of semiconductors. The $\psi_{n\mathbf{k}}(\mathbf{r})$ must also belong to the irreps of the point group of $\mathbf{k}$. In particular, at $\mathbf{k}=0$ the Bloch functions $\Sigma_{j\tau i}a_{nij}f_j(\mathbf{r}-\mathbf{R}_i)$ must have the symmetry of the entire point group of the lattice. Our j sum is now over the *ns*, *np* atomic orbitals. The symmetry is obeyed if the sum over each specific primitive cell has the point group symmetry. Concentrate on the pair of atoms which are contained in any chosen primitive cell in InSb. From the *s* and *p* states of these two atoms we anticipate constructing approximate eigenfunctions that are symmetry-adapted bases for the Γ-point **k**-group (T_d). Under these point group operations an *s* orbital centred on the indium nucleus (taken to be the origin of the point group) obviously always transforms into itself. A similar orbital on the antimony nucleus transforms either into itself or into another *s* orbital on an atom in one of the adjacent primitive cells. As far as the complete lattice of indium *s* orbitals or the complete primitive cell (containing portions of functions centred in different cells) is concerned, the T_d operations leave them invariant. Table 19.10 summarizes the characters for the eight *s* and *p* orbitals, analysed separately.

Table 19.10 *s* and *p* state orbital symmetries for InSb.

	E	$8C_3$	$6S_4$	$3C_2$	$6\sigma_d$	
Γ^s(In)	1	1	1	1	1	$=\Gamma_1$
Γ^s(Sb)	1	1	1	1	1	Γ_1
Γ^p(In)	3	0	−1	−1	1	Γ_{15}
Γ^p(Sb)	3	0	−1	−1	1	Γ_{15}

A general state of Γ_1 symmetry therefore has the form

$$\psi^{\Gamma_1} = \sum_{\text{cells}} [a_{\text{In}} f_s(\text{In}) + a_{\text{Sb}} f_s(\text{Sb})].$$

There are two solutions, one with a_{Sb} of the same sign as a_{In} and an orthogonal solution with opposing signs. In the former the electronic charge is concentrated in the region between the nearest neighbour In and Sb atoms. As in the hydrogen molecule example, this represents an s-bonding orbital and has lower energy than the second, s-antibonding, solution. Similarly there is a pair of bonding and anti-bonding orbitals of symmetry Γ_{15}. These four $\mathbf{k}=0$ states are expected to correspond to the four lowest Γ-point states of the empty lattice model (Figure 19.11).

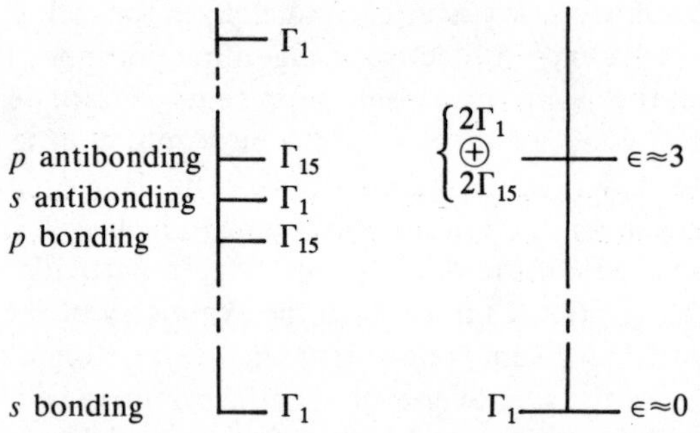

Figure 19.11 Comparison of bonding-orbital ideas with the empty-lattice model of the valence and condition states of InSb.

The implication is clear. The s-bonding state must correspond to the $\epsilon=0$ empty lattice state. Of the four states originating from $\epsilon=3$, one of the Γ_{15} states is likely to lie lowest in energy, corresponding to p-bonding. Next should come a Γ_1 and the Γ_{15} state (probably in that order) corresponding to the antibonding states. The other Γ_1 state is likely to be removed somewhat higher in energy. In Figure 19.10 we see that this is precisely what happens in practice.

We have discussed the need to form sp^3 hybridized bonding orbitals in the crystal and described the bonding nature of the $\mathbf{k}=0$ eigenfunctions. Intriguingly, though, symmetry properties forbid the mixture of atomic s and p states in a single eigenfunction. By a hybridized bond we are to understand (as in the molecular tetrahedral bond of CH_4, for example) that the electron distribution summed over *all* the valence states has such a bonding nature. This is confirmed by computation; Figure 19.12 shows the net electron distribution obtained through the pseudopotential calculation. The charge is concentrated between each pair of nearest-neighbour nuclei, being slightly more localized in the vicinity of the group V anion in the sphalerite case. In contrast the conduction levels would have an almost uniform charge distribution were they to be fully occupied by electrons.

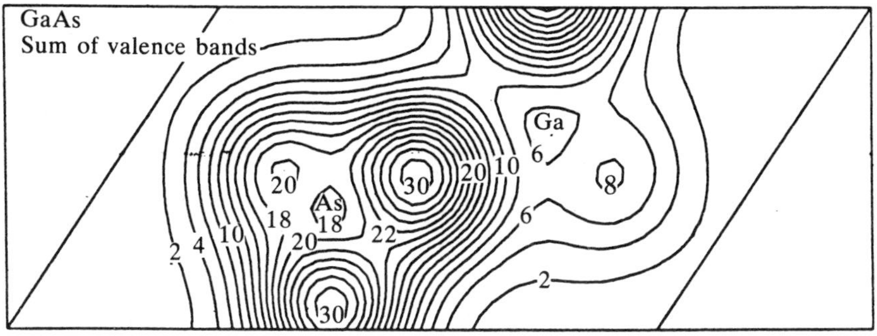

Figure 19.12 The valence charge density of GaAs, summed over all valence **k** states. [after J. P. Walter and M. L. Cohen, *Phys. Rev.* **4B,** 1877 (1971)].

(iv) ***The non-crossing rule***

Two or more empty-lattice states of the same symmetry might exist at some general point in **k**-space: for example, there are three Δ_1 states in the k_x direction, where $(\kappa_x-2)^2=(\kappa_x-1)^2+2$ (Figure 19.7). The empty lattice eigenfunctions at such crossing points are in general coupled by the periodic potential. This is the same statement as saying that the exact eigenfunctions are linear combinations of those approximate empty lattice functions of the same symmetry. If two otherwise degenerate states are coupled by some perturbation, then they repel each other and a gap forms. Similarly, in the vicinity of the cross-over the higher-energy empty lattice state is repelled to a higher energy yet, and vice versa. Thus no cross-overs in the band structure are expected for states of the same symmetry. This rule does not apply for different symmetry types.

The consideration of the four clues to band ordering – compatibility, time-reversal, bonding and non-crossing – lead one to construct for T_d face-centred-cubic crystals a nearly empty-lattice band structure is given in Figure 19.13. Apart from the distinction between the $X_{1,3}$ points, the ordering is well specified. Only the magnitudes of the various energy corrections due to the crystal potential need to be determined. We note that one is not always likely to be in such a good position prior to such calculations.

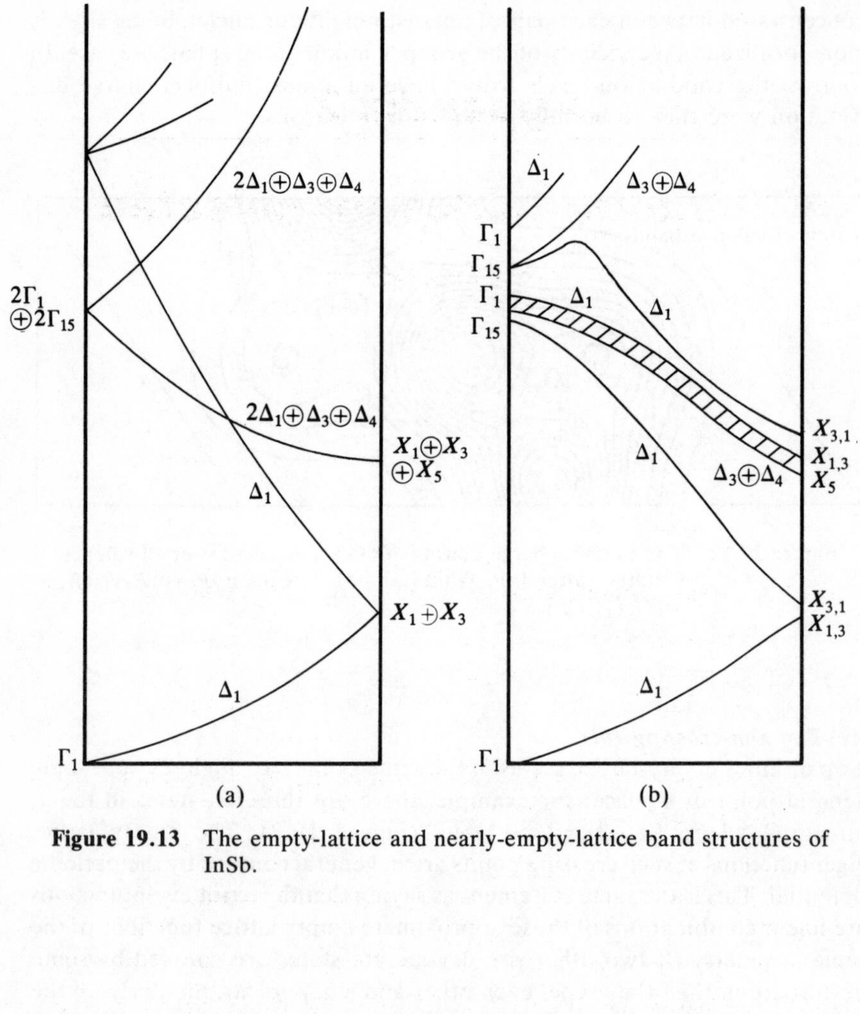

Figure 19.13 The empty-lattice and nearly-empty-lattice band structures of InSb.

19.5 Nearly-free-electron model for O_h^7 structures

Silicon and germanium are sufficiently important semiconductor materials for us to summarize the nearly empty lattice model for the non-symmorphic O_h^7 structure. Table 19.11 gives the character tables for the point group of $\mathbf{k}=0$, k_x and $2\pi/a$. The empty lattice band structure is of course the same as for the sphalerite materials. In order to use the character tables to reduce the representations based on the empty lattice eigenfunctions one must consider the operations $\{R/\boldsymbol{\tau}\}$ whenever indicated. Thus the function

$$\exp \mathrm{i}\, 2\pi(x+y+z)/a,$$

which is one of the eight degenerate $\epsilon=3$ functions, transforms under the operation $\{C_4/\boldsymbol{\tau}\}$ as follows:

$$(x+y+z)\xrightarrow{C_{4z}}(y-x+z)\xrightarrow{\boldsymbol{\tau}=\frac{a}{4}(1,1,1)}\left(y-\frac{a}{4},-x-\frac{a}{4},z-\frac{a}{4}\right),$$

so

$$\exp\left[\frac{\mathrm{i}2\pi}{a}(x+y+z)\right]\longrightarrow \mathrm{e}^{-\mathrm{i}3\pi/2}\exp\left[\frac{\mathrm{i}2\pi}{a}(-x+y+z)\right].$$

Table 19.11 Character tables for use with the O_h^7 group. Operations labels refer to the O_h point group operations. For details of the effective table for the X point see, for example, Landsberg (1969) (Bibliography, Section (d)).

(a) Γ point; $\mathbf{k}=0$. $\boldsymbol{\tau}=a(\frac{1}{4},\frac{1}{4},\frac{1}{4})$

	O_h	E	$8C_3$	$6C_4\boldsymbol{\tau}$	$3C_2$	$6C'_2\boldsymbol{\tau}$	$\mathrm{i}\boldsymbol{\tau}$	$8S_6\boldsymbol{\tau}$	$6S_4$	$3\sigma_h\boldsymbol{\tau}$	$6\sigma_d$	
Γ_1	A_{1g}	1	1	1	1	1	1	1	1	1	1	
Γ_2	A_{2g}	1	1	−1	1	−1						
Γ_{12}	E_g	2	−1	0	2	0	same block					
Γ'_{15}	T_{1g}	3	0	1	−1	−1						(R_x,R_y,R_z)
Γ'_{25}	T_{2g}	3	0	−1	−1	1						
Γ'_1	A_{1u}											
Γ'_2	A_{2u}	same block					(minus one)×(same block)					
Γ'_{12}	E_u											
Γ_{15}	T_{1u}											(x,y,z)
Γ_{25}	T_{2u}											

(b) Δ axis; $(k_x,0,0)$. $\Delta=\exp(\mathrm{i}\mathbf{k}_\Delta.\boldsymbol{\tau})$, $\boldsymbol{\tau}=a(\frac{1}{4},\frac{1}{4},\frac{1}{4})$

	C_{4v}	E	$2C_4\boldsymbol{\tau}$	C_2	$2\sigma_d$	$2\sigma_h\boldsymbol{\tau}$
Δ_1	A_1	1	$1.\Delta$	1	1	$1.\Delta$
Δ'_1	A_2	1	$1.\Delta$	1	−1	$-1.\Delta$
Δ'_2	B_1	1	$-1.\Delta$	1	1	$-1.\Delta$
Δ_2	B_2	1	$-1.\Delta$	1	−1	$1.\Delta$
Δ_5	E	2	0	−2	0	0

(c) X point; $\mathbf{k}=(2\pi/a,0,0)$. Effective character table.

	E	$C_{4x}\boldsymbol{\tau},C_{4x}^3\boldsymbol{\tau}$	C_{2x}	C_{2y},C_{2z}	$C'_{2\bar{y}z}\boldsymbol{\tau}$	$C'_{2yz}\boldsymbol{\tau}$	$\mathrm{i}\boldsymbol{\tau}$	S_{4x},S_{4x}^3	$\sigma_{hx}\boldsymbol{\tau}$	$\sigma_{hy}\boldsymbol{\tau},\sigma_{hz}\boldsymbol{\tau}$	$\sigma_{d\bar{y}z},\sigma_{dyz}$
X_1	2	0	2	0	0	0	0	0	0	0	2
X_2	2	0	2	0	0	0	0	0	0	0	−2
X_3	2	0	−2	0	2	−2	0	0	0	0	0
X_4	2	0	−2	0	2	2	0	0	0	0	0

In general the $\boldsymbol{\tau}$ translation will lead simply to an additional constant factor in the various non-zero matrix elements of the representation $\{R/\mathbf{O}\}$. In practice all of the eight Γ-point functions of greatest interest transform into different functions (amongst the set) under those operations in the $\mathbf{k}=0$ point group that contain the non-zero $\boldsymbol{\tau}$. The presence of the exp $(i\mathbf{K}_n.\boldsymbol{\tau})$ factor is therefore irrelevant, as $\chi=0$. Hence we obtain the characters for all ten classes:

	E	$8C_3$	$6C_4\boldsymbol{\tau}$	$3C_2$	$6C_2'\boldsymbol{\tau}$	$i\boldsymbol{\tau}$	$8S_6\boldsymbol{\tau}$	$6S_4$	$3\sigma_h\boldsymbol{\tau}$	$6\sigma_d$
$\Gamma(\epsilon=3)$	8	2	0	0	0	0	0	0	0	4

$$\Gamma(\epsilon=3) \Rightarrow \Gamma_1 \oplus \Gamma'_2 \oplus \Gamma_{15} \oplus \Gamma'_{25}$$

As before $\Gamma(\epsilon=1)=\Gamma_1$.

Along the Δ-axis we obtain, using Tables 19.3, 19.5 and 19.11, the characters shown in Table 19.12.

Table 19.12

	E	$2C_4\boldsymbol{\tau}$	C_2	$2\sigma_d$	$2\sigma_h\boldsymbol{\tau}$	
$\Gamma(0<\epsilon<1)$	1	Δ	1	1	Δ	$=\Delta_1$
$\Gamma(1<\epsilon<4)$	1	$-\Delta$	1	1	$-\Delta$	$=\Delta'_2$
$\Gamma(2<\epsilon<3)$	4	0	0	2	0	$\Rightarrow\Delta_1\oplus\Delta'_2\oplus\Delta_5$

To obtain the compatibility relations we must again account for $\boldsymbol{\tau}$. A set of functions transforming with character $\chi(G_r)$ in the $\mathbf{k}=0$ point group transforms as a reducible rep. or irrep. of character $\chi(G_r)$ exp $i\mathbf{k}_\Delta.\boldsymbol{\tau}$ in the subgroup, etc.

Table 19.13 Compatibility relations for the Δ-axis in the O_h^7 group.

Γ_1	Γ_2	Γ_{12}	Γ'_{15}	Γ'_{25}	Γ'_1	Γ'_2	Γ'_{12}	Γ_{15}	Γ_{25}
Δ_1	Δ_2	$\Delta_1\oplus\Delta_2$	$\Delta'_1\oplus\Delta_5$	$\Delta'_2\oplus\Delta_5$	Δ'_1	Δ'_2	$\Delta'_1\oplus\Delta'_2$	$\Delta_1\oplus\Delta_5$	$\Delta_2\oplus\Delta_5$

X_1	X_2	X_3	X_4
$\Delta_1\oplus\Delta'_2$	$\Delta'_1\oplus\Delta_2$	Δ_5	Δ_5

Finally the bonding considerations require us to recognize that whereas In and Sb were distinct before and whereas none of the T_d operations transformed an In atomic state into an Sb state, the opposite is true for the group IV crystals. For example, under the operation $\{i/\boldsymbol{\tau}\}$ an s-state on the atom at (0,0,0) transforms into itself under i and then into an s-state centred at the other nucleus $(\frac{1}{4},\frac{1}{4},\frac{1}{4})$ under $\boldsymbol{\tau}$.

The two s-states (combined over all cells) therefore form a basis for a two-dimensional representation of the group of $\mathbf{k}=0$. The reduction of this representation and of the six-dimensional p-state representation is as shown in Table 19.14.

Table 19.14

	E	$8C_3$	$6C_4\tau$	$3C_2$	$6C'_2\tau$	$i\tau$	$8S_6\tau$	$6S_4$	$3\sigma_h\tau$	$6\sigma_d$	
$\Gamma(s_1,s_2)$	2	2	0	2	0	0	0	2	0	2	$\Rightarrow\Gamma_1\oplus\Gamma'_2$
$\Gamma(p_1,p_2)$	6	0	0	−2	0	0	0	−1	0	1	$\Rightarrow\Gamma'_{25}\oplus\Gamma_{15}$

All τ operations transform s_1 into s_2 etc. and must give a zero character. The pure point group operations lead to characters equal to those of $\{\Gamma^p(\text{In})\oplus\Gamma^p(\text{Sb})\}$ of the compound case (Table 19.10). The bonding / antibonding states can be identified precisely in this case, because each irrep. in the above reduction appears just once. Thus the s-bonding state contains a combination

$$\sum_{\text{cells}} [f(s_1)+f(s_2)],$$

and has Γ_1 symmetry. In the p-bonding state the p-orbitals along a given bond must be directed towards each other, concentrating the change in this region; they have Γ'_{25} symmetry. (Note that these are the two irreps of even parity). The order of states at $\mathbf{k}=0$ is summarized in Figure 19.14.

It turns out that in this example the above order of the antibonding states is not necessarily taken up in the crystal. In silicon the Γ_{15} state lies below Γ'_2. For the present we note that the different symmetry types lead to different nonlinear optical properties for materials of the two structures.

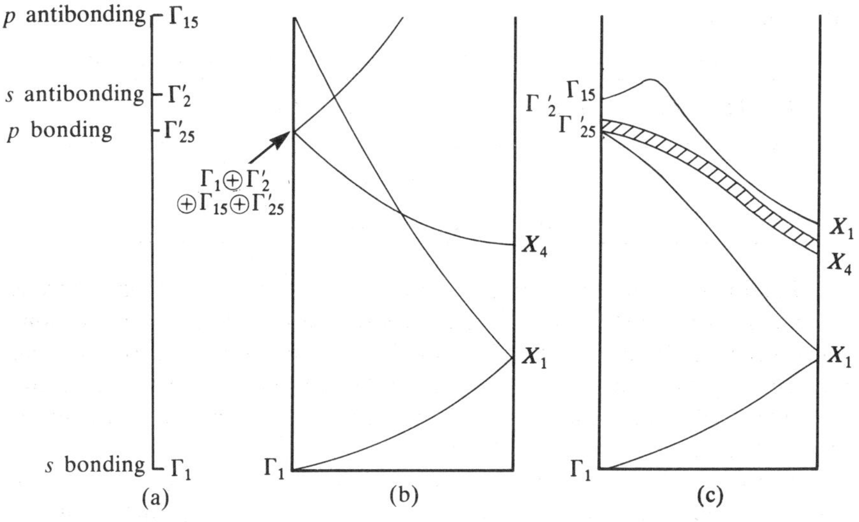

Figure 19.14 Summary of model band structures for the diamond lattice: (a) bonding; (b) empty-lattice; (c) nearly-free-electron model.

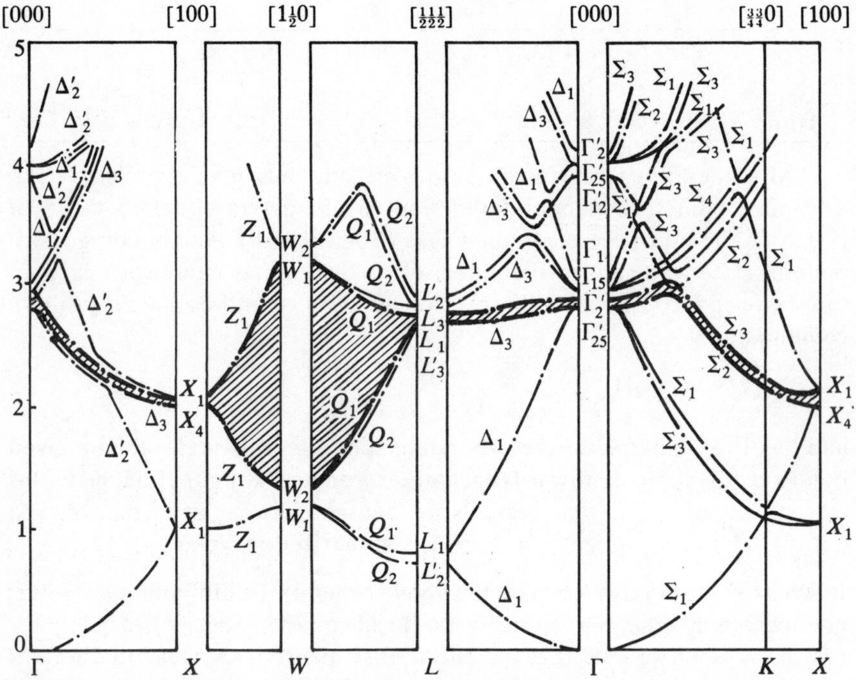

Figure 19.15 Nearly-empty-lattice band structure for germanium, after Herman (1953) (see Bibliography, Section (f)).

19.6 Energy band computations

We should remind ourselves of the approximations made in our considerations of energy band structure, and discuss briefly a few of the methods that may be used to derive the magnitude of the crystal field perturbation of the empty lattice results. No attempt will be made to follow through such calculations in detail as in general little extra understanding is achieved by so doing. Still ignoring spin effects, the many particle Hamiltonian for which we are in principle attempting a solution is:

$$\mathcal{H}_{\text{crystal}} = \sum_n \frac{-\hbar^2}{2M_n}\nabla_n^2 + \sum_i \frac{-\hbar^2}{2m}\nabla_i^2 + \sum_n V_{n-n} + \sum_{i,n} V_{e-n} + \sum_{i,j} V_{e-e}.$$

Respectively the terms in this sum are the nuclear and electronic kinetic energy terms, the nuclear–nuclear coulombic repulsion, electron–nuclear attraction and electron–electron repulsion. The nuclear kinetic terms are removed by application of the Born–Oppenheimer approximation as in the molecular case. The resultant equations can be placed in one-electron form only if the electron–electron interactions can be averaged in some manner:

$$\mathcal{H}_e \Rightarrow \sum_i \frac{-\hbar^2}{2m}\nabla_i^2 + \sum_n V_{e-n} + \bar{V}_{e-e}.$$

Given first guesstimates of the one-electron wave functions, ψ_j, then $\bar{V}_{e-e}$ is obtained by averaging over the occupied states:

$$\bar{V}_{e-e}(\mathbf{r}) = \sum_j{}' e^2 \frac{\int \psi_j^*(\mathbf{r}')\,\psi_j(\mathbf{r}')\,\mathrm{d}\mathbf{r}'}{|\mathbf{r}-\mathbf{r}'|}.$$

Within the Hartree approximation one can therefore attempt self-consistent solutions with

$$\mathcal{H} = \sum_i \mathcal{H}_i = \sum_i \left[\frac{-\hbar^2}{2m}\nabla_i^2 + V_i(\mathbf{r})\right]; \quad \Psi = \prod_i \psi_i.$$

As far as we are concerned it is sufficient that $V_i(\mathbf{r})$ has the symmetry of the crystal. The only account taken of the electron–electron interactions is that they contribute a portion of the periodic potential $V_i(\mathbf{r})$ and that the self-consistent one-electron states one would calculate are populated in accord with the Pauli exclusion principle (Fermi–Dirac statistics). Proper account of Pauli exclusion of course would require us to consider Slater determinant states and the resulting exchange-interaction terms. We shall not consider this problem. Clearly we need to have an expression for an initial $V(\mathbf{r})$ in order to proceed with computation of the energy levels. This $V(\mathbf{r})$ is provided by the placement of the nuclei in the primitive cell and the initial guess for the electronic eigenfunctions.

In the nearly empty lattice model a reasonable starting point for the ψ_i is that the core electrons should be supposed to be in the appropriate atomic orbitals and the valence electrons in the lowest energy empty-lattice states. This also gives an initial guess at $\bar{V}_{e-e}(\mathbf{r})$, and $V(\mathbf{r})$. Rather than attempt to solve the differential Hamiltonian equation to obtain the next iteration, we would then like to expand the eigenfunctions as *linear combinations of plane waves*. The point is that the translational symmetry determines that only states of the same $\mathbf{k}$ value mix. Thus for the next iteration construct

$$\psi_{n\mathbf{k}}(\mathbf{r}) = \{\sum_j a_{nj} \exp \mathrm{i}\mathbf{K}_j.\mathbf{r}\}\, \mathrm{e}^{\mathrm{i}\mathbf{k}.\mathbf{r}}.$$

A finite number of plane waves must be used if the solution is to be tractable. This solution takes the form

$$a^{-1}[\mathcal{H} - E]a = [\mathcal{E} - E]$$

where $\mathcal{E}$ is diagonal with $\mathcal{E}_{nn'} = E_n(\mathbf{k})\delta_{nn'}$. The matrix elements of $\mathcal{H}$ are

$$\mathcal{H}_{ij} = \int \exp[-\mathrm{i}(\mathbf{K}_i + \mathbf{k}).\mathbf{r}]\left(\frac{-\hbar^2}{2m}\nabla^2 + V(\mathbf{r})\right) \exp[\mathrm{i}(\mathbf{K}_j + \mathbf{k}).\mathbf{r}]\,\mathrm{d}\mathbf{r}.$$

The sum over j may be further reduced by consideration of the point group symmetry of **k**. Starting from the plane waves, one can in principle calculate all the resulting symmetry-adapted functions belonging to each irrep. of the group of **k**. The j summation should be only over symmetry-adapted functions belonging to the same row of any given irrep. Given a self-consistent solution for some finite number of terms in Σ_j, one should check that a larger number of terms give the same solution. Unfortunately over 100 waves are required for full self-consistency, necessitating a huge computational effort. The reason is that in the vicinity of the nuclei the valence states, which we understand to originate from the outer atomic orbitals, have several nodes. Their oscillatory behaviour in this region therefore requires an expansion over many plane waves.

In order to reduce this computational problem one can start by constructing plane-wave combinations that are orthogonal to the core atomic states, and therefore automatically contain the required oscillatory properties within the unit cell. These functions can then be combined in the above manner to solve self-consistently the finite matrix problem. This is called the *orthogonalized plane wave* method.

An alternative to adapting the plane wave functions is to divide the crystal potential problem into two regions. In spherical regions close to the nuclei (and core electrons) one considers the Coulombic terms, but $V(r)$ is taken to be constant in the regions away from the nuclei. The resulting potential is referred to as the *muffin-tin* potential. The advantage of its use is that a plane-wave expansion is appropriate where the potential is uniform and spherical functions may be combined in the nuclear regions. Boundary conditions must of course be satisfied. This is the *augmented plane wave* method.

Both potential and basis functions may be adapted in the various energy band calculations. The *pseudo-potential* method takes advantage of both techniques. This is the most successful method applied to date. Figure 19.16 shows the results of such calculations, for four semiconductors. Clearly the small shifts in energy discussed in the nearly empty lattice model at some Brillouin zone positions may become relatively large. Nevertheless the overall picture of the energy bands given by the simple model proves to be extremely useful. Further, the structure obtained for the different materials is clearly very similar. It is only because one is in a position to probe experimentally in great detail this structure that the differences between semiconductors can be established. Features to note in the bands schemed are that the top of the valence band appears at or very close to $\mathbf{k}=0$ in each material. The bottom of the conduction band however may lie at $\mathbf{k}=0$ as in the case of InSb, InAs and GaAs, or at the Brillouin zone boundary as in Ge or GaP, or at some general point in the zone, as in the case of silicon.

Metals and insulators

Before leaving this section a small amount of perspective needs to be made by commenting on the relation of the nearly empty lattice models to metals and insulators. Aluminium and copper are both face-centred-cubic crystals and therefore have the same empty lattice band structure as InSb, Ge etc. Of course the number of electrons to be contained in these bands differs and this is the reason for the differing conductivity and optical properties. Outside of closed shells aluminium has three electrons, there is one atom per primitive cell, and so the Fermi energy is expected to lie near $\epsilon = 1.3$ on the band diagram (see Figure 19.8). There is no band gap at this energy that can open up as the crystal potential is accounted for; the Fermi energy is at a position that has a high density of states available just above it. This is the requirement for a metal. The latter statements are also true for copper, although its band structure is complicated by the presence of $(n-1)$ d-electrons which form a band within the region of the free-electron valence band. For divalent metals the Brillouin zone structure is such that the Fermi energy lies below the top of a band in one **k**-space direction. Again no band gap exists between occupied and empty states.

The insulator condition is easier to describe; the band structure is more closely resembled by the tightly bound core electron states for all the occupied levels. The nearly free states now form an entirely empty 'conduction' band.

Problems

19.1 Show that the two functions

$$(\cos 2\pi x/a \quad \cos 2\pi y/a \quad \cos 2\pi z/a)$$

$$(\sin 2\pi x/a \quad \sin 2\pi y/a \quad \sin 2\pi z/a)$$

each belongs to the Γ_1 irrep. of the T_d group, and that the functions appropriately abreviated as

$$\begin{matrix} (\cos & \cos & \sin) & & (\cos & \sin & \sin) \\ (\cos & \sin & \cos) & \text{and} & (\sin & \cos & \sin) \\ (\sin & \cos & \cos) & & (\sin & \sin & \cos) \end{matrix}$$

form bases for Γ_{15}.

(You will need to consider the effect on x, y and z of the 24 operations in the T_d group.)

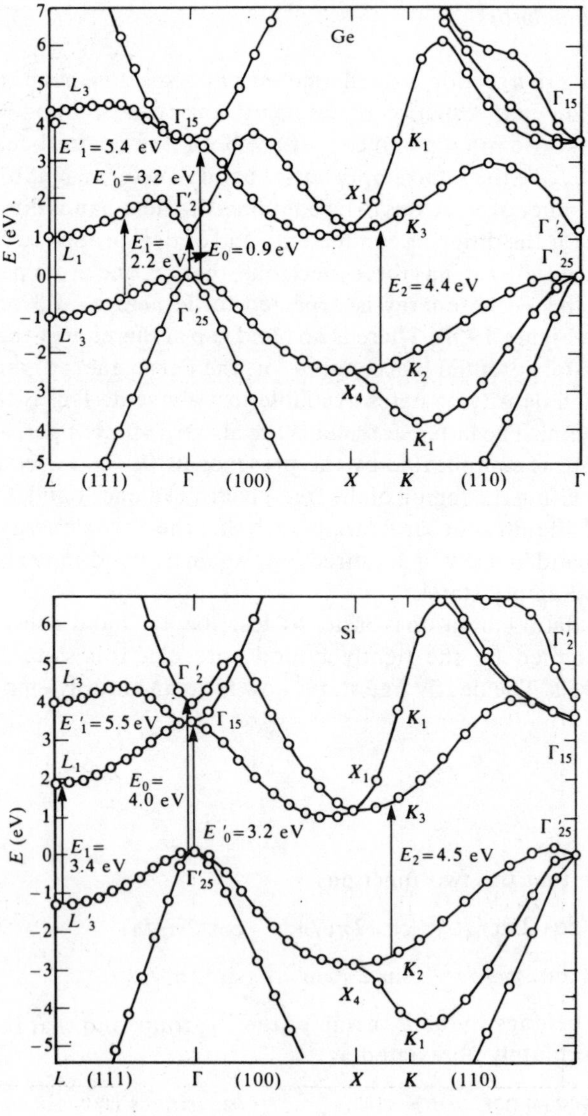

Figure 19.16 Pseudopotential calculations of the energy bands $E_n(\mathbf{k})$ of germanium, silicon, indium arsenide and gallium arsenide, near the energy gap. Optical transitions between the valence and conduction bands are indicated by arrows. The labelling of the transitions is conventional, but the inter-band energies quoted are the ones derived from experiment.

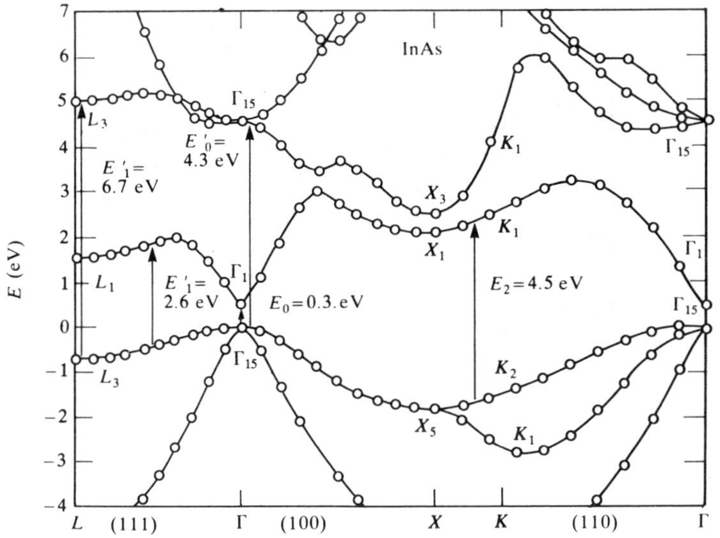

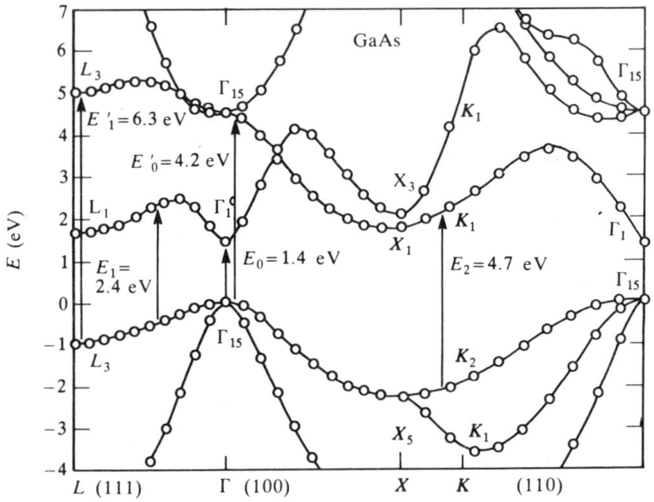

The open circles are simply the calculated values of $E_n(\mathbf{k})$, while the solid lines are curves drawn through the calculated points. In these calculations the spin–orbit interaction is omitted for simplicity, and the symmetries of the wave functions at symmetry points or along symmetry lines are indicated in some cases.

After M. L. Cohen and T. K. Bergstresser (1966), *Phys. Rev.* **141**, 789–96.)

19.2 *A model band structure

For a simple cubic lattice of lattice parameter a.

(a) Determine the primitive reciprocal lattice vectors.
(b) Hence construct the first Brillouin zone.
(c) Write down a few of the smaller reciprocal lattice vectors, in units of π/a, that is,

$$(0,0,0),\ (\pm 2,0,0),\ \text{etc.}$$

(d) For a **k** point in the first Brillouin zone, $(\kappa_x,0,0)\ \pi/a$, write down the free-electron energies corresponding to

$$E=(\hbar^2/2m)|\mathbf{k}+\mathbf{K}|^2,\ \text{in units of}\ (\hbar^2/2m)\ (\pi/a)^2.$$

(e) Hence for the Δ axis ((000)→(100)) sketch the free electron 'energy bands', $E(\kappa)$, and write down the degeneracies of the various bands and symmetry points.
(f) Put in the position of the Fermi energy, assuming eight electrons per primitive cell and two electrons per **k**-state.
(g) Write down the plane-wave functions for the following:
 (i) The $\mathbf{k}=0$ position at energies 0 and 4 on your scale.
 (ii) The $\mathbf{k}=\pi/a$ (1,0,0) position at energy 1.
 (iii) The k_x axis energies between 0 and 1; and between 1 and 4.
(h) The point-group of **k** for $\mathbf{k}=0$ is the O_h group. Draw up a table that shows how x, y and z each transform under one choice of operation from each class of O_h:

	E	C_3	...
x	x	y	
y	y	z	
z	z	x	

[*Hint*: For the improper rotations consider i times the corresponding rotation. (i always sends $x\rightarrow -x$ etc.)]
(i) Hence obtain the point-group symmetries of the eigenfunctions at the $\mathbf{k}=0$ (Γ) points above. (Use the chemical notation A_{1g} etc.)
(j) Obtain the point-groups at the X point (π/a) (1,0,0) and along the Δ-axis (k_x).
(k) Hence obtain the symmetries of the eigenfunctions at these positions.
(l) Do you have enough information to determine the order of splitting at $\epsilon=6$ due to the finite periodic lattice potential? [*Hint*: Consider the compatibility relations.]

(m) At the X point of energy 1 there are two empty-lattice states. Obtain the symmetry-adapted functions at the X point using the two appropriate plane waves as a basis. What are the eigenfunctions?

(n) Consider a model potential of simple cubic symmetry and period a:

$$V(r) = V_0 + V \cos\frac{2\pi x}{a} \cos\frac{2\pi y}{a} \cos\frac{2\pi z}{a}.$$

Hence find the energy gap at the position considered above and show that the eigenfunctions above are *not* coupled by the periodic potential.

19.3 *Radiation selection rules

Radiative transition between bands are essentially vertical in **k** space. In addition point-group selection rules must be obeyed between the initial and final states.

Using the T_d and O_h character tables, find the selection rules for electric-dipole transitions amongst the Γ-point states that originated from $\epsilon = 3$, for InSb and Ge respectively. For these semiconductors the fundamental band gaps lie respectively between (i) Γ_{15}(valence) and Γ_1(conduction) bands and (ii) Γ'_{25} (V) and Γ'_2 (C) bands. Verify that direct inter-band transitions are allowed in both cases.

19.4 *Second harmonic generation*

In certain materials it is possible to generate radiation at frequency 2ω from incident radiation at frequency ω. A particular process for the generation involves transitions between three electronic states each of which couple to both of the other two; that is, electric-dipole transitions should be allowed between all the states.

Consider the Γ-point states of InAs and demonstrate that second harmonic generation is possible. Is the same true of Si?

19.5 *Effective masses*

In the vicinity of band extrema the band curvatures in **k**-space, and hence the effective masses of the electrons and holes, are determined by the interaction between nearby bands. The Hamiltonian for the interaction takes the form

$$\mathcal{H}' = \frac{\hbar}{m}\mathbf{k}.\mathbf{p}.$$

Consider the InSb bands in the k_x direction. Explain why the Δ_1 conduction band interacts with the Δ_1 valence band but not the Δ_3, Δ_4 bands. (The Δ_3, Δ_4 bands near $k_x = 0$ are known as heavy-hole bands (ignoring spin effects) and the Δ_1 level is a light-hole, because the interaction reduces the effective mass of the interacting levels.)

19.6 ***Long wavelength lattice vibrations***
The analysis of long wavelength lattice vibrations follows closely that used for molecules (Part II and Chapter 16). In the crystal case, however, $\mathbf{k}=0$ vibrations involve identical displacements of equivalent atoms in each cell. A transformation that changes an atomic displacement to an entirely equivalent displacement in an adjacent cell must therefore be considered to contribute $(+1)$ to the trace of the transformation matrix.

Consider orthogonal displacements of In and Sb atoms in sites $(0,0,0)$ and $(\frac{1}{4},\frac{1}{4},\frac{1}{4})$. Use the T_d operations and, treating these atoms as a quasi-molecule (with the above proviso), follow through the normal-mode treatment to find the symmetry types, degeneracies and eigenvectors for the InSb long-wavelength lattice vibrations.

19.7 Analyse the nearly-free electron band structure for GaAs in the $[1,1,1]$ direction.

Appendix 1

Symmetry notations used in the text

(a) Symmetry operations

E	the identity operation
C_n	rotation by $2\pi/n$
i	inversion
σ	reflection
S_n	C_n followed by reflection in the plane normal to the C_n axis
σ_v	reflection in a vertical plane containing the major symmetry axis
σ_h	reflection in a horizontal plane normal to the major symmetry axis
σ_d	reflection in a diagonal plane containing the major axis and bisecting a pair of axes each perpendicular to the major axis
C'_n, C''_n	C_n axes physically distinguishable from the major axes
C_{2z}	C_2 rotation about a z-axis (a major axis)
σ_{xy}	σ reflection in an x–y plane, $\equiv \sigma_h$
σ_{xz}	σ reflection in an x–z plane, $\equiv \sigma_v$
$R(\alpha,\xi)$	rotation by an arbitrary angle α about an arbitrary axis in the ξ direction

(b) Symmetry groups: Schoenflies notation

The groups labelled here refer to symmetric systems that contain the symmetry operations indicated.

C_1	no symmetry other than E
C_n	a single, simple n-fold axis
C_{nv}	C_n plus n σ_v reflections
C_{nh}	C_n plus a σ_h reflection
D_n	C_n plus two n-fold axes perpendicular to the major axis
D_{nd}	D_n plus σ_d reflections
D_{nh}	D_n plus a σ_h reflection
S_n	a single S_n axis
T	the symmetry of a regular tetrahedron $(E,3C_2,8C_3)$
T_d	T plus reflections
T_h	T plus inversion
O	the rotational symmetry of a regular octahedron or a cube
O_h	O plus inversion
I	the rotational symmetry of a regular icosahedron
I_h	I plus inversion
C_∞	the group of rotations by any angle about a fixed axis
$C_{\infty v}$	C_∞ plus an infinite set of σ_v reflection planes
$D_{\infty h}$	$C_{\infty v}$ plus the σ_h operation
R_3	rotations by any angle about any axis
$R_3 \otimes S_2$	R_3 and the inversion operation

Several of the above groups may be constructed as direct-product groups (Chapter 13). S_2 and C_{1h} play a particular role in this respect as the elements of S_2 (E and i) commute with all rotations and the elements of C_{1h} (E and σ_h) commute with all rotations about an axis normal to the σ_h plane. The resulting character tables (Appendix A2) take a simple form as discussed in Chapter 13. In general, $\mathcal{G}_h \equiv \mathcal{G} \otimes C_{1h}$ for any of the rotational groups with h subscripts. However, as the product $C_2\sigma_h$ is equivalent to the inversion operation, then for those groups $\mathcal{G}$ that contain a C_2 operation (C_n or D_n with even n: T, O, I and R_3) one may write $\mathcal{G}_h \equiv \mathcal{G} \otimes S_2$.

(c) The 32 crystal point groups: comparison of Schoenflies and International notation

The crystal point groups are discussed in detail in Chapter 17. Schoenflies symbols have been used throughout this text to label the groups. However, the International notations are also commonly used in solid state physics. In the International system the following notation is used for symmetry axis and planes:

n	C_n axis
$\bar{n}$	S_n axis

m σ_v or σ_d plane
$/m$ σ_h plane

Thus, for example, the T_d group containing the operations E, $8C_3$, $3C_2$, $6\sigma_d$, $6S_4$ has the International notation $\bar{4}3m$, where $\bar{4}$ indicates the presence of S_4, S_4^2 ($\equiv C_2$) and S_4^3 operations, 3 indicates the 3-fold axes (because of the presence of $\bar{4}$ there must be 4 such axes) and m indicates the σ_d reflection planes.

The crystal point group symmetries

Crystal class	*Schoenflies symbol*	*International notation*
Triclinic:	C_1	1
	S_2	$\bar{1}$
Monoclinic:	C_{1h}	m
	C_2	2
	C_{2h}	$2/m$
Orthorhombic:	C_{2v}	$mm2$
	D_2	222
	D_{2h}	mmm
Tetragonal:	C_4	4
	C_{4v}	$4mm$
	C_{4h}	$4/m$
	D_4	422
	D_{2d}	$\bar{4}2m$
	D_{4h}	$4/mmm$
	S_4	$\bar{4}$
Rhombohedral:	C_3	3
	C_{3v}	$3m$
	D_3	32
	D_{3d}	$\bar{3}m$
	S_6	$\bar{3}$
Hexagonal:	C_6	6
	C_{6v}	$6mm$
	C_{3h}	$\bar{6}$
	C_{6h}	$6/m$
	D_6	622
	D_{3h}	$\bar{6}m2$
	D_{6h}	$6/mmm$
Cubic:	T	23
	T_d	$\bar{4}3m$
	T_h	$m3$
	O	432
	O_h	$m3m$

(d) Labels for irreducible representations: Mulliken symbols

One-dimensional irreps	A,B	; Σ
Two-dimensional	E	; $\pi,\Delta\ldots$
Three-dimensional	T	
Four-dimensional	G	icosahedral groups only
Five-dimensional	H	icosahedral groups only
$(2l+1)$-dimensional	D^l	full rotation goups ($l\geqslant 0$)

Arabic letters refer to all point groups, Greek letters are used in reference to molecular orbital states for the linear groups.

In addition the following modifications are used:

A	used if the characters for all rotations about major axes are unity; $\chi(C_n)=1$.
B	used if any of the above characters is equal to minus one; $\chi(C_n)=-1$.
Subscript 1	used if there is more than one A or B representation and if rotations about the first labelled non-major axis or vertical reflection in the first labelled plane give characters of unity.
Subscript 2	if the above character is negative, etc.
Subscripts 1,2	appear on E and T labels; the logic behind these subscripts is complex and can be ignored without detriment to symmetry analysis.
Primed superfixes	one prime if the group contains σ_h and the character is positive; two primes if it is negative.

Finally, and most importantly, for those groups that contain the inversion operation all irreps have defined parity.

g(subscript or superscript)	even parity (from the German *gerade*, meaning even). The character under i will be positive.
u	odd parity (*ungerade*).
$+$	symmetry with respect to σ_v.
$-$	anti-symmetry with respect to σ_v.

Additional notation for the solid state is described in Chapter 19.

Appendix 2

Character tables for selected groups

The tables in this section contain the following:

The group in question	top left-hand corner
The group operations, displayed in classes	top row
The group irreducible representations	left-hand column
The irrep. characters	body of the tables

The irreps for which various parameters form basis entities are indicated in the right-hand columns

x, y, z	polar vectors, translations, or linear functions
R_x, R_y, R_z	axial vectors or rotations
z^2 etc.	quadratic functions

Unbracketed parameters are alternative bases; bracketed parameters transform together. Transformation properties for cubic bases and additional group tables may be found for example in Harris and Bertolucci (1978) (see Bibliography, Section (b)).

The cyclic groups, C_n

C_n	E	C_n	C_n^2	$\cdots$	C_n^{n-1}
$\Gamma^0 = A$	1	1	1		1
Γ^1	1	ω	ω^2		ω^{n-1}
Γ^2	1	(ω^2)	$(\omega^2)^2$		$(\omega^2)^{n-1}$
$\vdots$					
Γ^{n-1}	1	ω^{n-1}	$(\omega^{n-1})^2$		$(\omega^{n-1})^{n-1}$

Here $\omega^n = 1$.

The general cyclic group is discussed in Chapter 18 with respect to translational symmetries of a periodically repeated crystal structure; it is appropriate to choose $\omega = \exp(-i2\pi/n)$ in that context.

In the following specific cyclic groups the convention is followed that ω is set equal to $\exp(+i2\pi/n)$. The order of the irreps is also changed. Irreps for which the characters for each operation are complex conjugates (Γ^m and Γ^{n-m}) are paired together because under time-reversal symmetry they are required to lead to twofold degeneracy. Various notations exist in the literature, for example ε is often used as an alternative to ω, complex conjugates are often written explicitly. Here the lowest powers of ω are used (thus in C_4, $(\omega^2)^2$ is written as ω) and where ω^m is equal to $\pm$i or ± 1 the latter are used. Bases for the *E* irreps are written in pairs (x,y) etc. For the individual rows the complex combinations $x+iy$ etc. form bases. This is spelt out for the group C_6 as an example; it is easy to see how it will apply to all such groups.

C_1	E	
A	1	all bases

C_2	E	C_2			
A	1	1	z	R_z	x^2, y^2, z^2, xy
B	1	-1	x, y	R_x, R_y	xz, yz

C_3	E	C_3	C_3^2				
A	1	1	1	z	R_z	x^2+y^2, z^2	
E	$\begin{cases}1\\1\end{cases}$	$\begin{matrix}\omega\\ \omega^2\end{matrix}$	$\begin{matrix}\omega^2\\ \omega\end{matrix}$	(x,y)	(R_x, R_y)	(x^2-y^2, xy)	(xz, yz)

$\omega = \exp(i2\pi/3)$, $\omega^2 = \omega^*$, $\omega^3 = 1$.
$\omega + \omega^2 = \cos(2\pi/3) = -1$.

C_4	E	C_4	C_2	C_4^3			
A	1	1	1	1	z	R_z	x^2+y^2, z^2
B	1	−1	1	−1			x^2-y^2, xy
E	$\begin{cases}1\\1\end{cases}$	$\begin{matrix}i\\-i\end{matrix}$	$\begin{matrix}-1\\-1\end{matrix}$	$\begin{matrix}-i\\i\end{matrix}$	(x,y)	(R_x,R_y)	(xz,yz)

Note: $\omega=\exp(i2\pi/4)=i$, $\omega^2=-1$, $\omega^3=-i$

C_5	E	C_5	C_5^2	C_5^3	C_5^4			
A	1	1	1	1	1	z	R_z	x^2+y^2, z^2
E_1	$\begin{cases}1\\1\end{cases}$	$\begin{matrix}\omega\\\omega^4\end{matrix}$	$\begin{matrix}\omega^2\\\omega^3\end{matrix}$	$\begin{matrix}\omega^3\\\omega^2\end{matrix}$	$\begin{matrix}\omega^4\\\omega\end{matrix}$	(x,y)	(R_x,R_y)	(xz,yz)
E_2	$\begin{cases}1\\1\end{cases}$	$\begin{matrix}\omega^2\\\omega^3\end{matrix}$	$\begin{matrix}\omega^4\\\omega\end{matrix}$	$\begin{matrix}\omega\\\omega^4\end{matrix}$	$\begin{matrix}\omega^3\\\omega^2\end{matrix}$			(x^2-y^2, xy)

$\omega=\exp(i2\pi/5)$
$\omega+\omega^4=2\cos(2\pi/5)$ $\omega^2+\omega^3=2\cos(4\pi/5)$

C_6	E	C_6	C_3	C_2	C_3^2	C_6^5			
A	1	1	1	1	1	1	z	R_z	$(x+iy)(x-iy), z^2$
B	1	−1	1	−1	1	−1			
E_1	$\begin{cases}1\\1\end{cases}$	$\begin{matrix}\omega\\-\omega^2\end{matrix}$	$\begin{matrix}\omega^2\\-\omega\end{matrix}$	$\begin{matrix}-1\\-1\end{matrix}$	$\begin{matrix}-\omega\\\omega^2\end{matrix}$	$\begin{matrix}-\omega^2\\\omega\end{matrix}$	$\begin{matrix}x+iy\\x-iy\end{matrix}$	$\begin{matrix}R_x+iR_y\\R_x-iR_y\end{matrix}$	$\begin{matrix}(x+iy)z\\(x-iy)z\end{matrix}$
E_2	$\begin{cases}1\\1\end{cases}$	$\begin{matrix}\omega^2\\-\omega\end{matrix}$	$\begin{matrix}-\omega\\\omega^2\end{matrix}$	$\begin{matrix}1\\1\end{matrix}$	$\begin{matrix}\omega^2\\-\omega\end{matrix}$	$\begin{matrix}-\omega\\\omega^2\end{matrix}$			$\begin{matrix}(x+iy)^2\\(x-iy)^2\end{matrix}$

$\omega=\exp(i2\pi/6)$, $\omega^3=-1$, $\omega^*=-\omega^2$, $\omega-\omega^2=2\cos\pi/3=-1$.

Note that $(x+iy)(x-iy)=x^2+y^2$ and as transforming pairs:

$(x+iy, x-iy)$ is equivalent to (x,y), $(R_x+iR_y, R_x-iR_y)\equiv(R_x, R_y)$, $((x+iy)z, (x-iy)z)\equiv(xz,yz)$, $((x+iy)^2, (x-iy)^2)\equiv(x^2-y^2, xy)$.

C_∞	E	$R(\alpha,z)$			
$A\equiv\Sigma$	1	1	z	R_z	x^2+y^2, z^2
$E_1\equiv\Pi$	$\begin{cases}1\\1\end{cases}$	$\begin{matrix}e^{i\alpha}\\e^{-i\alpha}\end{matrix}$	(x,y)	(R_x,R_y)	(xz,yz)
$E_2\equiv\Delta$	$\begin{cases}1\\1\end{cases}$	$\begin{matrix}e^{i2\alpha}\\e^{-i2\alpha}\end{matrix}$			(x^2-y^2, xy)
⋮	⋮	⋮			
E_m	$\begin{cases}1\\1\end{cases}$	$\begin{matrix}e^{im\alpha}\\e^{-im\alpha}\end{matrix}$			$\begin{matrix}(x+iy)^m z^p\\(x-iy)^m z^p;\ p=0,1,2,\ldots\end{matrix}$

The C_{nv} groups

(The group with no rotation axis, but a single reflection plane is conventionally called C_{1h} rather than C_{1v}.)

C_{2v}	E	C_2	σ_{xz}	σ_{yz}			
A_1	1	1	1	1	z		x^2,y^2,z^2
A_2	1	1	−1	−1		R_z	xy
B_1	1	−1	1	−1	x	R_x	xz
B_2	1	−1	−1	1	y	R_y	yz

C_{3v}	E	$2C_3$	$3\sigma_v$			
A_1	1	1	1	z		x^2+y^2,z^2
A_2	1	1	−1		R_z	
E	2	−1	0	(x,y)	(R_x,R_y)	$(x^2-y^2,xy),(xz,yz)$

C_{4v}	E	$2C_4$	C_2	$2\sigma_v$	$2\sigma_d$			
A_1	1	1	1	1	1	z		x^2+y^2,z^2
A_2	1	1	1	−1	−1		R_z	
B_1	1	−1	1	1	−1			x^2-y^2
B_2	1	−1	1	−1	1			xy
E	2	0	−2	0	0	(x,y)	(R_x,R_y)	(xz,yz)

C_{5v}	E	$2C_5$	$2C_5^2$	$5\sigma_v$			
A_1	1	1	1	1	z		x^2+y^2,z^2
A_2	1	1	1	−1		R_z	
E_1	2	$2\cos x$	$2\cos 2x$	0	(x,y)	(R_x,R_y)	(xz,yz)
E_2	2	$2\cos 2x$	$2\cos x$	0			(x^2-y^2,xy)

$x=2\pi/5=72°$

C_{6v}	E	$2C_6$	$2C_3$	C_2	$3\sigma_v$	$3\sigma_d$			
A_1	1	1	1	1	1	1	z		x^2+y^2,z^2
A_2	1	1	1	1	−1	−1		R_z	
B_1	1	−1	1	−1	1	−1			
B_2	1	−1	1	−1	−1	1			
E_1	2	1	−1	−2	0	0	(x,y)	(R_x,R_y)	(xz,yz)
E_2	2	−1	−1	2	0	0			(x^2-y^2,xy)

$C_{\infty v}$	E	$2R(\alpha,z)$	$\infty\sigma_v$			
$A_1 \equiv \Sigma^+$	1	1	1	z		x^2+y^2, z^2
$A_2 \equiv \Sigma^-$	1	1	-1		R_z	
$E_1 \equiv \Pi$	2	$2\cos\alpha$	0	(x,y)	(R_x,R_y)	(xz,yz)
$E_1 \equiv \Delta$	2	$2\cos 2\alpha$	0			(x^2-y^2,xy)
.	.	.	.			
.	.	.	.			
.	.	.	.			
E_M	2	$2\cos M\alpha$	0			$\{(x+\mathrm{i}y)^M z^p, (x-\mathrm{i}y)^M z^p\}$

The C_{nh} groups

C_{1h}	E	σ_h			
A'	1	1	x,y	R_z	x^2,y^2,z^2,xy
A''	1	-1	z	R_x,R_y	xz,yz

C_{2h}	E	C_2	i	σ_h			
A_g	1	1	1	1		R_z	x^2,y^2,z^2,xy
B_g	1	-1	1	-1	z		
A_u	1	1	-1	-1		R_x,R_y	xz,yz
B_u	1	-1	-1	1	x,y		

The top left-hand corner of the C_{2h} corresponds to the C_2 group, and is included in direct-product group tables to emphasize the form of the character tables:

A	A
A	$-A$

$C_{2h} \equiv C_2 \times S_2$

C_{3h}	E	C_3	C_3^2	σ_h	S_3	S_3^5			
A'	1	1	1	1	1	1		R_z	x^2+y^2,z^2
E'	$\begin{cases}1\\1\end{cases}$	$\begin{matrix}\omega\\\omega^2\end{matrix}$	$\begin{matrix}\omega^2\\\omega\end{matrix}$	$\begin{matrix}1\\1\end{matrix}$	$\begin{matrix}\omega\\\omega^2\end{matrix}$	$\begin{matrix}\omega^2\\\omega\end{matrix}$	(x,y)		(x^2-y^2,xy)
A''	1	1	1	-1	-1	-1	z		
E''	$\begin{cases}1\\1\end{cases}$	$\begin{matrix}\omega\\\omega^2\end{matrix}$	$\begin{matrix}\omega^2\\\omega\end{matrix}$	$\begin{matrix}-1\\-1\end{matrix}$	$\begin{matrix}-\omega\\-\omega^2\end{matrix}$	$\begin{matrix}-\omega^2\\-\omega\end{matrix}$		(R_x,R_y)	(xz,yz)

$\omega = \exp(\mathrm{i}2\pi/3)$, $\omega + \omega^2 = -1$.
$C_{3h} \equiv C_3 \times C_{1h}$. $(S_3^5 \equiv \sigma_h C_3^2)$

C_{4h}	E	C_4	C_2	C_4^3	i	S_4^3	σ_h	S_4			
A_g	1	1	1	1	1	1	1	1		R_z	x^2+y^2, z^2
B_g	1	−1	1	−1	1	−1	1	−1			x^2-y^2, xy
E_g	1	i	−1	−i	1	i	−1	−i		(R_x, R_y)	(xz, yz)
	1	−i	−1	i	1	−i	−1	i			
A_u	1	1	1	1	−1	−1	−1	−1	z		
B_u	1	−1	1	−1	−1	1	−1	1			
E_u	1	i	−1	−i	−1	−i	1	i	(x,y)		
	1	−i	−1	i	−1	i	1	−i			

$C_{4h} \equiv C_4 \otimes S_2$. $(S_4 \equiv iC_4^3)$
$C_{2nh} \equiv C_{2n} \otimes S_2$
$C_{(2n+1)h} \equiv C_{(2n+1)} \otimes C_{1h}$

$C_\infty \otimes S_2$	E	$R(\alpha,z)$	i	$iR(\alpha,z)$			
A_g	1	1	1	1		R_z	x^2+y^2, z^2
E_{+1g}	1	$e^{i\alpha}$	1	$e^{i\alpha}$		$R_x - iR_y$	$(x+iy)z$
E_{-1g}	1	$e^{-i\alpha}$	1	$e^{-i\alpha}$		$R_x - iR_y$	$(x-iy)z$
E_{+2g}	1	$e^{i2\alpha}$	1	$e^{-i2\alpha}$			$(x+iy)^2$
.	.	.	.	.			
.	.	.	.	.			
.	.	.	.	.			
E_{-Mg}	1	$e^{-iM\alpha}$	1	$e^{-iM\alpha}$			
A_u	1	1	−1	−1	z		
E_{+1u}	1	$e^{i\alpha}$	−1	$-e^{i\alpha}$	$x+iy$		
E_{-1u}	1	$e^{-i\alpha}$	−1	$-e^{-i\alpha}$	$x-iy$		
.	.	.	.	.			
.	.	.	.	.			
.	.	.	.	.			
E_{-Mu}	1	$e^{-iM\alpha}$	−1	$e^{-iM\alpha}$			

The dihedral groups, D_n

D_2	E	C_{2z}	C_{2y}	C_{2x}			
A	1	1	1	1			x^2, y^2, z^2
B_1	1	1	−1	−1	z	R_z	xy
B_2	1	−1	1	−1	y	R_x	xz
B_3	1	−1	−1	1	x	R_y	yz

D_3	E	$2C_3$	$3C'_2$			
A_1	1	1	1			x^2+y^2, z^2
A_2	1	1	−1	z	R_z	
E	2	−1	0	(x,y)	(R_x, R_y)	$(xz, yz), (x^2-y^2, xy)$

D_4	E	$2C_4$	C_2	$2C'_2$	$2C''_2$			
A_1	1	1	1	1	1			x^2+y^2, z^2
A_2	1	1	1	−1	−1	z	R_z	
B_1	1	−1	1	1	−1			x^2-y^2
B_2	1	−1	1	−1	1			xy
E	2	0	−2	0	0	(x,y)	(R_x,R_y)	(xz,yz)

D_5	E	$2C_5$	$2C_5^2$	$5C'_2$			
A_1	1	1	1	1			x^2, y^2, z^2
A_2	1	1	1	−1	z	R_z	
E_1	2	$2\cos x$	$2\cos 2x$	0	(x,y)	(R_x,R_y)	(xz,yz)
E_2	2	$2\cos 2x$	$2\cos x$	0			(x^2-y^2,xy)

$x=2\pi/5$

D_6	E	$2C_6$	$2C_3$	C_2	$2C'_2$	$2C''_2$			
A_1	1	1	1	1	1	1			x^2+y^2, z^2
A_2	1	1	1	1	−1	−1	z	R_z	
B_1	1	−1	1	−1	1	−1			
B_2	1	−1	1	−1	−1	1			
E_1	2	1	−1	−2	0	0	(x,y)	(R_x,R_y)	(xz,yz)
E_2	2	−1	−1	2	0	0			(x^2-y^2,xy)

The D_{nd} groups

D_{2d}	E	$2S_4$	C_2	$2C'_2$	$2\sigma_d$			
A_1	1	1	1	1	1			x^2+y^2, z^2
A_2	1	1	1	−1	−1		R_z	
B_1	1	−1	1	1	−1			x^2-y^2
B_2	1	−1	1	−1	1	z		xy
E	2	0	−2	0	0	(x,y)	(R_x,R_y)	(xz,yz)

D_{3d}	E	$2C_3$	$3C'_2$	i	$2S_6$	$3\sigma_d$			
A_{1g}	1	1	1	1	1	1			x^2+y^2, z^2
A_{2g}	1	1	−1	1	1	−1		R_z	
E_g	2	−1	0	2	−1	0		(R_x,R_y)	$(x^2-y^2,xy),(xz,yz)$
A_{1u}	1	1	1	−1	−1	−1			
A_{2u}	1	1	−1	−1	−1	1	z		
E_u	2	−1	0	−2	1	0	(x,y)		

$D_{3d} \equiv D_3 \otimes S_2$. $(iC_3 \equiv S_6)$

D_{4d}	E	$2S_8$	$2C_4$	C_2	$2S_8^3$	$4C'_2$	$4\sigma_d$			
A_1	1	1	1	1	1	1	1			x^2+y^2, z^2
A_2	1	1	1	1	1	−1	−1		R_z	
B_1	1	−1	1	1	−1	1	−1			
B_2	1	−1	1	1	−1	−1	1	z		
E_1	2	$\sqrt{2}$	0	−2	$-\sqrt{2}$	0	0	(x, y)		
E_2	2	0	−2	2	0	0	0			(x^2-y^2, xy)
E_3	2	$-\sqrt{2}$	0	−2	$\sqrt{2}$	0	0		(R_x, R_y)	(xz, yz)

$D_{(2n+1)d} \equiv D_{(2n+1)} \otimes S_2$

The D_{nh} groups

D_{2h}	E	C_{2z}	C_{2y}	C_{2x}	i	σ_{xy}	σ_{xz}	σ_{yz}			
A_g	1	1	1	1	1	1	1	1			x^2, y^2, z^2
B_{1g}	1	1	−1	−1	1	1	−1	−1		R_z	xy
B_{2g}	1	−1	1	−1	1	−1	1	−1		R_y	xz
B_{3g}	1	−1	−1	1	1	−1	−1	1		R_x	yz
A_u	1	1	1	1	−1	−1	−1	−1			
B_{1u}	1	1	−1	−1	−1	−1	1	1	z		
B_{2u}	1	−1	1	−1	−1	1	−1	1	y		
B_{3u}	1	−1	−1	1	−1	1	1	−1	x		

$D_{2h} \equiv D_2 \otimes S_2$

D_{3h}	E	$2C_3$	$3C'_2$	σ_h	$2S_3$	$3\sigma_v$			
A'_1	1	1	1	1	1	1			x^2+y^2, z^2
A'_2	1	1	−1	1	1	−1		R_z	
E'	2	−1	0	2	−1	0	(x, y)		(x^2-y^2, xy)
A''_1	1	1	1	−1	−1	−1			
A''_2	1	1	−1	−1	−1	1	z		
E''	2	−1	0	−2	1	0		(R_x, R_y)	(xz, yz)

$D_{3h} \equiv D_3 \otimes C_{1h}$

D_{4h}	E	$2C_4$	C_2	$2C'_2$	$2C''_2$	i	$2S_4$	σ_h	$2\sigma_v$	$2\sigma_d$			
A_{1g}	1	1	1	1	1	1	1	1	1	1			x^2+y^2, z^2
A_{2g}	1	1	1	−1	−1	1	1	1	−1	−1		R_z	
B_{1g}	1	−1	1	1	−1	1	−1	1	1	−1			xy
B_{2g}	1	−1	1	−1	1	1	−1	1	−1	1			x^2-y^2
E_g	2	0	−2	0	0	2	0	−2	0	0		(R_x, R_y)	
A_{1u}	1	1	1	1	1	−1	−1	−1	−1	−1			
A_{2u}	1	1	1	−1	−1	−1	−1	−1	1	1	z		
B_{1u}	1	−1	1	1	−1	−1	1	−1	−1	1			
B_{2u}	1	−1	1	−1	1	−1	1	−1	1	−1			
E_u	2	0	−2	0	0	−2	0	2	0	0	(x,y)		

$D_{4h} = D_4 \otimes S_2$

D_{6h}	E	$2C_6$	$2C_3$	C_2	$3C'_2$	$3C''_2$	i	$2S_3$	$2S_6$	σ_h	$3\sigma_v$	$3\sigma_d$			
A_{1g}	1	1	1	1	1	1	1	1	1	1	1	1			x^2+y^2, z^2
A_{2g}	1	1	1	1	−1	−1	1	1	1	1	−1	−1		R_z	
B_{1g}	1	−1	1	−1	1	−1	1	−1	1	−1	1	−1			
B_{2g}	1	−1	1	−1	−1	1	1	−1	1	−1	−1	1			
E_{1g}	2	1	−1	−2	0	0	2	1	−1	−2	0	0		(R_x, R_y)	(xz, yz)
E_{2g}	2	−1	−1	−2	0	0	2	−1	−1	−2	0	0			(x^2-y^2, xy)
A_{1u}	1	1	1	1	1	1	−1	−1	−1	−1	−1	−1			
A_{2u}	1	1	1	1	−1	−1	−1	−1	−1	−1	1	1	z		
B_{1u}	1	−1	1	−1	1	−1	−1	1	−1	1	−1	1			
B_{2u}	1	−1	1	−1	−1	1	−1	1	−1	1	1	−1			
E_{1u}	2	1	−1	−2	0	0	−2	−1	1	2	0	0	(x,y)		
E_{2u}	2	−1	−1	−2	0	0	−2	1	1	2	0	0			

$D_{6h} = D_6 \otimes S_2$
$D_{(2n)h} = D_{2n} \otimes S_2$
$D_{(2n+1)h} = D_{(2n+1)} \otimes C_{1h}$

$D_{\infty h}$	E	$2R(\alpha,z)$	$\infty C'_2$	i	$2iR(\alpha,z)$	$\infty\sigma_i$		
$A_{1g}\equiv\Sigma_g^+$	1	1	1	1	1	1		x^2+y^2,z^2
$A_{2g}\equiv\Sigma_g^-$	1	1	−1	1	1	−1		R_z
$E_{1g}\equiv\Pi_g$	2	2 cos α	0	2	2 cos α	0		(R_x,R_y) (xz,yz)
$E_{2g}\equiv\Delta_g$	2	2 cos 2α	0	2	2 cos 2α	0		(x^2-y^2,xy)
.	.	.	.	.	.	.		
.	.	.	.	.	.	.		
.	.	.	.	.	.	.		
E_{Mg}	2	2 cos $M\alpha$	0	2	2 cos $M\alpha$	0		
$A_{1u}\equiv\Sigma_u^+$	1	1	1	−1	−1	−1		
$A_{2u}\equiv\Sigma$	1	1	−1	−1	−1	1	z	
$E_{1u}\equiv\Pi_u$	2	2 cos α	0	−2	−2 cos α	0	(x,y)	
.	.	.	.	.	.	.		
.	.	.	.	.	.	.		
.	.	.	.	.	.	.		
E_{Mu}	2	2 cos $M\alpha$	0	−2	−2 cos $M\alpha$	0		

$D_{\infty h}\equiv D_\infty\otimes S_2$

The S_n groups

S_2	E	i			
A_g	1	1		R_x,R_y,R_z	x^2,y^2,z^2,xy,xz,yz
A_u	1	−1	x,y,z		

S_4	E	S_4	C_2	S_4^3			
A	1	1	1	1		R_z	x^2+y^2,z^2
B	1	−1	1	−1	z		
E	$\begin{cases}1 & \mathrm{i} & -1 & -\mathrm{i}\\ 1 & -\mathrm{i} & -1 & \mathrm{i}\end{cases}$				(x,y)	(R_x,R_y)	(xz,yz), (x^2-y^2,xy)

$S_6\equiv C_3\otimes S_2$
S_n only exists if n is even
$S_{2n}=C_m\otimes S_2$ if n is odd

For even n, S_{2n} is isomorphic with C_{2n}. A useful tabulation is:

S_n	E	S_n	S_n^2		S_n^{n-1}			
A	1	1	1	...	1		R_z	x^2+y^2, z^2
B	1	-1	1	...	-1	z		
E_1	$\begin{cases}1\\1\end{cases}$	ω ω^{n-1}	ω^2 $(\omega^{n-1})^2$	... ...	ω^{n-1} $(\omega^{n-1})^{n-1}$	(x,y)	(R_x,R_y)	
E_2	1	ω^2	ω^4	...				(x^2-y^2, xy)
E_3	1	ω^3	...					(xz, yz)

$\omega = \exp(i2\pi/n)$

The cubic groups

T	E	$4C_3$	$4C_3^2$	$3C_2'$			
A	1	1	1	1			
E	1	ω	ω^2	1			$(3z^2-r^2,\ x^2-y^2)$
	1	ω^2	ω	1			
T	3	0	0	-1	(x,y,z)	(R_x,R_y,R_z)	(xy, yz, zx)

$\omega = \exp(i2\pi/3)$. $\omega + \omega^2 = 2\cos(2\pi/3) = -1$. $r^2 = x^2+y^2+z^2$.

It is useful to consider the C_2' axis as in the x, y and z directions; the C_3 axes will lie along $x=y=z$, etc. It is conventional to ignore the prime referring to physically distinct axes, without loss of ambiguity in the cubic groups, as below.

T_d	E	$8C_3$	$6S_4$	$3C_2$	$6\sigma_d$			
A_1	1	1	1	1	1			r^2
A_2	1	1	-1	1	-1			
E	2	-1	0	2	0			$(3z^2-r^2,\ x^2-y^2)$
T_1	3	0	1	-1	-1		(R_x,R_y,R_z)	
T_2	3	0	-1	-1	1	(x,y,z)		(xy, yz, zx)

T_h	E	$4C_3$	$4C_3^2$	$3C_2$	i	$4S_6$	$4S_6^5$	$3\sigma_h$		
A_g	1	1	1	1	1	1	1	1		r^2
E_g	1	ω	ω^2	1	1	ω	ω^2	1		$(3z^2-r^2,\ x^2-y^2)$
	1	ω^2	ω	1	1	ω^2	ω	1		
T_g	3	0	0	−1	3	0	0	−1		(R_x,R_y,R_z) (xy,yz,zx)
A_u	1	1	1	1	−1	−1	−1	−1		
E_u	1	ω	ω^2	1	−1	$-\omega$	$-\omega^2$	−1		
	1	ω^2	ω	1	−1	$-\omega^2$	$-\omega$	−1		
T_u	3	0	0	−1	−3	0	0	1	(x,y,z)	

$T_h \equiv T \otimes S_2$

O	E	$8C_3$	$6C_4$	$3C_2$	$6C_2'$		
A_1	1	1	1	1	1		r^2
A_2	1	1	−1	1	−1		
E	2	−1	0	2	0		$(3z^2-r^2,\ x^2-y^2)$
T_1	3	0	1	−1	−1	(x,y,z) (R_x,R_y,R_z)	
T_2	3	0	−1	−1	1		(xy,yz,zx)

O_h	E	$8C_3$	$6C_4$	$3C_2$	$6C_2'$	i	$8S_6$	$6S_4$	$3\sigma_h$	$6\sigma_d$			
A_{1g}	1	1	1	1	1	1	1	1	1	1			r^2
A_{2g}	1	1	−1	1	−1	1	1	−1	1	−1			
E_g	2	−1	0	2	0	2	−1	0	2	0			$(3z^2-r^2,\ x^2-y^2)$
T_{1g}	3	0	1	−1	−1	3	0	1	−1	−1		(R_x,R_y,R_z)	
T_{2g}	3	0	−1	−1	1	3	0	−1	−1	1			
A_{1u}	1	1	1	1	1	−1	−1	−1	−1	−1			
A_{2u}	1	1	−1	1	−1	−1	−1	1	−1	1			
E_u	2	−1	0	2	0	−2	1	0	−2	0			
T_{1u}	3	0	1	−1	−1	−3	0	−1	1	1	(x,y,z)		
T_{2u}	3	0	−1	−1	1	−3	0	1	1	−1			

$O_h \equiv O \otimes S_2$

The icosahedral groups

I	E	$12C_5$	$12C_5^2$	$20C_3$	$15C_2$	
A	1	1	1	1	1	
T_1	3	$\frac{1}{2}(1+\sqrt{5})$	$\frac{1}{2}(1-\sqrt{5})$	0	-1	(x,y,z) (R_x,R_y,R_z)
T_2	3	$\frac{1}{2}(1-\sqrt{5})$	$\frac{1}{2}(1+\sqrt{5})$	0	-1	
G	4	-1	-1	1	0	
H	5	0	0	-1	0	$(3z^2-r^2,x^2-y^2,xy,yz,zx)$

$I_h \equiv I \otimes S_2$

(x,y,z) form the basis of T_{1u}, (R_x,R_y,R_z) etc. form bases, as above, for g irreps.

The full rotation and rotation–inversion groups

R_3	E	$R(\alpha,\xi)$		
D^0	1	1		r^2
D^1	3	$\sin(3\alpha/2)/\sin(\alpha/2)$	(x,y,z) (R_x,R_y,R_z)	
D^2	5	$\sin(5\alpha/2)/\sin(\alpha/2)$		$(3z^2-r^2, x^2-y^2, xy, yz, zx)$
.	.	.		
.	.	.		
.	.	.		
D^l	$2l+1$	$\sin(l+\frac{1}{2})\alpha/\sin(\alpha/2)$		

$R_3 \otimes S_2$	E	$R(\alpha,\xi)$	i	$iR(\alpha,\xi)$			
D^{0g}	1	1	1	1			r^2
D^{1g}	3	$\sin(3\alpha/2)/\sin(\alpha/2)$	3	sin/sin		(R_x,R_y,R_z)	
D^{2g}	5	$\sin(5\alpha/2)/\sin(\alpha/2)$	5	sin/sin			$(3z^2-r^2, x^2-y^2, xy, yz, zx)$
.	.	.					
.	.	.					
.	.	.					
D^{lg}	$2l+1$	$\sin(l+\frac{1}{2})\alpha/\sin(\alpha/2)$	$2l+1$	sin/sin			
D^{0u}	1	1	−1	1			
D^{1u}	3	sin/sin	−3	−sin/sin	(x,y,z)		
D^{2u}	5	sin/sin	−5	−sin/sin			
.	.	.					
.	.	.					
.	.	.					
D^{lu}	$(2l+1)$	sin/sin	$-(2l+1)$	−sin/sin			

Appendix 3

Glossary of notation

Arabic, Italic, Bold and Script

a; A; $\mathbf{a}$; $\mathbf{A}$; $\mathcal{A}$

a	matrix	$(a,b,\ldots,f)$
$[a]$	matrix relating eigenfunctions to basis functions	
a_{ij}	matrix element	
a_j	coefficient in the reduced form of a representation	
a_l	number of occurrences of Γ^l in a reduced representation	
a	bond label	(a,b,c)
$\mathbf{a}$	Bravais real-lattice vector	$(\mathbf{a},\mathbf{b},\mathbf{c})$
a	magnitude of $\mathbf{a}$	(a,b,c)
$\mathbf{a}_1$	primitive real-lattice vector	$(\mathbf{a}_1,\mathbf{a}_2,\mathbf{a}_3)$
a_1	magnitude of $\mathbf{a}_1$	(a_1,a_2,a_3)
a	reference-frame axis	$(\sigma_a,\sigma_b,\sigma_c)$, $(0a,0b,0c)$
a	subscript, position of a nucleus in a molecule	$(\mathbf{R}_a,\mathbf{R}_b)$
a	superscript, an irrep. in the group $\mathcal{A}$	(Γ^a,Γ^b)
A	arbitrary matrix	(A, B, C, D)
A	Coulomb energy coefficient in the H_2 problem	
A^{III}	group three element	(A,B)
$A_1, \ldots.$	Irreducible representation label	$(A_1,A_2;B_1,B_2)$
A_g	Irreducible representation label	(A_g,A_u,B_g,B_u)
$A', \ldots.$	Irreducible representation label	$(A',A'';B',B'')$
$A^{L,l_1,l_2}_{m1,M-m1}$	Clebsch–Gordon coefficient	

A	electromagnetic vector potential	
A	atomic position in diamond structure	
$\mathcal{A}$	a group	$(\mathcal{A},\mathcal{B})$
$\mathcal{A}_i$	elements of a group	$(\mathcal{A}_i,\mathcal{B}_i)$

b; B; **b**; **B**; $\mathcal{B}$

$[b]$	matrix relating eigenfunctions to symmetry adapted functions	
$b_{nn'}$	matrix elements	
$\mathbf{b}_1$	reciprocal-lattice vector	$(\mathbf{b}_1,\mathbf{b}_2,\mathbf{b}_3)$
b_1	magnitude of $\mathbf{b}_1$	
B	Coulomb exchange energy	
B	magnetic field vector	

See also nomenclature of a, A, $\mathcal{A}$

c; C

c	velocity of light in vacuum	
c	separation constant for the Schrödinger equation	
c	subscript, class	
cm	subscript, centre of mass	
c_{nj}	coefficients in the expansion of basis functions in terms of eigenfunctions	
$c_{kl}(\mathbf{Q})$	coefficient in the expansion for symmetry-adapted vectors	
$c\text{–}j,k$	subscript, combination vibration level	
C	total direct integral in the H_2 problem	
C	general normalization constant	(C,C')
C_n	n-fold rotation axis	(C_n,C'_n,C''_n)
C_n	symmetry-group labels	(C_n,C_{nv},C_{nh})

d; D; **d**; **D**; $\mathcal{D}$; $\boldsymbol{\mathcal{D}}$

d	subscript, diagonal	
d	differential operator	
$\mathbf{d}_i$	unit displacement vector	
$\mathbf{d}'_i$	transformed displacement	
d	row vector of elements $\mathbf{d}_i$	$(\mathbf{d},\mathbf{D},\boldsymbol{\mathcal{D}})$
disp	arbitrary displacement vector	
d	superscript, referred to **d**-displacements	
D	exchange integral in the H_2 problem	
D_e	dissociation energy for diatomic molecules	
D_n	symmetry-group labels	$(D_n,D_{nv},D_{nh},D_{nd})$
2–D	two dimensional	

$\mathbf{D}_i$	symmetry-adapted displacement vector
D	superscript, referred to **D**-displacements
$\mathbf{D}_k^j(\mathbf{Q})$	symmetry-adapted displacement vector belonging to the kth row of the jth irrep. of a group
$\mathfrak{D}_i$	normal-mode eigenvector
$\mathfrak{D}$	superscript, referred to $\mathfrak{D}$-displacements

e; E; $\mathbf{e}$; $\mathbf{E}$; $\mathcal{E}$; $\boldsymbol{\varepsilon}$; E

e	electron charge
$e,elect$	subscript for electrons
$\mathbf{e}_i$	alternative set of basis displacements
e–d	electric–dipole
eq	equilibrium
E	the identity operation
E	energy E_{H}, E_{atom}, E_{mol}, E_{elect}, E_{motion} E_i, E_{cm}, E_{nlm}, E_{nl} E_e, $E_{e\mathbf{R}}$, E_n, E_s E_e', $E_n(\mathbf{k})$, $E_e(R_0)$ E_T, E_R, E_V as defined in the text
E_n	energy eigenvalue
E_F	Fermi energy
$\mathbf{E}$	electromagnetic electric-field vector
$\mathcal{E}$	diagonal, energy matrix
$\mathcal{E}_{nn}$	matrix element
$\boldsymbol{\varepsilon}$	uniform electric field vector
E	two-dimensional irrep. (E, E_1, E_2, E_g, E_u)
E^N	two-dimensional irrep. eqivalent to E

f; F; $\mathcal{F}$

f	fraction
f_i	basis functions or eigenfunctions of $\mathcal{H}_0$
$f_s(r)$	s-state basis function ($f_s(r)$,$f_p(r)$)
f	superscript, refers to basis functions
$f_j(\mathbf{r}-\mathbf{R}_i)$	atomic function for an atom at $\mathbf{R}_i$
f–j	subscript, jth fundamental vibrational level
F_i	symmetry-adapted functions
F^j	symmetry-adapted function belonging to Γ^j
F_k^j	symmetry-adapted function belonging to the kth row of Γ^j
F^E	symmetry-adapted function belonging to E
$F_{n'}^{\mathbf{k}}(I)$	n'th symmetry-adapted function belonging to $\Gamma^{\mathbf{k}}$
F	subscript, Fermi
$\mathcal{F}$, $\mathcal{F}_i$	arbitrary functions

g; G; $\mathcal{G}$

g	number of elements in a group (order of a group)	
$g_{\mathcal{A}}$	order of group $\mathcal{A}$	
g	number of elements in a class	(g_r, g_σ)
g	*gerade* (even)	
g	subscript, ground-state	
G	group	
G_e	entire point group	
G_r	element of a group	(G_r, G_s, G_t)
$\mathcal{G}$	direct-product group	
$\mathcal{G}_{ij}$	element of a direct-product group	

h; H; $\mathcal{H}$

h	Planck's constant
$\hbar$	$(h/2\pi)$
h	horizontal
H	hydrogenic
$H_{n_i}(\alpha_i R_i)$	Hermite polynomial
$\mathcal{H}$	Hamiltonian
$[\mathcal{H}]$	Hamiltonian matrix
$[\mathcal{H}^f]$	Hamiltonian matrix based on functions f_i
$\mathcal{H}_0$	unperturbed Hamiltonian (a soluble Hamiltonian)
	$\mathcal{H}_{\text{atom}}$, $\mathcal{H}_{\text{mol}}$, $\mathcal{H}_{\text{H}}$, $\mathcal{H}_{\text{elect}}$, $\mathcal{H}_{\text{crystal}}$
$\mathcal{H}_i$	one-electron or single mode Hamiltonian
$\mathcal{H}_{ij}^f$	Hamiltonian matrix element
$\mathcal{H}'$	perturbation Hamiltonian

i; i; I; $\mathbf{i}$

i	$\sqrt{(-1)}$	
i	the inversion operation	
$\mathbf{i}$	orthogonal unit vector	$(\mathbf{i},\mathbf{j},\mathbf{k})$
$\mathbf{i}_1,\mathbf{i}',\mathbf{i}_N$	alternative sets of unit vectors	
$\hat{\mathbf{i}}$	alternative sets of unit vectors	$(\hat{\mathbf{i}},\hat{\mathbf{j}},\hat{\mathbf{k}})$
	carets are used where necessary to avoid ambiguity, particularly for the reciprocal lattice	
ij	matrix element subscripts	
i	one-electron or single-mode subscript	
i	various dummy variables as subscripts or superscripts	
irrep.	inequivalent irreducible representation	
I	unit diagonal matrix of arbitrary dimension	

I	the icosahedral symmetry group	(I, I_h)
I	superscript, the totally symmetric irreducible representation for any group	

j; $\mathbf{j}$

j	superscript for the jth irreducible representation of a group	
j	dummy variable, subscript or superscript	
$\mathbf{j}, \hat{\mathbf{j}}$	unit vectors	

K; k; K; $\mathbf{k}$; $\mathbf{K}$

k	subscript for the kth row of an irrep.	
k	dummy variable subscript	
k, k_{bond}	force constant for harmonic bond	
$\mathbf{k}, \hat{\mathbf{k}}$	unit vectors	
$\mathbf{k}$	vector in reciprocal space, vector within the first Brillouin zone	
$\mathbf{k}'$	general vector in reciprocal space	
k, k'	magnitudes of $\mathbf{k}$, $\mathbf{k}'$	
k_F'	Fermi wavevector	
$\mathbf{k}$	subscript referring to the group of $\mathbf{k}$	
$\mathbf{k}$	superscript referring to the $\mathbf{k}$ irrep.	
$\mathbf{K}$	reciprocal lattice vector	
K	magnitude of $\mathbf{K}$	
$\mathbf{K}_1$	nearest neighbour reciprocal lattice vector	
$\mathbf{K}_n$	nth reciprocal lattice vector	
K^d	force matrix referred to **d**-displacements	$(K^d, K^D, K^{\mathfrak{D}})$
KE	kinetic energy	

L; l; L; $\mathbf{L}$; $\mathcal{L}$

l	dimension of a matrix representation	
l	degeneracy of a normal mode	
l	repeat distance for lattice vibrations	
l_j	dimension of the jth irrep of a group	
l	total angular momentum quantum number for one particle	
L	total angular momentum quantum number for two or more particles	
$\mathbf{L}$	orbital angular momentum vector	
L	(1,1,1) point of a reciprocal lattice	
L	size of a crystal in one dimension	
L_x	crystal dimensions	(L_x, L_y, L_z)

LCAO	linear combination of atomic orbitals	
$\mathcal{L}$	Lagrangian	

m; M; $\mathbf{M}$

m	mass	
m	quantum number of component of angular momentum	
mol	molecular	
$m\text{–}d$	magnetic dipole	
m_{atom}	atomic mass	
m_e	electron mass	
m_1	integer in the definition of primitive lattice sites	(m_1,m_2,m_3)
M	quantum number for component of angular momentum for two or more particles	
M	total mass	
M^d	mass matrix referred to **d**-displacements	$(M^d,M^D,M^{\mathcal{D}})$
M_n	nuclear mass	
M_T	total mass of a molecule	
$M\text{–}O$	molecular orbital	
$\mathbf{M}$	molecular electric dipole	
$M(R_0),M_0$	permanent dipole moment	

n; N; $\mathcal{N}$

n	quantum number for energy eigenvalues	
n	nth energy band in a solid or nth harmonic oscillator level	
n	n-fold rotation	
n	number of atoms in a crystal unit cell	
n_c	number of classes in a group or subgroup	
n	nuclear	
n_1	integer used for defining positions within the Bravais lattice	(n_1,n_2,n_3)
n_1	integer for defining reciprocal lattice points	(n_1,n_2,n_3)
N	orthogonal matrix for relating similar bases	
N	subscript referring to transformed axes	
N	orthogonal matrix relating symmetry-adapted displacements to basis displacements	
N	number of particles or electrons in a system	
N_1	number of primitive translations in one direction in a crystal	(N_1,N_2,N_3)

N	number of primitive cells in a lattice	
N_F	number of electronic states beneath the Fermi energy	
$\mathcal{N}$	number of cells in a crystal	

o; O

o	subscript for unperturbed Hamiltonian	
o–j	subscript, overtone vibration level	
O	point group centre, reference frame origin	
O	symmetry group label	(O, O_h)

P; p; P; $\mathcal{P}$

p	parity	
p	atomic p-orbital	
$\mathbf{p}$	scaled reciprocal lattice vector	
P	point on an object	(P, P')
PE	potential energy	
$P^l_{\|m\|}(\cos\theta)$	Legendre polynomial	
$\mathcal{P}$	crystal point group	
$\mathcal{P}'$	subgroup of the crystal point group	
$\mathcal{P}_\mathbf{k}$	the point group of the vector $\mathbf{k}$	
p_x	integer used to define reciprocal lattice points	(p_x, p_y, p_z)

q; Q; $\mathbf{q}$; $\mathbf{Q}$; $\mathcal{q}$; $\mathcal{Q}$

$\mathbf{q}$	radiation wavevector	
q_i	generalized coordinates	
Q	arbitrary function	
$\mathbf{Q}$	arbitrary displacement vector	
$\mathcal{q}_i$	normal mode coordinates	
$\mathcal{Q}_i$	normal mode coordinate eigenvectors	

r; R; $\mathbf{r}$; $\mathbf{R}$; $\mathcal{R}$

r	subscript for the rth group element	(r, s, t)
r_i	amplitude coefficient for the basis displacement $\mathbf{d}_i$	
$\mathbf{r}$	column vector	
$\tilde{\mathbf{r}}$	row vector	
$\mathbf{r}, \mathbf{r}_i$	vector coordinate of an electron	
r	spherical polar coordinate	(r, θ, ϕ)
$\mathbf{r}$	relative coordinate for internal motion of an atom	

$\mathbf{r}_e$	electronic coordinates
r	radial (atomic orbital)
r_i	electron–nucleus separation
r_{in}	electron–nucleus separation
r_{ij}	electron–electron separation
R	rotational eigenfunction
R_i	amplitude coefficient for symmetry-adapted displacement $\mathbf{D}_i$
$\mathbf{R}$	centre-of-mass coordinates or nuclear configuration
$\mathbf{R}$	lattice site vector
$\mathbf{R}_n$	nuclear coordinates
$R(r)$,$R_{ne}(r)$	radial part of a wavefunction
R_{np}	separation of nuclei in a molecule
R_{eq}	equilibrium separation of a diatomic molecule
R_x	rotational function for the x-axis (R_x,R_y,R_z)
R_3	the rotational group
R	general point group operation
$R(\alpha)$	rotation by angle α
$R(\alpha,z)$	rotation by angle α about the z axis
$R(\alpha,\xi)$	rotation about an arbitrary axis
$\mathcal{R}_i$	amplitude coefficient for normal mode eigenvectors $\mathfrak{D}_i$
$\mathcal{R}_\mathbf{k}$	space subgroup for which $\mathbf{k}$ is unaltered

s; S; $\mathcal{S}$

s	subscript for s state
s	sth group operation
sabd	symmetry-adapted basis displacement
S_n	overlap integral
S_n	symmetry group label
S	overlap integral
$\mathcal{S}$	symmorphic space group
$\mathcal{S}'$	subgroup of space group
$\mathcal{S}_\mathbf{k}$	space group of the vector $\mathbf{k}$

t; T; $\mathbf{t}$; $\mathbf{T}$; $\mathcal{T}$

t	time	
t	tangential (orbitals)	
$\mathbf{t}$	arbitrary translation	
T	irrep. label	(T_1,T_2)
T	symmetry group label	(T,T_d,T_h)
T	translational eigenfunction	
T_1	primitive vector translation	(T_1,T_2,T_3)
$\mathbf{T}$	lattice translation vector	

$\mathcal{T}$	the translation group	
$\mathcal{T}_1$	cyclic group of order N_1	$(\mathcal{T}_1, \mathcal{T}_2, \mathcal{T}_3)$

u; U

u	*ungerade* (uneven)	
$U_{n\mathbf{k}}(\mathbf{r})$	Bloch function, or lattice periodicity	
$U(\mathbf{r})$	arbitrary Bloch function	

v; V

v	vertical	
V	volume, dV integrals over all space	
V	total crystal volume	
V	vibrational eigenfunction	
$V(\mathbf{r})$	arbitrary potential energy	
$\bar{V}_e(\mathbf{r}_i)$	average potential energy of electron i, in the presence of additional electrons	
V_0	constant potential energy	
V_{n-n}	Coulomb potential energy	$(V_{n-n}, V_{e-n}, V_{e-e})$

x; X

x	x coordinate in fixed frame of reference	(x,y,z)
x',x''	altered coordinates	(x',y',z') (x'',y'',z'')
$x-y$	plane containing x and y axes	
X	(1,0,0) point of a reciprocal lattice	
$X(\mathbf{R})$	wavefunction for nuclear motion	

y; Y

y	y coordinate	
Y^l_m	spherical harmonic function	
$Y(\theta,\phi)$	general description of angular part of a wavefunction	
$Y^{Ll_1l_2}_M$	two-particle spherical harmonic function	

z; Z

z	z coordinate	
z	z directed orbital	
Z	charge number of a nucleus	

Greek

α	angle of rotation about a specified axis	
α	spin-up state	(α,β)
α	one-electron energy in the benzene problem	(α,β)
α	Morse coefficient	
$\alpha,[\alpha]$	matrix relating normal mode eigenvectors to basis vectors	
α_i	harmonic oscillator parameter $(M_i\omega_i/\hbar)^{\frac{1}{2}}$	
α_{ij}	matrix element	
α_1	primitive lattice angle	$(\alpha_1,\alpha_2,\alpha_3)$
α	Bravais lattice angle	(α,β,γ)
β	angle of rotation about a secondary axis	
β	Bravais lattice angle	
β	spin-down state	
β	adjacent-atom interaction energy in the benzene problem	
$\beta;[\beta]$	matrix relating eigenvectors to symmetry-adapted vectors	
γ	Bravais lattice angle	
γ_n	coefficients in an expansion of vectors belonging to the same row of an irrep.	
γ_{kj}	coefficients in the expansion of $\mathscr{H}'_{fj}$	
Γ	arbitrary matrix representation of a group of operations	
$\Gamma(G_r)$	arbitrary matrix representation of the operation G_r	
$\Gamma^j(G_r)$	arbitrary matrix irrep. of the operation G_r	
Γ^d	representation based on **d** displacements	$(\Gamma^d,\Gamma^D,\Gamma^{\mathscr{D}},\Gamma^f,\Gamma^F,\Gamma^{\mathscr{F}})$
Γ^a	irrep. of the group	(Γ^a,Γ^b)
Γ^I	the symmetric irrep.	
$\Gamma^{\mathscr{H}'f_j}$	matrix rep. based on $\mathscr{H}'f_j$ and its partner vectors	
Γ	representation based on an arbitrary set of functions	
$\Gamma^j(G_r)_{kl}$	matrix element of $\Gamma^j(G_r)$	
Γ_N	representation matrix obtained from Γ by a similarity transformation	
Γ_f	representation based on f states	$(\Gamma_s,\Gamma_p,\Gamma_f,\ldots)$
Γ_i	subscripted Γ irreps for use at $\mathbf{k}=0$ in solids	
∂	partial derivative operator	
δ_{ij}	Dirac delta	
Δ	the [100] axis	
ΔE	energy change due to a perturbation	$(\Delta E,\Delta V)$
$\Delta l,\Delta m$	transition selection rule for change in l or m	
Δ_k	volume per $\mathbf{k}$ state in reciprocal space	
Δ_{sr}	unity if s and r refer to the same class, zero otherwise	

ϵ_0	permittivity of free space	
ϵ_j	energy eigenvalues for the $\mathcal{H}_0$ problem	
ϵ_k	dimensionless, scaled energies for use in the semiconductor problem	
ϵ_F	scaled Fermi energy	
κ	dimensionless, scaled vectors in reciprocal space	
θ	spherical polar angle	(r,θ,ϕ)
$\Theta(\theta)$	function of θ	
$\Theta^l_{\vert m\vert}(\theta)$	θ part of spherical harmonic function	
λ	radiation wavelength	
μ	reduced H-atom mass	
ν	frequency in Hz	
$\bar{\nu}$	radiation wavenumber	
π	pi	
π	π-orbitals in molecules	
π	π-states in diatomics, irrep. label	
Π	many-electron states in diatomics, irrep. label	
Π	product	
ξ	arbitrary rotation axis	
σ	reflection operation	$(\sigma,\sigma_a,\sigma_v,\sigma_h,\sigma_d)$
σ	one-electron state for axial molecules, irrep. label	
Σ	summation	
Σ	many-electron state for molecules, irrep. label	
τ	translation by a vector smaller than a primitive lattice vector	
ϕ	spherical polar angle	
ϕ	phase	
Φ	function of ϕ	
χ	character (trace)	
$\chi^j(G_r)$	character for the jth irrep. for the operation G_r	
$\chi^d(G_r)$	character for G_r, based in **d**-displacements	
χ^d	set of characters for each G_r	(χ^D,χ^f,χ^F)
$\chi(\Gamma_s)$	characters for the representation with s states as basis	

$\chi^l(G_r)$	characters for the symmetric irrep.
χ^l	characters for the lth irrep. of the rotation group
ψ	arbitrary eigenfunction
ψ_n	specific eigenfunction
$\psi(r,\theta,\phi)$	internal motion eigenfunction
$\psi_{nlm}(r,\theta,\phi)$	H-atom eigenfunction
ψ_{elect}; $\psi_{\mathrm{e}\mathbf{R}}$, $\psi_{\mathrm{n}\mathbf{k}}(\mathbf{r})$	
Ψ	many-particle eigenfunction
$\Psi(\mathbf{r}_e,\mathbf{R}_n)$	two-particle H-atom eigenfunction
ψ_{elect}	many-electron wavefunction
$\psi_{e\mathbf{R}}(\mathbf{r})$	electron wavefunction for nuclear configuration $\mathbf{R}$
ω	vibration angular frequency or radiation angular frequency
ω_i	natural frequency for ith normal mode
ω	$\exp(-2\pi i/N_1)$ in cyclic groups
Ω	diagonal matrix of elements $\Omega_{ii}=\omega_i^2$
Ω	primitive cell volume
Ω_k	Brillouin zone volume in reciprocal space

Miscellaneous

∇	nabla
$\equiv$	is equivalent to
$\approx$	is approximately equal to
$\propto$	is proportional to
$\Rightarrow$	reduces to
$\rightarrow$	to or tend to
$\in$	belongs to
$\oplus$	matrix direct addition
$\otimes$	matrix direct product or direct product of two groups
$\mathbf{r}$	vector-r
$\tilde{\boldsymbol{\alpha}}$	transpose of a matrix α
$\bar{V}$	average value of V
$\lvert\mathbf{r}\rvert$	magnitude of $\mathbf{r}$
$\hat{\mathbf{k}}$	unit vector in $\mathbf{k}$-direction
$\mathbf{a}\wedge\mathbf{b}$	cross product of $\mathbf{a}$ and $\mathbf{b}$
$\dot{r}$	derivative of r with respect to time
χ^*	complex conjugate of χ
†	footnote citation

(1,0,0)	real-space Bravais unit cell vector (**a**)
(1,0,0)	first Brillouin zone boundary point in the **a** direction
[1,0,0]	the **a** direction, in **k**-space
⟨1,0,0⟩	translation by **a** in **k**-space
$\{R/\mathbf{t}\}$	Seitz vector for a space group operation
$\{E/\mathbf{T}\}$	lattice translation operation

Appendix 4

Bibliography

(a) Symmetry

Two texts which describe the diversity of symmetry in nature are

Shubnikov, A. V. and Koptsik, V. A. (1974) *Symmetry in Science and Art*, Plenum Press, New York;

Weyl, H. (1952) *Symmetry*, Princeton University Press, New Jersey.

Two-dimensional space groups are presented as artistic mosaics in

MacGillavry, C. H. (1965) *Symmetry Aspects of M.C. Escher's Periodic Drawings*, A. Oosthoek's Uitgeversmaatschappij N.V., Utrecht.

(b) Introductory texts on the application of symmetry

A brief survey covering many of the group-theoretical ideas used in the present text is

Leech, J. W. and Newman, D. (1969) *How to Use Groups*, Science Paperbacks; Chapman and Hall, London.

Two texts close in content and at a level similar to the present text are

Boardman, A. D., O'Connor, D. E. and Young, P. A. (1973) *Symmetry and its Applications in Science*, McGraw-Hill, London;

Burns, G. (1977) *Introduction to Group Theory with Applications*, Academic Press, New York.

A step-by-step approach to using basic group theory ideas in chemistry is
Vincent, A. (1977) *Molecular Symmetry and Group Theory*, Wiley, Chichester.
Three texts aimed at chemists and containing sections on introductory group theory are
Atkins, P. W. (1970) *Molecular Quantum Mechanics*, Parts I, II; Clarendon Press, Oxford.
Cotton, F. A. (1936) *Chemical Applications of Group Theory*, Wiley, New York.
Harris, D. C. and Bertolucci, M. D. (1978) *Symmetry and Spectroscopy*, Oxford University Press, New York.

(c) Intermediate text

The book to refer to having read the present text, for extensions into atomic and molecular physics, is
Tinkham, M. (1964) *Group Theory and Quantum Mechanics*, McGraw-Hill, London.

(d) Solid-state group theory

Bouckaert, L. P., Smoluchowski, R. and Wigner, E. P. (1936) Theory of Brillouin Zones and Symmetry Properties of Wave Functions in Crystals, *Phys. Rev.*, **50,** 58.
Bradley, C. J. and Cracknell, A. P. (1972) *The Mathematical Theory of Symmetry in Solids*, Oxford University Press.
Burns, G. and Glazer, A. M. (1978) *Space Groups for Solid State Scientists*, Academic Press.
Landsberg, P. T. (ed.) (1969) *Solid State Theory*, Wiley-Interscience, London.
Koster, G. F. (1957) Space Groups and their Representations, in *Solid State Physics*, Vol. 5, Seitz, F. and Turnbull, D. (eds.) Academic Press, New York.
Kovalev, O. V. (1965) *Irreducible Representations of the Space Groups*, Gordon and Breach, New York.

(e) Advanced texts

Eliot, J. P. and Dawber, P. G. (1979) *Symmetry in Physics*, Vols. I, II, MacMillan, London.

Hamermesh, M. (1962) *Group Theory and its Applications to Physical Problems*, Addison-Wesley, Reading, Mass.

Heine, V. (1960) *Group Theory with Applications*, Pergamon Press, London.

Lax, M. (1974) *Symmetry Principles in Solid State and Molecular Physics*, Wiley, New York.

Slater, J. C. (1963) *Quantum Theory of Molecules and Solids*, Vol. 1, McGraw-Hill, New York.

Wigner, E. P. (1959) *Group Theory and its Applications to the Quantum Mechanics of Atomic Spectra*, Academic Press, New York.

(f) Background reading

Condon, E. O. and Shortley, G. H. (1951) *The Theory of Atomic Spectra*, Cambridge University Press, New York.

Herman, F. (1953) Theoretical Investigation of the Electronic Energy Band Structure of Solids, *Rev. Mod. Phys.*, **30,** 102 and cited references.

Jones, H. (1960) *The Theory of Brillouin Zones and Electronic States in Crystals*, North-Holland, Amsterdam.

Long, D. (1968) *Energy Bands in Semiconductors*, Wiley-Interscience, New York.

Philips, J. C. (1973) *Bonds and Bands in Semiconductors*, Academic Press, New York.

Wilson, E. B., Decius, J. C. and Cross, P. (1955) *Molecular Vibrations*, McGraw-Hill, New York.

(g) Tables for semiconductor compounds

Landolt–Börnstein, *Numerical Data and Functional Relationships in Science and Technology*, Group III, Vols. 17a,b, Springer-Verlag, Berlin (1982).

Index